Hydrological Modelling and Integrated Water Resources Management in Ungauged Mountainous Watersheds

Hydrological Modelling and Integrated Water Resources Management in Ungauged Mountainous Watersheds

Edited by:

WEI-LIN XU, TIAN-QI AO & XIN-HUA ZHANG

State Key Laboratory of Hydraulics & Mountain River Engineering
Sichuan University, Chengdu, China

Proceedings of a symposium held at Chengdu, China, November 2008

IAHS Publication 335
in the IAHS Series of Proceedings and Reports

Published by the International Association of Hydrological Sciences 2009

IAHS Publication 335

ISBN 978-1-907161-07-0

British Library Cataloguing-in-Publication Data.
A catalogue record for this book is available from the British Library.

The papers included in this volume have been reviewed and some were extensively revised by the Editors, in collaboration with the authors, prior to publication.

IAHS is indebted to the employers of the Editors for the invaluable support and services provided that enabled them to carry out their task effectively and efficiently.

Publications in the series of Proceedings and Reports are available from:
IAHS Press, Centre for Ecology and Hydrology, Wallingford, Oxfordshire OX10 8BB, UK
tel.: +44 1491 692442; fax: +44 1491 692448; e-mail: jilly@iahs.demon.co.uk

Printed by Information Press

Preface

In recent years, China has been increasingly exposed to water-related problems such as flood, drought, water shortage, water pollution and soil loss. These problems have already become a major constraint to China's economic development and a big challenge for the protection of natural ecosystems and environments, especially in western China, centred on Sichuan, Yunnan and Guizhou provinces, where high mountains and plateaus are located and many rivers originate. Water resources and hydro-power in these regions are therefore very rich and many hydraulic projects have been developed. During the implementation of the National Strategies of China Western Development in recent years, and also under the pressure of the increasing population, water-related problems have become even worse due to complex natural changes and human activities in these mountainous watersheds, where ground observations are almost unavailable. In addition, China is now changing from traditional water resources development to water resources management for sustainable development. It is against such a background that hydrological modelling and integrated water resources management strongly call for new innovations in approaches, techniques and theories. The key issue is how to make predictions in ungauged or poorly-gauged mountainous basins with a major reduction in predictive uncertainties.

The China PUB organization was formed in 2004 and focuses on the new methodology of hydrological simulation and prediction under natural and human-induced global changes. To promote exchanges between China and IAHS, and with help from the international IAHS-PUB Scientific Steering Group (SSG), the China PUB National Steering Committee organized the first International Symposium on Flood Forecasting and Water Resources Assessment for IAHS-PUB in September 2006*. The second International Symposium on Hydrological Modelling and Integrated Water Resources Management in Ungauged Mountainous Watersheds for IAHS-PUB was successfully held in November 2008. The Symposium organizers would like to express their heartfelt thanks to Prof. Pierre Hubert, the Secretary General of IAHS, Prof. Kuniyoshi Takeuchi (ICHARM, Japan), Prof. Jeffrey McDonnell (Oregon State University, USA), and Prof. E. Foufoula-Georgiou (University of Minnesota, USA) for their enthusiastic and valuable assistance.

We received a huge response to the second Symposium; more than 170 abstracts were submitted. After an initial review, we accepted 125 papers and invited the authors to attend the Symposium. After the Symposium and a further peer-review process, we accepted 101 papers in total, of which 41 papers are published in this IAHS volume and 60 papers were published in a special issue of the Journal of Sichuan University (Engineering Science Edition).

The editors gratefully acknowledge the assistance of the reviewers who made valuable contributions to this volume. We also thank Penny Perrins and Cate Gardner of IAHS Press for their professional approach and help with the processing of the manuscripts.

Wei-lin Xu, Tian-qi Ao & Xin-hua Zhang
State Key Laboratory of Hydraulics & Mountain River Engineering
Sichuan University, Chengdu, China

---

* Selected papers from the event were published in 2008 in *Hydrological Research in China* (edited by D. Yang *et al.*). IAHS Publ. 322.

Contents

3 Integrated Water Resources Management

4 Eco-environmental Protection

1 Hydrological Modelling and Predictive Uncertainty

A study on calibration of the Nash model parameters in no-data or scarce-data regions

YINGNA SUN[1,3], XIAOFANG RUI[2], QIANG FU[1] & NING JIANG[1,3]

1 *College of Water Conservancy and Civil Engineering, Northeast Agricultural University, Harbin 150030, China*
ynsunchina@yahoo.com.cn

2 *College of Water Resources and Environment, Hohai University, Nanjing 210098, China*

3 *College of Water Conservancy, Heilongjiang University, Harbin 150086, China*

Abstract Hydrologic simulation for regions without observed data or with scarce data is one of the main interests of study in hydrological forecasting. In this paper, the theory of stochastic differential formulae was used to find the relationship between the parameters of the watershed flow concentration models and characterization of a flow concentration system with a random input and output process. Based on output of a flow concentration system conditioned on an input process with white noise and Markov noise, the linear regression method was used to estimate the Nash model parameters, and then a method to calculate flow concentration progress in ungauged basins theoretically was proposed. Case studies of the Xiang-yang-Huang-zhuang channel of Han-jiang River and the Yan-du-he basin in the Three Gorges area showed the method worked well. For channel routing, calibration of Nash model parameters should be based on an input process with Markov noise, and for basin confluence, Nash model parameter calibration should be based on a rain process with white noise; this is consistent with the stochastic character of the input and rainfall processes.

Key words Nash Model; stochastic differential equation; correlation function; Markov noise

INTRODUCTION

Hydrological simulation without empirical data, or for scarce data regions, is one of the hot topics in hydrological forecasting. As China is one of the largest developing countries in the world, there are only 3400 hydrological stations. And of these most are distributed in several economically developed eastern areas, where the population is large. Observed data are scarce for most other regions, and cannot reflect the time–space distribution law of basin rainfall for those regions. Therefore, it is very important to establish a theoretical method for determining parameters of the watershed flow concentration models in without-data or data-scarce regions. The Nash model is widely used in watershed concentration. The key in parameter determination for the Nash model is to estimate n and k. Through case studies, a relationship between parameters of the Nash model and the random structure of the inflow or outflow of a flow concentration system was found, and the hydrological forecasting in data-scarce regions was done based on the output of the flow concentration system with a known stochastic structure of the input process.

THEORETICAL BASIS

Suppose the flow concentration system is a linear time-invariant system, and the input and output processes meet the convolution formula as below:

$$Q(t) = \int_0^{+\infty} u(s) I(t-s) \mathrm{d}s \tag{1}$$

where $u(t)$ is the instantaneous unit hydrograph of confluence, $I(t)$ is the input process and $Q(t)$ is the output process of the hydrological model.

According to equation (1), the cross-correlation function of the input process and the output process is:

$$\begin{aligned}R_{QI}(\tau) &= \mathbf{E}[Q(t)I(t-\tau)] \\ &= \mathbf{E}\left[\int_0^{+\infty} u(s)I(t-s)I(t-\tau)\mathrm{d}s\right] \\ &= \int_0^{+\infty} u(s)\mathbf{E}[I(t-s)I(t-\tau)\mathrm{d}s \\ &= \int_0^{+\infty} u(s)R_I(\tau-s)\mathrm{d}s \\ &= u(\tau) * R_I(\tau)\end{aligned} \tag{2}$$

According to equation (1), the autocorrelation function of the outflow process is:

$$\begin{aligned}R_Q(\tau) &= \mathbf{E}\left[Q(t)Q(t+\tau)\right] \\ &= \mathbf{E}\left[\int_0^{+\infty} u(s)I(t-s)Q(t+\tau)\mathrm{d}s\right] \\ &= \int_0^{+\infty} u(s)\mathbf{E}[I(t-s)Q(t+\tau)]\mathrm{d}s \\ &= R_{QI}(\tau) * u(-\tau)\end{aligned} \tag{3}$$

where $R_{QI}(\tau)$ is the cross-correlation function of the combined inflow outflow process, $R_I(\tau)$ is the autocorrelation function of the inflow process, $R_Q(\tau)$ that of the outflow process, and * stands for convolution operation.

Fourier transformation is applied to equations (1) and (2) and the following equations can be obtained:

$$S_{QI}(\omega) = U(\omega)S_I(\omega) \tag{4}$$

$$S_Q(\omega) = S_{QI}(\omega)U^*(\omega) \tag{5}$$

where $S_{QI}(\omega) = \mathbf{F}[R_{QI}(\tau)]$; $U(\omega) = \mathbf{F}[u(\tau)]$; $S_I(\omega) = \mathbf{F}[I(\tau)]$; $S_Q(\omega) = \mathbf{F}[R_Q(\tau)]$ and F is the symbol of Fourier transformation.

Equation (5) expressed in frequency area meets the relationship between the stochastic process correlation function of equation (1).

The redistribution effect that the Nash Model makes on the net rain divided the regulation function in which linear linked reservoir made a series of same effects. The expression of the instantaneous unit hydrograph of the Nash model took this form:

$$u(t) = \frac{1}{k(n-1)!}\left(\frac{t}{k}\right)^{n-1} e^{-t/k} \tag{6}$$

where n and k are parameters of the Nash model.

Fourier transformation is applied to equation (6):

$$U(\omega) = \frac{1}{(1+j\omega k)^n} \tag{7}$$

Situation of input process with white noise

Suppose an input process with white noise and the autocorrelation function has the following form:

$$R_I(\tau) = \sigma^2\delta(\tau) \qquad (-\infty < \tau < \infty) \tag{8}$$

where σ is the mean square deviation of input data and the meanings of the other symbols is as above.

The Fourier transform is applied to equation (6):

$$S_I(\omega)=\sigma^2 \qquad -\infty<\omega<+\infty \tag{9}$$

By substitution of equation (7) and equation (9) into equation (5):

$$S_Q(\omega)=\frac{1}{(1+\omega^2k^2)^n}\sigma^2 \tag{10}$$

This is the autocorrelation function of the outflow process in the frequency domain. Then the autocorrelation function of the outflow procession time domain was converted by using the theorem of residues:

$$R_Q(\tau)=\mathrm{Re}\,s[S_Q(s)\mathrm{e}^{s\tau};s=-1/k]$$
$$=\frac{1}{(n-1)!}\cdot\frac{\mathrm{d}^{n-1}[S_Q(s)\mathrm{e}^{s\tau}\cdot(s+1/k)^n]}{\mathrm{d}s^{n-1}}\Big|s=-1/k$$

and equation (11) was obtained:

$$R_Q(\tau)=\sigma^2\frac{\exp(-\tau/k)}{k(n-1)!2^n}\sum_{i=0}^{n-1}\frac{(n+i-1)!}{(n-i-1)!2^i i!}(\tau/k)^{n-i-1} \tag{11}$$

Situation of input process with Markov noise

The input process with Markov noise, and autocorrelation function has the form:

$$R_I(\tau)=\sigma^2\exp(-\beta\tau) \qquad (-\infty<\tau<\infty) \tag{12}$$

where β is the attenuation coefficient. In same way as with white noise:

$$S_Q(\omega)=\frac{1}{(1+\omega^2k^2)^n}\cdot\frac{2\sigma^2\beta}{(\beta-j\omega)(\beta+j\omega)} \tag{13}$$

Inverse Fourier transform is applied to equation (12) with the residue theorem:

$$R_Q(\tau)=\frac{\sigma^2\mathrm{e}^{-\beta\tau}}{(1-\beta k)^n(1+\beta k)^n}+\frac{1}{(n-1)!}\cdot\frac{\mathrm{d}^{n-1}}{\mathrm{d}s^{n-1}}\left[\frac{1}{(1-sk)^n}\cdot\frac{2\sigma^2\beta}{(\beta+s)(\beta-s)}\cdot\mathrm{e}^{s\tau}\cdot\frac{1}{k^n}\right]_{s=-\frac{1}{k}}$$

Then equation (14) is obtained:

$$R_Q(\tau)=\frac{\sigma^2\mathrm{e}^{-\beta\tau}}{(1-\beta k)^n(1+\beta k)^n}+\frac{\sigma^2e^{-\tau/k}}{(n-1)!}\sum_{i=0}^{n-1}\left[\frac{1}{(1+k\beta)^{n-i}}-\frac{1}{(1-k\beta)^{n-i}}\right]\sum_{j=0}^{i}\frac{(n+j-1)!(\tau/k)^{i-j}}{(i-j)!j!2^{n+j}} \tag{14}$$

STEPS OF APPLICATION

(1) Calculate the correlation function of the measured output process with equation (15):

$$\hat{\rho}_Q(\tau)=\sum_{i=1}^{m-\tau}(Q_i-\bar{Q})(Q_{i+\tau}-\bar{Q})\left[\sum_{i=1}^{m}(Q_i-\bar{Q})^2\right]^{-1} \tag{15}$$

(2) Determine β of the correlation function of the input process. Then, calculate the input autocorrelation function with equation (16):

$$\rho_{II'}(\tau)=\sum_{i=1}^{p-\tau}(I_i-\bar{I})(I_{i+\tau}-\bar{I})\left[\sum_{i=1}^{p}(I_i-\bar{I})^2\right]^{-1} \tag{16}$$

where I_i is the inflow rate at time $i\Delta t$; P is integral times of total time Δt of the input process; Δt is calculation interval; and $\bar{I}$ is the sample mean of the inflow process.
Then determine β according to equation (12).

(3) Determine the parameters n and k in the Nash Model. Firstly, the $\rho_{QQ}(\tau)$ was calculated according to equation (11) or (14), and then the correlation function peaks $\rho_{QQ}(\tau)\sim\tau$ were drawn with parameter k, and called the theoretical correlation function peaks. Secondly, for a series of measured flows, each order correlation function was calculated according to equation (15), the correlation function peaks, $\rho_{QQ'}(\tau)\sim\tau$, which were called experience correlation function peaks, was obtained. Theoretically, if the inflow process follows a certain distribution, it must have a certain k to make the theoretical correlation function equal to the practical correlation function; the theoretical curve $\rho_{QQ}(\tau)\sim\tau$ basically coincides with the experience curve $\rho_{QQ'}(\tau)\sim\tau$ and the value of k is the Nash model parameter.

CASE STUDIES

Application in the channel confluence

The length of the Xiang-yang–Huang-zhuang reach is 158 km, and the velocity of flow is 1.83 m/s. According to the concept of character length of river, the value of n determined preliminarily for this channel is 3. The calculation period is Δt = 6 h. Six floods were chosen from 1974 to 1985, and were used to calibrate the parameters of the Nash model.

After the white noise was input, the parameters were obtained (Table 1). For input with Markov noise, the range of input process parameter β is 0.029–0.035. The set mean, corresponding to parameter k was obtained (Table 1). Table 1 shows that the precision of Markov noise input is higher than that of white noise input. When the Markov noise is input to two simulated floods from 1980 to 1982, the parameters mean k = 6.43 h (Table 2), and the forecasts were satisfactory.

Table 1 Parameters of Nash model in the river calculated by the proposed method, Xiang-yang–Huang-zhuang reach.

Flood number	Q_m (m^3/s)	Input white noise			Input Markov noise			
		k	QE (%)	QC	β	k	QE (%)	QC
741002	18900	8.9	–3.05	0.98	0.03	6.2	–0.95	0.99
810820	12564	11	–17.0	0.74		7.0	–7.96	0.97
810901	10867	15	–13.19	0.38		6.2	–7.98	0.94
830730	11900	15.9	–10.85	--		6.5	–8.76	0.70
831005	25922	15.5	–17.59	0.04		6.5	–8.43	0.93
850603	4295	16	–8.05	--		6.2	–4.17	0.42

Table 2 Results of flood forecast calculated with the parameters calibrated by the method presented in the paper, Xiang-yang–Huang-zhuang reach.

Flood no.	Q_m (m^3/s)	β	K (h)	QE (%)	Peak time (h)	QC
800701	10386	0.003	6.43	–6.67	–2	0.91
820728	10553			–3.80	0	0.93

Q_m: flood peak discharge; QE: flood peak fitting relative error; QC: deterministic coefficient; β: Markov noise parameters.

Application in watershed concentration

The area of the Yandu River basin is 601 km^2. According to the method of Sun Yingna & Ren Xiaofang (2005), parameters $n = 3$, the set calculation period is 6 h, six floods were chosen from 1981 to 1983 to determine the parameters of the model. When the white noise was input, the parameters were obtained (Table 3). When the Markov noise is input, we obtain the range of input process parameters β is 0.5–0.9. The set mean $\beta = 0.7$, the corresponding parameters k were obtained (Table 3). Table 3 shows that the precision of white noise input is higher than the precision of Markov noise input. We can obtain the parameters mean $k = 2.1$ h when the Markov noise was input to simulate three flood events from 1984 to 1986 (Table 4).

Table 3 Parameter of Nash model in the watershed calculated by the proposed methods, Yan-du-he basin.

Flood number	Q_m (m^3/s)	Input white noise			Input Markov noise			
		k	QE (%)	QC	β	k	QE (%)	QC
810625	1046	1.70	13.7	0.78	0.7	1.25	25.8	0.28
810714	540	2.20	6.80	0.94		1.90	12.03	0.85
810812	552	1.95	22.8	0.41		1.50	48.55	0.18
820720	929	1.70	10.11	0.63		1.30	12	0.58
820821	489	2.40	2.02	0.87		2.00	12	0.89
830909	799	2.30	11.8	0.78		1.90	21.4	0.73

Table 4 Results of flood forecast calculated by the method presented in the paper with the parameters calibrated above; Yan-du River basin.

Flood no.	Q_m (m^3/s)	β	K (h)	QE (%)	Peak time (h)	QC
840613	484	2.1	–3.3	+1	0.88	840613
840725	948		–15	+1	0.93	840725
860615	451		–4.2	+2	0.79	860615

CONCLUSION

Tables 1 and 3 show that, for channel routing, because the rainfall process has a Markov noise character, based on the theory of equation (14), the parameter k of the Nash Model was defined. For watershed concentration, we can see that the rainfall process has the character of white noise, and the k parameter was defined according to the method proposed by Rui (1993). This conclusion is consistent with the character of the stochastic structure of the input process and rainfall process.

Acknowledgements This research was supported by the Foundation of Heilongjiang Educational Bureau of Science and Technology (11531267), the Research Foundation of Northeast Agricultural University, and the Foundation of the Heilongjiang Water Conservancy Bureau (HSKY2007–06).

REFERENCES

Rui Xiaofang (1993) Determining parameters of Nash Model only according to outflow data of the concentration flow system. *Water Science Progress,* 6, 141–146 (in Chinese).

Rui Xiaofang (1999) A study on determining the parameters of the Nash Model using physiognomy data. *Hydrology* **3**, 6–10 (in Chinese).

Sun Yingna & Rui Xiaofang (2005) Study of Nash unit hydrograph based on random theory. *Hydroelectric Energy* **6**, 18–20 (in Chinese).

Wang Guangde (1991) An auto/cross covariance approach for parameter estimation of the discrete linear transfer function nodel. *Water Science Progress* **6**, 113–119 (in Chinese).

An NDVI-based precipitation interpolation to improve hydrology simulation in the Upper Reaches of the Yangtze River

XUAN YU[1,2], HAO WANG[1], DENGHUA YAN[1], YANGWEN JIA[1] & CUNWEN NIU[1]

1 *Department of Water Resources China Institution of Water Resources and Hydropower Research, Beijing 100038, China*
akyuxuan@126.com

2 *Department of Civil and Environmental Engineering, Pennsylvania State University, University Park, Pennsylvania 16802, USA*

Abstract Spatially interpolated rainfall estimates from raingauges are widely used as input to hydrological models, but there is no universally suitable interpolation method. In southwest China there are very few raingauges. Therefore, geographical information related to rainfall needs to be used in spatial interpolation methods. Among elevation, longitude, latitude, slope, and the normalized difference vegetation index (NDVI), NDVI is found to be the most optimum geographical information indicator for precipitation in the upper reaches of the Yangtze River in China, after carefully analysis. Accordingly, a gradient plus reverse distance squared method based on NDVI was developed and applied in the upper reaches of the Yangtze River from 1956 to 2000. The interpolation result was compared with the result of Countrywide Water Resources Planning. The difference was not too much and the result of the interpolation showed a reasonable precipitation distribution in the mountainous areas, which was greatly improved compared to the direct use of the inverse-distance weighted method (IDW). And then the interpolation result was input to the Water and Energy transfer Process (WEP) model to simulate the hydrological cycle of the basin. The simulation of the hydrological processes show that water quantity balance errors are less than 5% and the Nash-Sutcliffe efficiency coefficient is above 0.7, which prove that the interpolation method presented in this paper is credible.

Key words precipitation; spatial interpolation; NDVI

INTRODUCTION

Precipitation is a fundamental input of the hydrological cycle. It is widely known that the inhomogeneous spatial and temporal distribution of precipitation has a great influence on basin streamflow (Dawdy & Bergmann, 1969). Therefore, the spatial and temporal variability of precipitation should always be taken into consideration. However, China and many other developing countries have very few raingauges. Also, advanced techniques, such as radar, or satellites, have not been generally used. The available high resolution rainfall data is limited. Interpolation is therefore the only way to derive spatial rainfall from point measurements.

As the original input of rainfall–runoff modelling, the spatial precipitation data are usually interpolated by various methods, such as: Thiessen polygon method, spline interpolations, inverse-distance weighted method (IDW), gradient plus reverse distance squared (GRDS), and kriging method. There is not a universal method fitting all areas. Thus, a comparison between different methods should be carried out to choose a favourable one (Dubrule, 1983; Philips *et al.*, 1992; Price *et al.*, 2001). The variable-order inverse distance method (Chia-Ling Chang *et al.*, 2006) has been proved more accurate, and the reliability of hydrological modelling, and hence of general water resource management, can be significantly improved with it.

With the increasingly broad application of remote sensing, vegetation information can be easily obtained from NOAA/AVHRR data (Defries *et al.*, 1994). A popular method is to derive the normalized difference vegetation index (NDVI, NDVI = (NIR – Red)/(NIR + Red), NIR denotes the reflection of near infrared and Red denotes the reflection of red light from visible light (Howard *et al.*, 1991) and it is an efficient indicator of growth status (Purevdorj *et al.*, 1998; Sun *et al.*, 1998; Liu *et al.*, 1999). In addition, the growth processes of vegetation are immensely affected by temperature, precipitation, etc. (Nicholson *et al.*, 1990; Li *et al.*, 2000; Schmidt *et al.*, 2000; Zhang *et al.*, 2003). Therefore, incorporation of vegetation information into hydrological modelling has the potential to improve our capability to constrain uncertainty in rainfall inputs and extend water resource simulations.

This paper reports on the potential of using NDVI data to help interpolate precipitation in the upper reaches of Yangtze River, where there are now too few data to allow reliable estimates of spatial rainfall, which would finally extend the simulation of the hydrological model.

STUDY AREA

The study area extends from 24°27′41″N to 35°45′10″E and 90°32′8″N to 104°55′36″E, which is the upper reaches of the Yangtze River (Fig. 1). Generally speaking, the upper reaches refers to the section above the Yichang City in Hubei Province. Here, the section above Yibin City in Sichuan Province is the focus of study, as the West-route of the South-to-North Water Transfer Project may cause significant changes to the runoff in the area. In winter, the climate of the area is controlled by the Siberian high and the Mongolian high, and it is very cold and arid, with little rain and strong winds. In summer, the climate is controlled by the Indian low, and is warm and moist (Zhao, 1998). The mountainous area forms various habitats which offer favourable conditions for most plants. From southeast to northwest, there are seven types of vegetation: subtropical evergreen broad-leaved forest, deciduous broad-leaved forest, alpine brush, alpine meadow, alpine grassland and high and cold desert (Chen *et al.*, 1999).

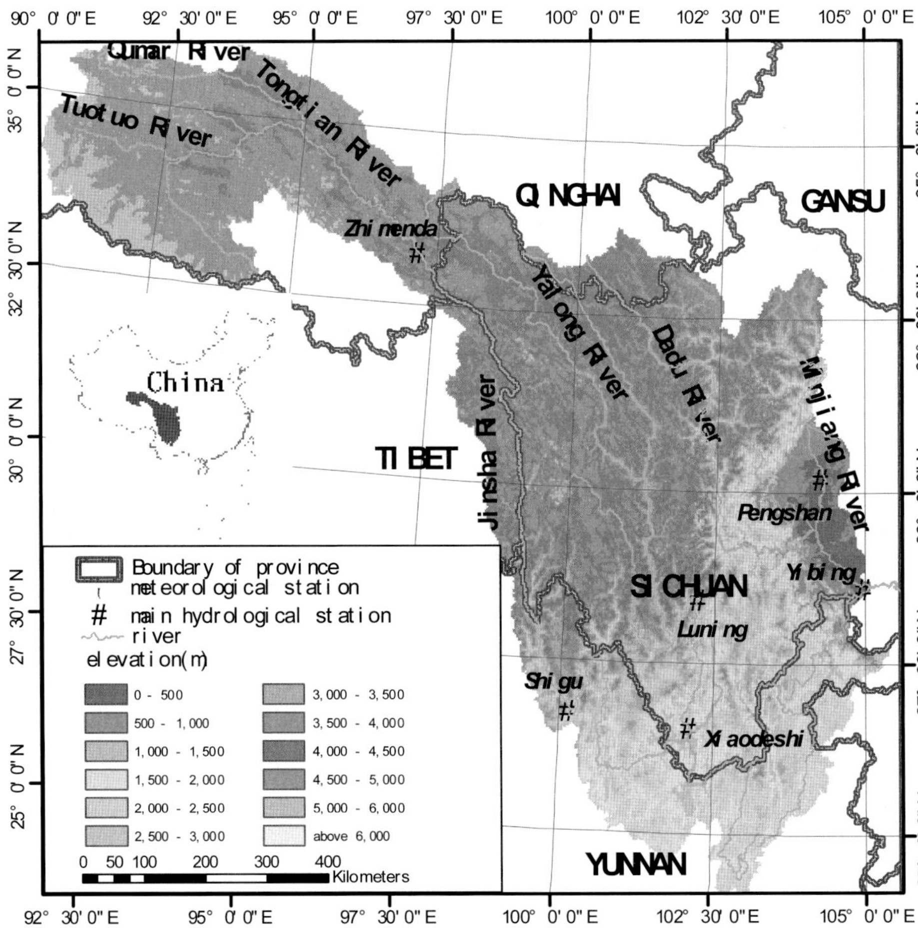

Fig. 1 Research region.

DATA AND METHODS

The historical precipitation records are available from databases of the China Meteorology Administration. Daily rainfall records from 1956 to 2001 of 73 stations (Fig. 2) over the study area and its surrounding areas were collected. The NDVI data were derived from free data of the EOS Data Gateway (EDG). To remove the effect of clouds, the Maximum Value Composite (MVC) was used in data processing, which presumes that the position of clouds varies during a period of time, such as 10 days, and any point has a day without cloud sheltering. The value of NDVI in this day at the point is higher than those of the cloudy days, and this study selects this value as the value of the point for the 10 days, then it selects the highest value of the point as the value of the given month. In this study, we convert the primitive projection of these images into Albers Projection and select the area of upper reaches of Yangtze River in China.

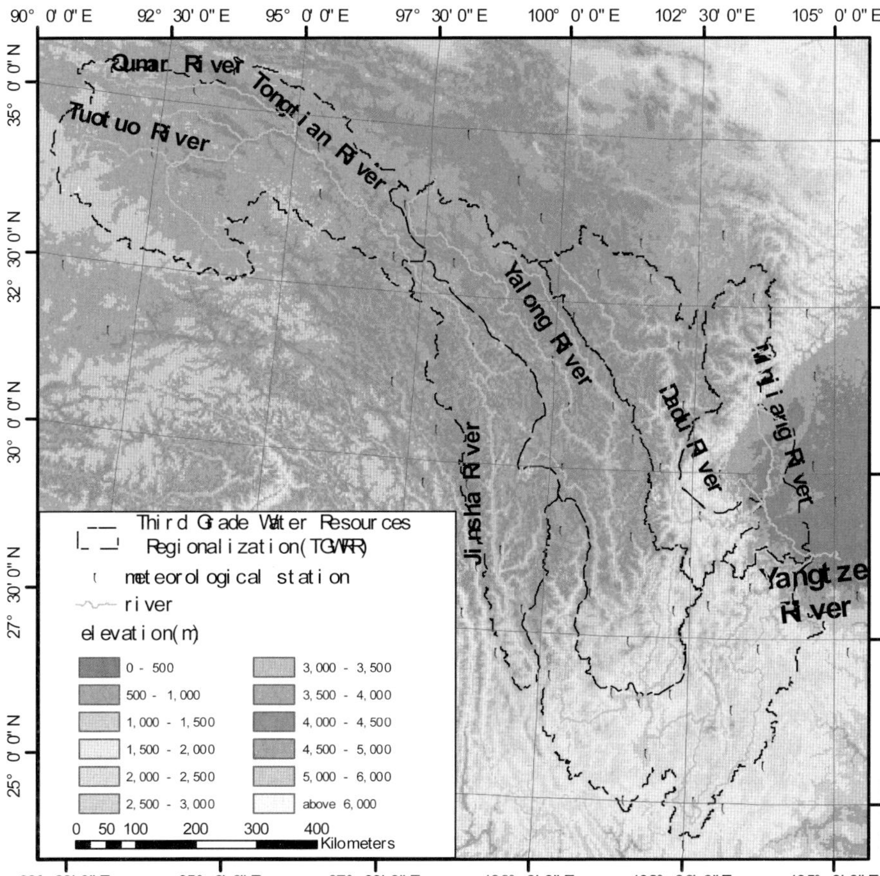

Fig. 2 Locations of the meteorological stations.

Relationship of geographical factors and precipitation

The correlation between mean monthly maximum NDVI and monthly precipitation from 1982 to 1999 is very weak in the western, northern and southern plateau, and very strong in the central and eastern plateau (Ding *et al.*, 2007), which is part of the study area. Therefore the relationship of geographical factors (elevation, longitude, latitude, slope, and NDVI) and precipitation were compared to choose auxiliary information for rainfall interpolation. The annual change of NDVI of

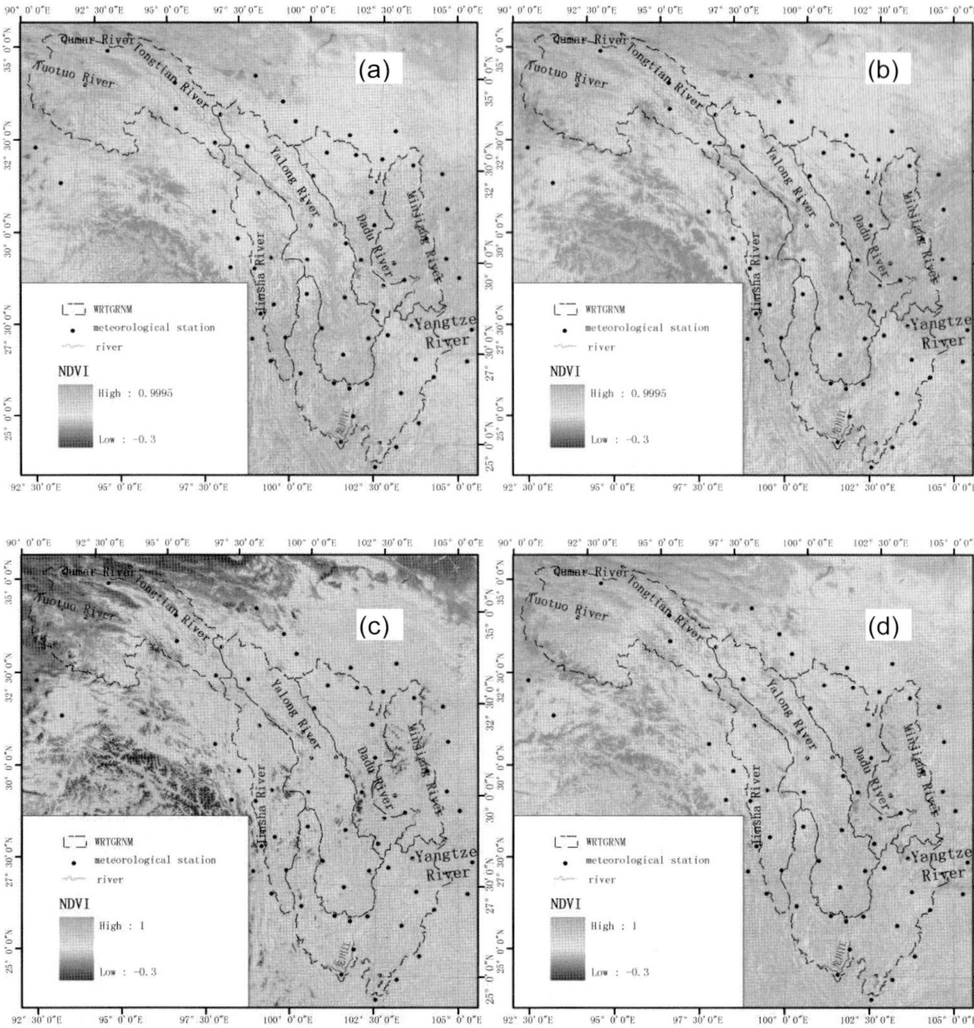

Fig. 2 Four typical months NDVI in 1985: (a) January, (b) April, (c) July, (d) October.

southwest China showed an increase, but it was not significant, from 1982 to 1999 (Fang *et al.*, 2003). In addition, the candidate factors are used to reflect the spatial rainfall distribution. So, only the NDVI data of 1985 are chosen (Fig. 2).

The linear regression method was used to compare the relationship of geographical factors with precipitation. Table 1 shows the regression coefficients. Monthly precipitation had notable negative correlation with elevation, significant positive correlation with longitude, and negative correlation with latitude. This is consistent with previous studies (Ding *et al.*, 2007). Because of control by the southwest warm and humid air flow, the southeast monsoon and Siberian high, the amount of precipitation decreases from southeast to northwest (Ding *et al.*, 2007). Monthly precipitation also has appreciable positive correlation with NDVI. As a result of sufficient moisture supply during rainy days, the plants may grow fast, which can be reflected by NDVI. Steeper terrain is more favourable to larger orographic rainfall amounts. However, the study area is extensive and mountainous, and the wind direction changes frequently. So, it is probable that the windward side of a mountain may be turned into a leeward side, which makes the relationship between precipitation and slope insignificant.

Table 1 The correlation coefficients between monthly precipitation and some geographical factors.

Month	Elevation		Longitude		Latitude		Slope	NDVI	
1	–0.514	***	0.321	***	–0.409	***	–0.086	0.220	*
2	–0.386	***	0.188	*	–0.295	***	–0.027	0.284	**
3	–0.310	***	0.149		–0.200	*	0.026	0.182	
4	–0.494	***	0.421	***	–0.172		0.079	0.504	***
5	–0.630	***	0.724	***	–0.360	***	0.011	0.589	***
6	–0.564	***	0.570	***	–0.616	***	0.071	0.249	**
7	–0.753	***	0.560	***	–0.584	***	0.078	0.353	***
8	–0.718	***	0.499	***	–0.516	***	0.114	0.166	
9	–0.675	***	0.605	***	–0.492	***	0.071	0.171	
10	–0.678	***	0.590	***	–0.689	***	0.033	0.350	***
11	–0.785	***	0.597	***	–0.676	***	–0.031	0.446	***
12	–0.657	***	0.472	***	–0.512	***	–0.046	0.472	***

Notes: * represents a significant difference at $P < 0.1$, ** represents a significant difference at $P < 0.05$, *** represents a significant difference at $P < 0.01$. (T-test)

Interpolation methods

The IDW was implemented to compare with the NDVI-based interpolation method, which incorporating NDVI by GRDS. So, the predicted value for a site is given by:

$$P(x,y,v)=[\sum_{i=1}^{n}\frac{P_i+(v-v_i)\times R}{d_i^2}]/[\sum_{i=1}^{n}\frac{1}{d_i^2}]$$

where P is the estimated precipitation, P_i is the value at the climate station i, d_i is the distance from the site to climate station i, and v and v_i are the NDVI of the estimate site and the climate station i, respectively, and R is the regression coefficient for NDVI, and n is the number of climate stations used for the interpolation.

Assessment criteria

The outcome of the Chinese Countrywide Water Resource Planning (CWRP) includes the annual rainfall of the Third Grade Water Resources Regionalization (TGWRR) intersected by the prefectural-level regions, which are regarded as the authorities in the Chinese water resources sector. So, the data are compared with results of each interpolation method for assessment.

Hydrological model response to rainfall inputs

The method was applied in the hydrological simulation on the source water area of the West-route of the South-to-North Water Transfer Engineering. A module for NDVI-based GRDS interpolation method was added to the WEP (Jia, Y. *et al.*, 2001).

RESULTS

Both IDW and NDVI-based GRDS methods were used for the precipitation interpolation from 1956 to 2000. At first, the study area was divided into 5 km × 5 km units; there was a total of 24 378 units. And then the centre sites' precipitation were estimated by the two methods, which presented each whole unit. Finally, the average rainfalls of each TGWRR intersected by the prefectural-level region were calculated as the average of its units' rainfalls. The average rainfalls from 1956 to 2000 are listed in Table 2.

We can easily differentiate between the interpolation methods according to the errors. The difference between the results of NDVI-based GRDS method and the outcome of the CWRP is not too great, which shows that the NDVI-based GRDS method is credible.

Table 2 The result of interpolation and Countrywide Water Resources Planning.

Region number	Results of CWRP (mm)	Interpolation results(mm)		Relative error (%)	
		RDS	NDVI-based GRDS	RDS	NDVI-based GRDS
1	775.5	713.2	801.2	–8.0	3.3
2	940.4	738.7	815.1	–21.5	–13.3
3	1519.7	1418.3	1537.1	–6.7	1.1
4	1561.7	1093.5	1205.3	–30.0	–22.8
5	1200.7	1045.3	1160.9	–12.9	–3.3
I Dadu River	961.0	805.0	893.2	–16.2	–7.1
6	946.1	837.5	924.1	–11.5	–2.3
7	1254.3	1149.6	1260.4	–8.3	0.5
8	1488.0	1289.5	1406.7	–13.3	–5.5
II Qingyi River and Minjiang River	1258.1	1100.7	1205.5	–12.5	–4.2
9	936.2	854.6	937.1	–8.7	0.1
10	871.4	791.1	886.8	–9.2	1.8
11	900.7	935.8	1030.5	3.9	14.4
12	993.0	939.4	1038.0	–5.4	4.5
13	1020.0	927.6	1030.2	–9.1	1.0
14	969.6	971.1	1071.0	0.2	10.5
15	937.8	889.9	982.5	–5.1	4.8
III Lower reach under Shigu	960.4	894.3	992.5	–6.9	3.3
16	340.0	322.8	323.2	–5.1	–5.0
17	357.7	363.7	382.9	1.7	7.1
IV Tongtian River	353.7	354.5	369.6	0.2	4.5
18	825.8	721.2	803.3	–12.7	–2.7
19	1144.1	985.8	1091.0	–13.8	–4.6
V Yalong River	848.2	739.8	823.6	–12.8	–2.9
20	617.0	567.0	637.2	–8.1	3.3
21	945.8	783.5	882.7	–17.2	–6.7
22	722.1	579.8	651.8	–19.7	–9.7
VI Reach from Zhimenda to Shigu	723.2	617.7	694.7	–14.6	-3.9
Whole basin	794.8	710.8	783.3	–10.6	–1.5

Based on the reasonable rainfall interpolation method, the hydrological simulation by the WEP model also showed accurate results.

DISCUSSION AND CONCLUSION

The study area was divided into 22 regions according to TGWRR and the prefecture-level regions, as shown in Table 2. Focusing on the results of NDVI-based GRDS interpolation, we analysed the errors. Among the 22 regions, there were eight regions where NDVI-based GRDS interpolation errors were above 5%. Number 4 region had the highest error, –22.8%, because there is only one meteorological station within it and sharp spatial variation nearby. Number 2 region also has a high error because few stations are present. These two large errors caused a small estimation in Dadu River TGWRR. The rainfall estimation in Qingyi and Minjiang rivers TGWRR showed favourable results. The estimation values were much higher than the results of CWRP in the lower reach under Shigu TGWRR, which may be caused by the locally developed irrigation agriculture. The dependent relation between NDVI and precipitation changed for human irrigation areas. Good results were also obtained in the Tongtian River TGWRR and Yalong River TGWRR. The errors were remarkably decreased by substituting NDVI-based GRDS for RDS in reach from

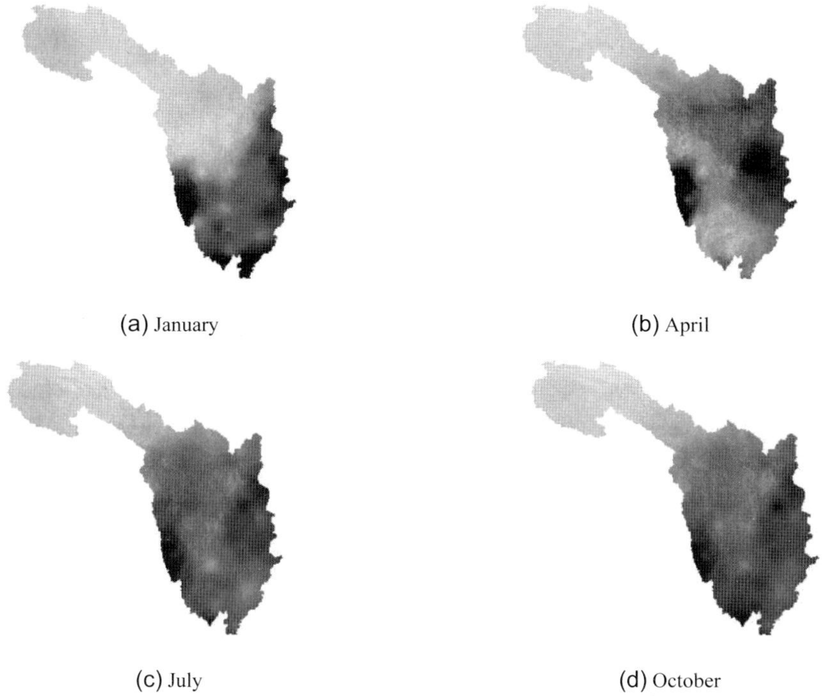

Fig. 3 Four typical month precipitation distributions in 2000.

Zhimenda to Shigu TGWRR. In sum, the results of NDVI-based GRDS were more accurate than that of RDS.

The spatial distribution of precipitation is shown in Fig. 3 for different months. The results were reasonable in the upper reaches of the Yangtze River, which reflects the trend of the rainfall distribution. So the precipitation NDVI-based GRDS interpolation method can be input to the WEP hydrological model with confidence. The verification of simulated monthly discharges at Xiaodeshi station is shown in Fig. 4. Because of limited input data, the WEP hydrological model was run with a daily time step from 1 January 1985 to 31 December 2000. The data from 1985–1995 are used for the calibration of the model, and 1996–2000 for verification.

Firstly, the relative error was determined, which is given by:

$$\delta_B = \left| \frac{\sum_{i=1}^{n} (Q_{si} - Q_{oi})}{\sum_{i=1}^{n} Q_{oi}} \right| \times 100\% \tag{1}$$

where n equals the total number of observation, Q_{oi} is the observed value on time step i, and Q_{si} is the modelled value on time step i.

Secondly, the Nash-Sutcliffe model efficiency (Nash & Sutcliffe, 1970) was determined, which is given by:

$$R^2 = 1.0 - \frac{\sum_{i=1}^{n} (Q_{si} - Q_{oi})^2}{\sum_{i=1}^{n} \left(Q_{oi} - \overline{Q_o}\right)^2} \tag{2}$$

The values range between $-\infty$ and 1, and the higher the value, the more efficient the calibration.

The results in Fig. 4 indicate that, during the calibration period, the relative error is 0.94%, the Nash-Sutcliffe efficiency coefficient is 0.78, and during the verification period the relative error is –3.20%, and the Nash-Sutcliffe efficiency coefficient is 0.86.

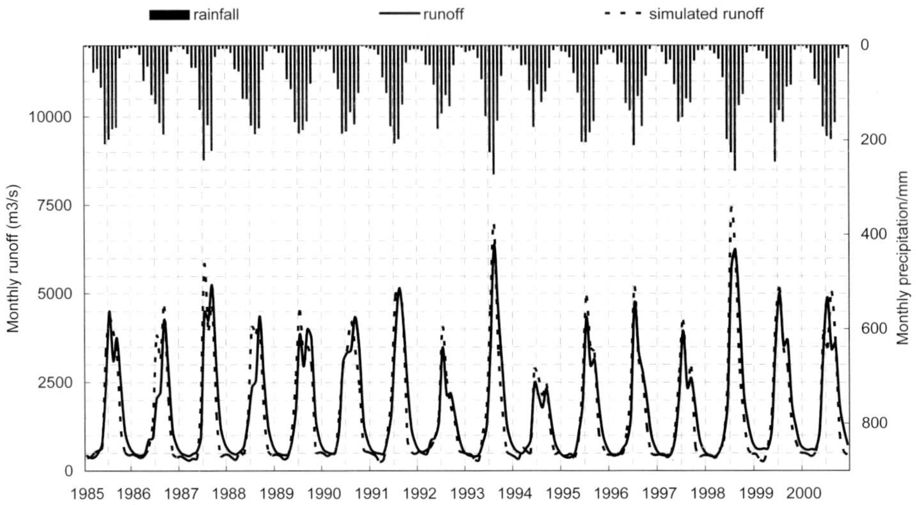

Fig. 4 Verification of simulated monthly discharges at Xiaodeshi station.

As GIS technology develops further and is applied more widely, it is possible to use the relevant information properly for precipitation interpolation. This study incorporated NDVI into GRDS interpolation, and then applied the method to the hydrological model, WEP; the results proved to be satisfactory.

The growth processes of vegetation have some relation to precipitation and temperature. So the distribution of vegetation can, to some extent, reflect the precipitation of a given district. Finding a more accurate relationship between them may be useful and practical. For example, it can be used in Predictions in Ungauged Basins (PUB) in the future.

Acknowledgement This paper was supported by the Chinese National Key Science and Technology Supporting Project "Assessment and regulation on the effects of the West route of the South-to-North water transfer project on its water source area" under Grant no. 2006BAB04A08 and the Social Common-wealth Research Program of the Sciences and Techniques Ministry "Quantified water resources assessment techniques on the source water area of West Route of the South-to-North Water Transfer Engineering" under Grant no. 2005DIB3J057. The authors wish to thank all collaborators.

REFERENCES

Chen Linzhi & Wang Zuwang (1999) *The Impact of Human Alteration on Ecosystem Diversity*, 267–277. Zhejiang Science and Technology Press, Hangzhou, China (in Chinese).

Chia-Ling Chang, Shang-Lien Lo & Shaw-L Yu (2006) The parameter optimization in the inverse distance method by genetic algorithm for estimating precipitation. *Environ. Monitor. Assess.* **117**, 1–528.

Dawdy, D. R. & Bergman, J. M. (1969) Effect of rainfall variability on stream flow simulation. *Water Resour. Res.* **5**(5), 958–966.

Defries R. S. & Townsend J. R. G. (1994) NDVI-derived land cover classification at a global scale. *Int. J. Remote Sensing*, **15**, 3567–3586.

Ding Mingjun, Zhang Yili & Liu Linshan (2007) The relationship between NDVI and precipitation on the Tibetan Plateau. *J. Geogr. Sci.* **17**(3), 259–268.

Dubrule, O. (1983) Two methods with different objectives: splines and kriging. *Math. Geology* **15**(2),245–257

Fang Jingyun, Piao Shilong, Field, Christopher B. *et al.* (2003) Increasing net primary production in China from 1982 to 1999. *Frontiers in Ecol. Environ.* **6**(8), 293–297.

Goward, S. N., Markham, B. *et al*. (1991) Normalized difference vegetation index measurements from the Advanced Very High Resolution Radiometer. *Remote Sens. Environ.*, **35**,257–277

Jia. Y., Ni. G., Kawahara. Y. *et al.* (2001) Development of WEP model and its application to an urban watershed. *Hydrol. Processes* **15**(11), 2175–2194.

Li Xiaobing &Shi Peijun (2000) Sensitivity analysis of variation in NDVI, temperature and precipitation in typical vegetation types across China. *Acta Phytoecologica Sinica* **24**(3), 379–382 (in Chinese).

Liu Shuzhen, Zhou Lin, Chou Chongshan *et al.* (1999) *Studies on Grassland Degradation and Desertification of Naqu Prefecture in Tibet Autonomous Region*, 73–79. Tibet People's Press, Lhasa, Tibet (in Chinese).

Nash, J. E. & Sutcliffe, J. E. (1970) River flow forecasting through conceptual model. *J. Hydrol*. **10**, 282–290.

Nicholson, S. E., Davenport, M. L. & Malo, A. R. A. (1990) Comparison of the vegetation response to rainfall in the Sahel and East Africa, using NDVI from NOAA AVHRR. *Climatic Change* **17**, 209–241.

Phillips, D. L., *et al*. (1992) A comparison of geostatistical procedures for spatial analysis of precipitation in mountainous terrain. *Agric. For. Met.* **58**,119–141.

Price, D. T., *et al*. (2000) A comparison of two statistical methods for spatial interpolation of Canadian monthly mean climate data. *Agric. For. Met.* **101**, 81–94.

Purevdorj, T. S., Tateishi, R., Ishiyam, T. *et al*. (1998) Relationships between percent vegetation cover and vegetation indices. *Int. J. Remote Sens.* **19**, 3519–3535.

Schmidt, H. & Gitelson, A. (2000) Temporal and spatial vegetation cover changes in lsraeli transition zone: AVHRR-based assessment of rainfall impact. *Int. J. Remote Sens.* **21**, 997–1010.

Sun Hongyu, Wang Changyao, Niu Zheng *et al*. (1998) Analysis of the vegetation cover change and the relationship between NDVI and environmental factors by using NOAA time series data. *J. Remote Sensing*, **2**(3), 205–210 (in Chinese).

Zhang Yuandong, Xu Yingtao, Gu Fengxue *et al*. (2003) Correlation analysis of NDVI climate and hydrological factors in oasis and desert. *Acta Phytoecologica Sinica* **27**(6), 816–821 (in Chinese).

Zhao Ji (1998) *Physical Geography of China*. Higher Education Press, Beijing, China (in Chinese).

The Bayesian probabilistic flood forecasting model based on AM-MCMC and BP

ZHENG-XIANG XING[1,2], XIAO-FANG RUI[3], FANG-GUI LIU[4] & JIE FENG[3]

1 *College of Water Conservancy and Civil Engineering, Northeast Agriculture University, Harbin 150030, China*
xingzhenxiang01@163.com

2 *Center for Postdoctoral Studies in Agricultural Engineering, Northeast Agriculture University, Harbin 150030, China*

3 *College of Hydrology and Water Resources, Hohai University, Nanjing 210098, China*

4 *College of Civil Engineering and Architecture, Guangxi University, Nanning 530004, China*

Abstract Under the general theoretical framework of the Bayesian Forecasting System (BFS), models of the prior density and likelihood function based on the Back Propagation of error Artificial Neural Net (BP ANN) were built in the paper. The posterior density of flood discharge was obtained by the Markov Chain Monte Carlo simulation method based on the Adaptive Metropolis algorithm (AM-MCMC), and then probabilistic forecasting of flood discharge was made. A Genetic Algorithm (GA) was used to optimise the weights and bias of BP ANN and the initial samples in AM. The results of a case study show that the accuracy of BFS was higher than that of the Xin'anjiang model. Not only the mean of forecast discharge, but also the variance of forecast discharge was given by the BFS based on the AM-MCMC and BP ANN model. The variance of the predictand qualified the uncertainty of forecast, and the uncertainty of the predictand could be considered in the decision-making for prevention of floods.

Key words BFS; adaptive metropolis algorithm; MCMC; BP; probabilistic forecasting

INTRODUCTION

The Bayesian Forecasting System (BFS) is a general theoretical framework for probabilistic forecasting via any deterministic hydrological model (Krzsyztofowicz, 1999). Research results of probabilistic hydrological forecasting are reported in many articles written by Krzsyztofowicz (1999), Kelly & Krzsyztofowicz (2000), and Krzsyztofowicz & Maranzano (2004). Fuzzy Bayesian forecasting was proposed by Huang (1995). The fuzzy prior distribution and the fuzzy likelihood function were used in BFS. Zhang (2006) used a linear perturbation model to describe the likelihood function and an autoregressive model to describe the prior distribution in BFS, then the complexity of solution of Bayesian posterior distribution was decreased to some extent. The complexity was further decreased through Li's (2005) use of the Artificial Neural Network (ANN) to describe the prior distribution and the likelihood function.

In the BFS proposed by Krzysztofowicz (1999), empirical discharge and predictand were all needed for meta-Gauss model and normal-linear assumption to obtain a nonlinear relationship of prior density and likelihood function, then a posterior density in transformed space was obtained. Therefore, the complexity of the meta-Gauss model and normal-linear assumption made it hard to apply widely. Li (2005) used ANN to describe prior density and likelihood function successfully in flood forecasting of the Shuangpai reservoir in Hunan province, China. But there were still two disadvantages: (i) a convergence rate of traditional ANN, related to initial weights and bias, was usually slow, (ii) a certain distribution and sample interval of parameters were specified in traditional Metropolis-Hasting algorithm, which was not able to ensure ergodicity and rapid convergence of the sampling process. So in the paper, the Markov Chain Monte Carlo based on the Adaptive Metropolis algorithm (AM-MCMC) and the BP ANN based on Genetic Algorithm (GA) were used in BFS to research the probabilistic flood forecast. The GA was used to optimise weights and bias of BP ANN, which made convergence rate of BP train rapidly. The AM-MCMC did not depend on a distribution specified in advance and prior intervals of parameters, which ensured ergodicity and fast sampling efficiency of MCMC.

BAYESIAN PROBABILISTIC FORECASTING THEORY

Under the hypothesis that a discharge process at the outlet of a basin is considered as a Markov process, set $Q_0 = \{Q_t, Q_{t-1}, \cdots, Q_{t-p+1}\}$ is a previous observed discharge process at time *t*; Q_n is a discharge process to forecast; S_n is a discharge process of output of a certain hydrological model (in the paper the certain deterministic hydrological model is Xin'anjiang Model); q_0, q_n, s_n are the realization of Q_0, Q_n, S_n; *n* is a forecast period, and $n = 1, 2, \ldots, N$.

The posterior density of q_n can be obtained via the Bayesian formula as follows:

$$\phi(q_n \mid s_n, q_0) = \frac{f(s_n \mid q_n, q_0) g(q_n \mid q_0)}{l(s_n \mid q_0)} = \frac{f(s_n \mid q_n, q_0) g(q_n \mid q_0)}{\int_{-\infty}^{\infty} f(s_n \mid q_n, q_0) g(q_n \mid q_0) dq_n} \tag{1}$$

where $\phi(q_n \mid s_n, q_0)$ is a posterior density of q_n; $g(q_n \mid q_0)$ is a prior density of q_n, which is given at forecast time and related to q_0; $f(s_n \mid q_n, q_0)$ is a likelihood function of q_n when s_n was given, which indicates the forecast capacity of the hydrological model.

The prior density and the posterior density of discharge in formula (1) can be described with BP ANN as follows (Li, 2006):

The BP ANN structure of the prior density of discharge is described in the following:

$$Q_n = G(Q_n \mid Q_0) + \Xi_n \tag{2}$$

where *G* is a nonlinear reflection of a prior density of discharge, Ξ_n is residual, which obeys a normal distribution $N(0, \xi^2)$ with mean zero and variance ξ^2.

A prior density of discharge was usually described as an ANN with one-hidden-layer (Fig. 1). Then the following equations (3)–(4) are derived according to (2), and take the form:

$$E(Q_n \mid Q_0 = q_0) = G(q_0) \tag{3}$$

$$Var(Q_n \mid Q_0 = q_0) = \varepsilon^2 \tag{4}$$

Then a prior density of discharge is described by the following normal distribution:

$$g(q_n \mid q_0) = \frac{1}{\sqrt{2\pi}\varepsilon} \exp(-\frac{(q_n - G(q_0))^2}{2\varepsilon^2}) \tag{5}$$

The BP ANN structure of the likelihood function of discharge is described in the following:

$$S_n = F(S_n \mid q_n, q_0) + \Theta_n \tag{6}$$

Where *F* is a nonlinear reflection of likelihood function as BP net, the distribution of Θ_n is a normal distribution $N(0, \theta^2)$ with mean zero and variance θ^2.

The likelihood function of discharge can usually be described as an ANN with three layers (Fig. 2). The conditional mean and variance of output S_n are obtained from (6):

$$E(S_n \mid Q_n = q_n, Q_0 = q_0) = F(q_n, q_0) \tag{7}$$

$$Var(S_n \mid Q_n = q_n, Q_0 = q_0) = \theta^2 \tag{8}$$

Then the likelihood function of discharge is described with following normal distribution:

$$f(s_n \mid q_n, q_0) = \frac{1}{\sqrt{2\pi}\theta} \exp(-\frac{(s_n - F(q_n, q_0))^2}{2\theta^2}) \tag{9}$$

Incorporating prior density $g(q_n \mid q_0)$ and likelihood function $f(s_n \mid q_n, q_0)$ into (1), the posterior density of discharge $\phi(q_n \mid s_n, q_0)$ could be solved. But it is usually difficult to get its analytic formula because the normalization constant *l* is hard to get without being given the interval of variate q_n. Therefore, the MCMC based on AM (AM-MCMC) is used to solve the problem above.

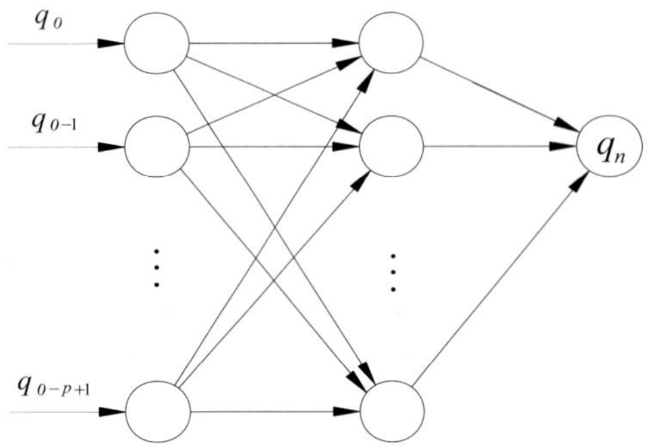

Fig. 1 BP ANN structure of the prior density.

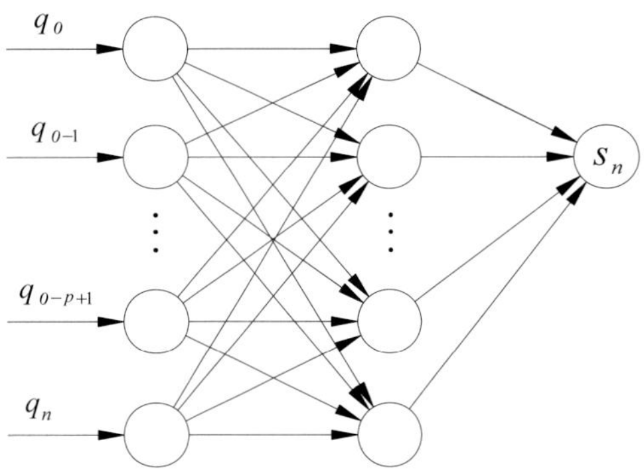

Fig. 2 BP ANN structure of the likelihood function.

AM-MCMC ALGORITHM

As a simulation method, MCMC had been used in several fields such as physics, astronomy, meteorology, communication. The key of MCMC is how to choose a transfer distribution to make sampling more effective. Generally, sampling methods are Metropolis-Hastings (M-H), Gibbs and Adaptive Metropolis (AM). Among these methods, AM does not depend on a specified transfer distribution and has a faster convergence speed, and is used in this paper.

AM algorithm

AM was one of the sampling methods for MCMC, which was put forward by Haario (2001). Compared with M-H and Gibbs, AM did not depend on a specified transfer distribution of variate, but on covariance of initial samples. In AM, the transfer distribution is defined as a multi-dimension normal distribution in parameter's space, and the initial covariance is determined by prior information. During the course of sampling, the transfer density (covariance matrix) is adjusted adaptively according to sample information, so the calculation speed and the convergence rate of MCMC are improved. The initial samples play an important part in sampling capability of AM, so GA was taken to optimize the initial samples.

A sample q^t can be obtained from a transfer distribution $q_t(\cdot \mid q^0, q^1, \cdots, q^{t-1})$ which is obtained according to the given samples q^0, q^1, …, q^{t-1} at time t. And the sample q^t is judged whether to be accepted or not by an acceptance probability described in equation (10):

$$\alpha(q^t, q^{t-1}) = \min\{1, \frac{f(s_n \mid q_n^{t}, q_0) g(q_n^{t} \mid q_0)}{f(s_n \mid q_n^{t-1}, q_0) g(q_n^{t-1} \mid q_0)}\} \tag{10}$$

The transfer distribution in AM algorithm is defined as a normal distribution with mean q_{t-1} and covariance $Cov(q^0,q^1,\cdots,q^{t-1})$. And the covariance is calculated through equation (11). The C_i is set as a constant as C_0 when the iteration number $i \le t_0$, then C_i is updated adaptively.

$$C_i = \begin{cases} C_0\text{iiiiiiiiiiiiiiiiiiii} \qquad \text{iiiiiii} & i \le t_0 \\ s_d(Cov(q^0,q^1,\cdots,q^{t-1})) + s_d\varepsilon\, I_d\text{iiii} & i > t_0 \end{cases} \tag{11}$$

where ε, a lesser positive number, ensuring C_i is not a singular matrix; s_d is a scaling factor, which depends on dimension d, ensuring the acceptance probability within a reasonable range and I_d is a unit matrix and t_0 is iteration times in initial sampling period. Gelman (1992) suggested s_d would be 2.4^2/day.

The covariance C_{i+1} is deprived from equation (12):

$$C_{i+1} = \frac{i-1}{i}C_i + \frac{s_d}{i}(i\bar{q}_{i-1}\bar{q}_{i-1}^{\ T} - (i+1)\bar{q}_i\bar{q}_i^{\ T} + q_i q_i^{\ T} + \varepsilon I_d) \tag{12}$$

where $\bar{q}_{i-1}$ and $\bar{q}_i$ are mean of samples before iteration i – 1 and i, respectively; $q_i^{\ T}$ is the transpose matrix of q_i.

In this way, the sampling mechanism of AM is related to historical information q_0, q_1, …, q_{t-1}. The convergence and the ergodicity of AM algorithm were proved by Haario (2001).

CASE STUDY

There are empirical data of 21 floods in the Yanduhe basin of the Three Gorges of the Yangtze River. The AM-MCMC was used to get the solution of the Bayesian posterior distribution of discharge. Twenty-one floods were selected for training and testing of the BP net of prior distribution and likelihood function, the other 10 floods were used for flood forecasting.

Prior density and likelihood function of discharge

The order of autocorrelation for the Markov process of floods data p, was determined by the partial correlations method. Then it was known that $p = 3$ for the Yanduhe basin. The number of hidden-layers of BP ANN was set to 5, which was optimized by the Genetic Algorithm, so the structure of ANN of prior density was 3-5-1; for the same reason, the structure of ANN of likelihood function of likelihood function was 4-5-1.

Solution to posterior density of discharge

For the forecast period $n = 1$, 11 floods were forecasted by BFS based on AM-MCMC. Ten thousand samples at each moment for one flood were simulated by AM-MCMC, the mean of samples was treated as the prediction of discharge at each forecast time (including the peak of discharge) at each moment. The flood peak, the 80% confidence interval, the error of forecasting flood peak, the deterministic coefficient and the flood peak delay-time are shown in Table 1 (XAJ stands for Xin'anjiang Model).

Shown as Table 1, the accuracy of flood peak, the deterministic coefficient and the flood peak delay-time of floods forecasted by BFS were higher than that of floods forecasted by the Xin'anjiang Model. The errors of flood peak were less than 10% except for three floods (870622, 870627, 870719); the deterministic coefficients were greater than 90% except for only one flood (870622). The cause of the forecasts of floods 870622 and 870627 being a little greater than that of the Xin'anjiang Model might be a result of the poor posterior information provided by forecasts of the Xin'anjiang Model, which proved the posterior information was very important to the forecast of BFS.

Table 1 Results of forecast flood discharge by BFS.

Flood code	Empirical discharge (m^3/s)	Forecast discharge (m^3/s)		Confidence interval of 80%	Error of flood peak (%)		Deterministic coefficient		Peak delay-time (h)	
		XAJ	BFS	BFS	XAJ	BFS	XAJ	BFS	XAJ	BFS
840612	632	524	606	(588, 626)	−17.1	−4.06	0.88	0.98	−1	0
840723	1060	990	992	(731, 1010)	−6.65	−6.44	0.94	0.97	0	1
850603	235	257	256	(234, 278)	9.32	8.87	0.76	0.98	−4	−4
850621	476	499	517	(496, 538)	4.81	8.63	0.93	0.99	−1	−1
860714	227	233	246	(224, 268)	2.9	8.5	0.74	0.96	−1	−1
860909	844	848	883	(861, 907)	0.51	4.69	0.95	0.99	0	0
870511	341	251	406	(384, 428)	−26.4	0.19	0.86	0.95	−1	0
870622	316	398	431	(409, 453)	25.89	36.43	0.60	0.81	−1	1
870627	367	288	453	(431, 475)	−21.4	23.5	0.86	0.96	0	0
870719	819	945	903	(879, 926)	15.4	10.2	0.92	0.98	1	1
870821	556	388	545	(523, 567)	−30.2	−1.9	0.79	0.96	0	1

VALIDATION OF ERRORS

The purpose of validation was to check whether the residual of (2) and (6) fits a normal distribution or not. The chi-square test method was used for the residual error serials of the two models (2) and (6). For forecast period $n = 1$, the null hypothesis was accepted at a 95% confidence level, i.e. the residuals Ξ and Θ fit normal distributions well, Ξ¡«$N(0,19.5^2)$, Θ¡«$N(0,22.2^2)$. The fit of the residual and theoretical distribution are illustrated in Fig. 3.

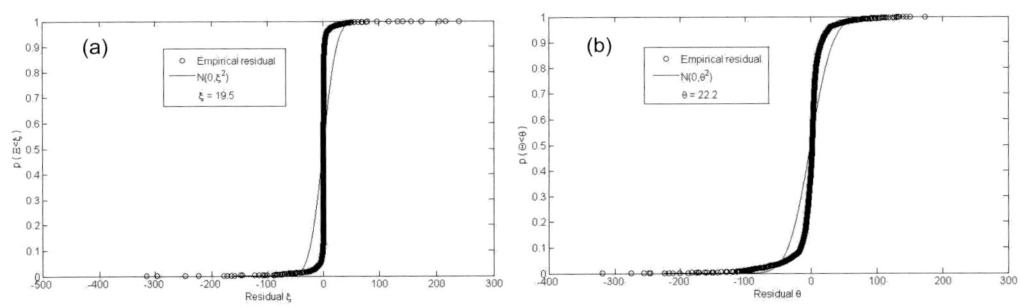

Fig. 3 Validation of residuals: (a) normality of residuals of prior model; (b) normality of residuals of likelihood function model.

CONCLUSION

A Bayesian probabilistic flood forecasting model based on AM-MCMC and BP ANN is presented. The purpose of use of AM-MCMC is to obtain a posterior distribution of the hydrological variate. BP ANN was used to construct a prior density and a likelihood function of BFS. The Bayesian probabilistic flood forecasting model based on AM-MCMC and BP ANN was tested on data for the Yanduhe basin in the Three Gorges of Yangtze River. The following conclusions were drawn on the test:

(i) The BP ANN can capture the nonlinear reflection of hydrological process well. The BP based on GA can make the rate of training convergence of BP of prior density and likelihood function speeded, and the efficiency of BP is improved.

(ii) A transfer distribution for parameters is not needed to be specified in advance in AM algorithm, which did not depend on experience and had a good ergodicity. The AM-MCMC could obtain the posterior mean process of flood discharge as forecast process. On the whole, the accuracy of probabilistic forecast is higher than that of Xin'anjing Model.

(iii) Not only discharge but also variance of the discharge at each moment is given by the Bayesian probabilistic flood forecasting model. The variance of predictand can be used to estimate the uncertainty of flood forecast, which can help decision-makers to consider the risk of flood control determination.

Acknowledgements Research leading to this article was funded from: (1) The Chinese Postdoctoral Science Foundation (20094511116), (2) the Postdoctoral Foundation of Heilongjiang province (LBH-Z08255), (3) Foundation of Heilongjiang Province Educational Committee (11451022), (4) the National Natural Science Foundation of China (50609005), (5) the Fok Ying-Tong Education Foundation for Young Teachers in the Higher Education Institutions of China (101075), (6) the Science Research Funds of Northeast Agricultural University (190125).

REFERENCES

Gelman, A. & Rubin, D. B. (1992) Inference from iterative simulation using multiple sequences. *Statistics Sci.* **7**(4), 457–511.

Haario, H. Saksman, E. & Tamminen, J. (2001) An adaptive metropolis algorithm. *Bernoulli* **7**(2), 223–242.

Huang, W. J. (1995) Application of grey fuzzy Bayesian theory on hydrology and water resources. PhD dissertation, Sichuan University, China.

Kelly, K. S. & Krzysztofowicz, R. (2000) Hydrologic uncertainty processor for probabilistic river stage forecasting. *Water Resour. Res.* **36**(11), 3265–3277.

Krzysztofowicz, R. (1985) Bayesian models of forecasted time series. *Water Resour. Bull.* **21**(5), 805–814.

Krzysztofowicz, R. (1999) Bayesian theory of probabilistic forecasting via deterministic hydrologic model. *Water Resour. Res.* **35**(9), 2739–2750.

Krzysztofowicz, R. (2002) Bayesian system for probabilistic river stage forecasting. *J. Hydrol.* **268**(1–4), 16–40.

Krzysztofowicz, R. & Herr, H. D. (2001) Hydrologic uncertainty processor for probabilistic river forecasting: precipitation dependent model. *J. Hydrol.* **249**(1–4), 46–48.

Krzysztofowicz, R. & Maranzano, C. J. (2004) Hydrologic uncertainty processor for probabilistic stage transition forecasting. *J. Hydrol.* **293**(1–4), 57–73.

Krzysztofowicz, R., Drzal, W. J., Drake, T. R., Weyman, J. C. *et al.* (1993) Probabilistic quantitative precipitation forecasts for river basin. *Weather Forecasting* **8**(4), 424–439.

Li, X. Y., Chen, C. T. & Liu, J. Y. (2006) Bayesian probabilistic hydrologic forecast model based on BP ANN. *ShuiLi XueBao* **37**(3), 354–359.

Zhang, H. G. (2006) Application and study on Bayesian probabilistic forecast system. PhD dissertation, Wuhan Uinversity, China.

Impacts of climate change on water resources in the Weihe River basin

XIANGYI DING[1], YANGWEN JIA[1], YAQIN QIU[1], ZUHAO ZHOU[1], CUNWEN NIU[1] & GE GAO[2]

1 *Department of Water Resources, China Institute of Water Resources and Hydropower Research, Beijing, China*
dingxiangyi840318@163.com

2 *National Climate Center, China Meteorological Administration (CMA), Beijing, China*

Abstract The Weihe River is the largest tributary of the Yellow River (with an area of 135 000 km^2), and both water resources shortage and flood problems exist in the basin. On the basis of analysing relevant domestic and international research on impacts of climate change on hydrology and water resources, this paper simulates the impacts of climate change in the next 30 years (2021–2050) in the Weihe River basin using the multi-model average dataset provided by National Climate Center (NCC) based on the World Climate Research Programme's (WCRP's) Coupled Model Intercomparison Project phase 3 (CMIP3) multi-model dataset under the A1B,A2 and B1 scenarios according to the Special Report on Emissions Scenarios (SRES) provided by the Intergovernmental Panel on Climate Change (IPCC) in 2000. Furthermore, after data scale conversion using the spatial interpolation method combining RDS (distance square inverse method) with the Theissen Polygon Method, by applying a distributed hydrological model – Water and Energy transfer Processes model in Large river basins (WEP-L), the change processes of hydrological components both under historical conditions in 1961–1990 and under future scenarios in 2021–2050 in the Weihe River basin are simulated, and the impacts of climate change on water resources are analysed. The research results illustrate: (1) from the view of average annual variation, comparing with the historical average (1961–1990), although the precipitation in the next 30 years (2021–2050) will appreciably increase, due to the temperature increase, the evapotranspiration will generally increase to a larger extent than the precipitation, resulting in a decrease in runoff, and it is unlikely that floods exceeding the maximum historical scale will occur, while it is possible that extreme drought may occur; (2) from the view of average monthly variation, compared with the historical average (1961–1990), the evapotranspiration in every month will increase, the precipitation and runoff will appreciably decrease in flood seasons, while in the remaining seasons, although the precipitation will appreciably increase, the runoff will decrease.

Key words climate change; CMIP3 data; WEP-L; spatial interpolation; water resources

INTRODUCTION

Climate change associated with global warming induced by the increase of carbon dioxide and other radiative trace gases in the atmosphere has been the focus of many scientific investigations for the past several decades. Climate change has significant implications for the environment, ecosystems, water resources and virtually every aspect of human life. Since the climate system is interactive with the hydrocycle, climate change has many impacts on water resources, which include the magnitude and timing of runoff, the frequency and intensity of floods and droughts, rainfall patterns, extreme weather events, and the quality and quantity of water availability. Moreover, evaporation and soil moisture are also affected due to climate change.

The general method of assessing the impact of global climate change is using General Circulation Models (GCMs) to predict the impacts of climate change through the establishment of a variety of scenarios. Benioff (1996) pointed out that the use of GCMs is the only credible way to simulate climate change scenarios. Due to the complexity of climate change, as well as the uncertainty of the structure of various models, the prediction results of different GCMs are not the same, but the trend forecasted by different GCMs is basically consistent (IPCC, 1992). Therefore, in this study we use the change trend estimated by GCMs under future scenarios, i.e. the change between a certain value and the average value, instead of directly using the value estimated.

The general research of climate change impacts on hydrology and water resources is to predict climate change impacts on river runoff by coupling GCM and water balance models. Arnell (1999) simulated the inflow and water consumption change of the world's different basins under different climate change patterns using two climate forecasting models and macro-hydraulic simulation models – HadCM2 and HadCM3. Mimikou (2000) simulated climate change impacts on the average runoff and water quality of central regions in Greece using two climate change models, HadCM2 and UKHI, as well as the hydrological balance model, WBUDG. Matondo (2004) simulated the conditions of water resources and hydrological changes of three basins in Switzerland using 11 GCMs and the hydrological model CLIRUN. Cao (2004) studied the atmospheric water change in north China and its impacts on water resources on the basis of NECP/NCAR's grid data. Wang (2000) analysed the sensitivity of river runoff response to climate change in the major runoff-area of the Yellow River under the climate scenarios assumed, and estimated the greenhouse effects on water resources according to the precipitation and temperature output results of GCMs. Liu (1997) studied possible changes of runoff and evaporation in some basins in China using a monthly water balance model taking the balanced GCMs output as the climate scenarios.

The aim of our research is to estimate the impacts of climate change on future water resources in the Weihe River basin. To achieve this, using the multi-model average dataset under different climate scenarios SRES-A1B, SRES-A2 and SRES-B1 provided by the National Climate Center (NNC) on the base of the World Climate Research Programme's (WCRP's) Coupled Model Intercomparison Project phase 3 (CMIP3) multi-model dataset, after data scale conversion using spatial interpolation method, we apply a distributed hydrological model WEP-L to simulate the change processes of hydrological components both under historical conditions in 1961–1990 and under future scenarios in 2021–2050 in the Weihe River basin, and furthermore, estimate the impacts of climate change on water resources.

DESCRIPTION OF STUDY AREA

Weihe River is the largest tributary of the Yellow River. It originates from Wushu Hill in Weiyuan County of Gansu Province, flows through Gansu Province, Ningxia Province, Shanxi Province, and finally flows into the Yellow River at Shanxi Tongguan. The total area of Weihe River basin is 135 000 km^2, the total length of the main stream is 818 km, and the whole basin is divided into five level-3 water resources areas (Fig. 1).

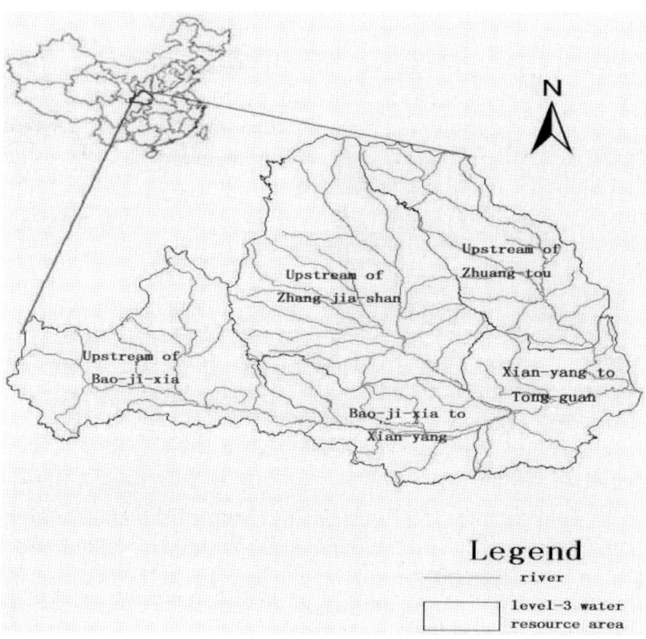

Fig. 1 Map of the Weihe River basin.

The Weihe River basin has a continental monsoon climate, there is more northwest and west wind in cold winter while more southeast wind in hot summer and more east wind in moderate spring and autumn. The annually temperature is 7.8–13.5°C with an extreme maximum temperature 42.8°C and an extreme minimum temperature –28.1°C. The annually average precipitation depth of the Weihe River basin is 572 mm, the annual average evaporation depth is about 700–1200 mm.

METHODOLOGY

Hydrological Model WEP-L

WEP-L (Water and Energy transfer Processes in Large river basins) is a distributed model developed in a National 973 Program of China entitled Evolutionary Laws and Renewable Mechanism of Water Resources in the Yellow River basin.

The horizontal structure of WEP-L model is shown in Fig. 2. Runoff routing on slopes is carried out by applying a one-dimensional kinematic wave approach or dynamic wave approach to trace the overland flow from upstream to downstream. Numerical simulation is made for groundwater flows in mountain and plain areas separately with the consideration of groundwater exchange with surface water, soil moisture and streamflow. The vertical structure of WEP-L model is shown in Fig. 2. From top to bottom, the vertical structure includes nine layers. The "Mosaic" method is applied to take unevenness of land use in the calculation unit into consideration by classifying the land into five groups.

The detailed content of the model is referred to in Jia (2005).

(a)

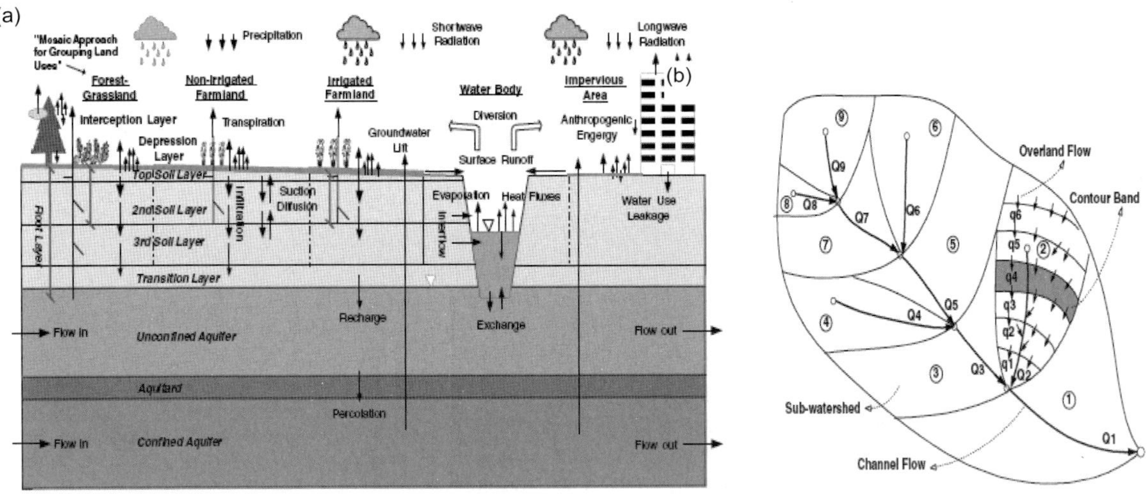

Fig. 2 Schematic illustration of WEP-L model structure: (a) vertical structure within a contour band, and (b) horizontal structure within a sub-basin

WEP-L Application to the Weihe River basin

In WEP-L the Weihe River basin is divided into 1523 sub-basins, each of which is assigned a Pfafstetter code (Fig. 3(a)). Each sub-basin in hilly and tableland areas is further divided into 1–10 contour bands, but no further division is performed for sub-basins in plain areas because of few topographic effects. According to the contour band, the whole Weihe River basin is further discretized into 7440 hydrological response units (HRU) (Fig. 3(b)).

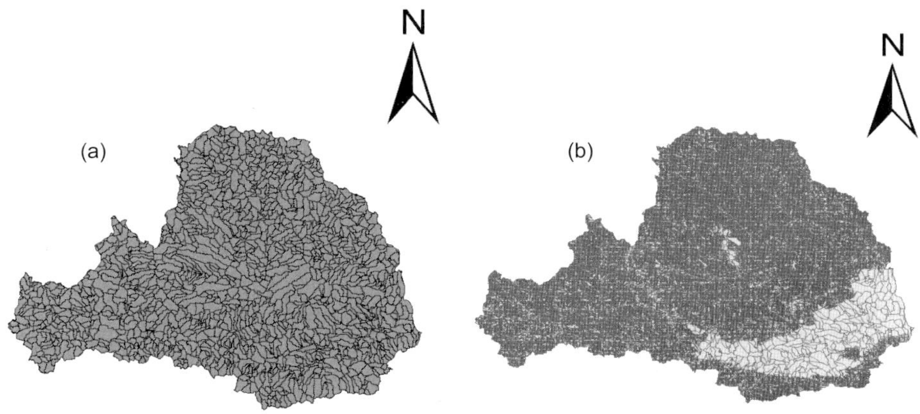

Fig. 3 Subdivision of hydrological response units: (a) subdivision of sub-basins; and (b) subdivision of contour bands in mountainous area.

Model calibration and validation

The model calibration period is 1980–2000, and the time step is 1 day. The model is calibrated using monthly discharge processes observed at three main hydrological stations, Huaxian, Linjiacun and Zhangjiashan. The main calibration parameters include saturated soil moisture hydraulic conductivity, groundwater aquifer hydraulic conductivity and specific yield, infiltration coefficient of riverbed material as well as the Manning roughness. The parameters are adjusted using the trial-and-error method.

After model calibration, keeping all model parameters unchanged, model validation for monthly discharge process is conducted from 1956 to 1979. Only model calibration and validation results at Huaxian station are listed here due to the page limit.

The monthly discharge process at the Huaxian hydrological station is well simulated with a high accuracy (Fig. 4). The simulation results are shown in Table 1. It could be found that the

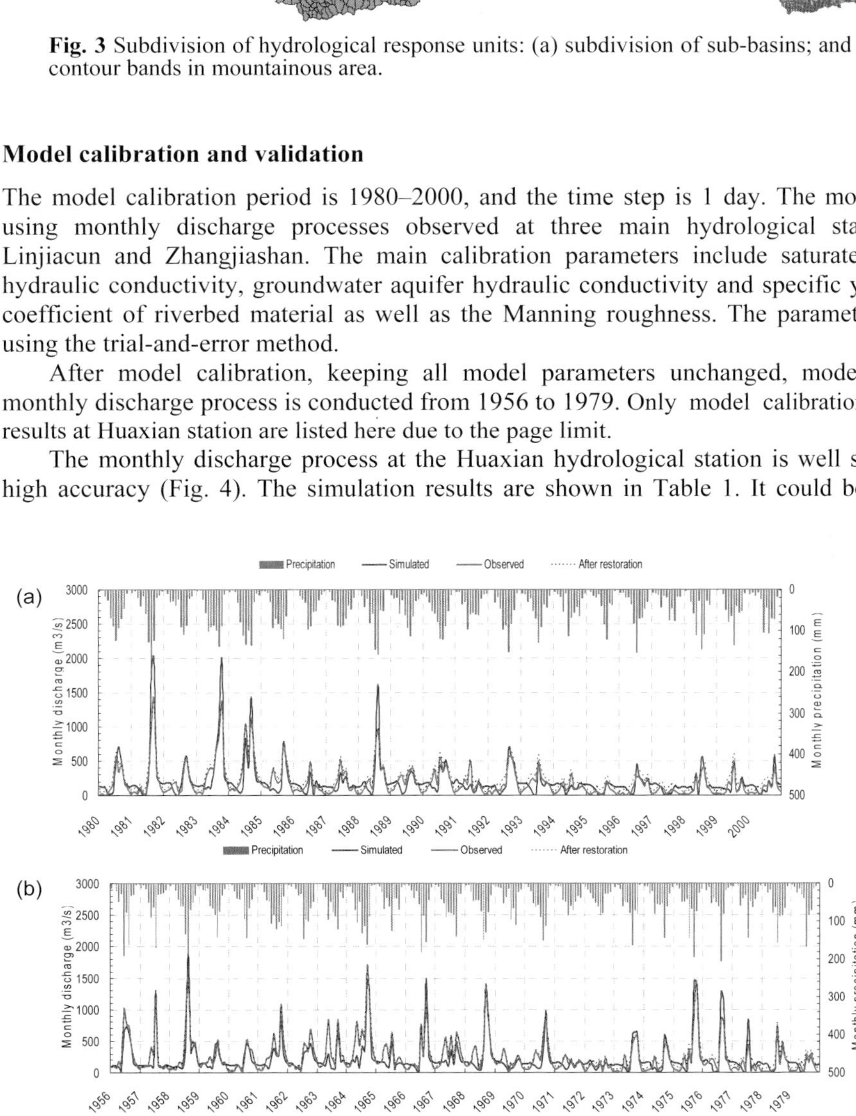

Fig. 4 Comparison of simulated results with observed values at Huaxian station: (a) calibration period, (b) validation period.

Table 1 Simulation results at Huaxian station.

	Observed annual average discharge (after restoration)($10^8 m^3$)	Simulated annual average discharge ($10^8 m^3$)	Relative difference	Nash-Sutcliffe efficiency coefficient
Calibration period (1980—2000)	78.4	79.9	1.9%	0.68
Validation period (1956—1979)	90.6	82.6	–8.8%	0.78

model is well validated in the validation period based on the relative error of annual average discharge and the Nash-Sutcliffe efficiency coefficient, which is calculated using the following equation:

$$NS = 1 - \frac{\sum_{i=1}^{N}[Q_s(i) - Q_o(i)]^2}{\sum_{i=1}^{N}\left[Q_o(i) - \overline{Q_o}\right]^2} \tag{1}$$

where NS is the Nash-Sutcliffe efficiency coefficient, i is the time step, N is the number of time step, Q_s is the simulated discharge, Q_0 is the observed discharge, and $\overline{Q_o}$ is the average discharge.

GCM data and scenarios

The GCM data used in this study are from the World Climate Research Programme's (WCRP's) Coupled Model Intercomparison Project phase 3 (CMIP3) multi-model dataset including estimation results of more than 20 GCMs provided by the Program for Climate Model Diagnosis and Intercomparison (PCMDI). On the basis of the multi-model dataset, NNC provides the multi-model average dataset through interpolation and downscaling calculation under the A1B, A2 and B1 scenarios according to the Special Report on Emissions Scenarios (SRES) provided by the Intergovernmental Panel on Climate Change (IPCC) in 2000. For more details about the dataset and scenarios, please refer to reference NCC (2008).

The specific data used in this study is the precipitation series and temperature series in 1961–1990 and 2021–2050 under the SRES-A1B, SRES-A2 and SRES-B1 scenarios.

Spatial interpolation of meteorological elements

There is currently a lot of research about spatial interpolation methods. In this study, considering interpolation precision, calculation efficiency and other factors, we use the integrated interpolation method combining RDS (distance square inverse method) with the Theissen polygon method. This method is introduced by taking the interpolation method of precipitation as an example, and the interpolation method of other meteorological elements is the same.

Before spatial interpolation, it is necessary to find out the reference sites for each sub-basin. Firstly, calculate the correlation coefficient between each of two sites, the sites that have larger correlation coefficients than the threshold (0.7 in this study) are considered correlated. Secondly, calculate the distance between each site and the most far correlated site as the maximum correlation distance of the site. Thirdly, calculate the distance between each site and sub-basin, and the sites could be selected as the reference site of the sub-basin if the distance is less than the maximum correlation distance. Finally, if correlated sites exist for a sub-basin and there is supporting data, use the RDS method written as in equation (1) to calculate:

$$P = \sum_{i \in I_R} P_i \frac{1/d_i^{\,n}}{\sum_i 1/d_i^{\,n}} \tag{2}$$

where P is the precipitation amount of each sub-basin, P_i is the precipitation amount of site I, d_i is the distance between the site and sub-basin, and n is the coefficient and usually values 2. If there is no correlated site for a sub-basin or there is no data at correlated sites, then the Theissen polygon method is used. The idea is: divide the study area into grids, for each grid, calculate the distance between each site and the centre of the grid, and take the value of the nearest site as the value of the grid.

RESULTS AND DISCUSSION

Average annual change rule

In accordance with the methods above, the annual change of the main hydrological elements including precipitation, evapotranspiration and runoff in the future 30 years (2021–2050) under three scenarios, compared with those under historical conditions (1961–1990), were obtained as shown in Fig. 5. The average annual characteristic values of the main hydrological elements under historical conditions as well as three scenarios in the future are shown in Table 2, including average annual value, maximum annual value and minimum annual value of precipitation, evapotranspiration and runoff.

From Table 2 we can see that, from the view of average annual change, in the future 30 years (2021–2050) under three scenarios, comparing with the historical average, the precipitation will appreciably increase, and the extreme will not obviously vary; the evaporation will generally increase 2–3%; whereas, the runoff will generally decrease 6–16%, and maximum annual runoff in the future will not exceed the historical level. Therefore, in these future 30 years, due to the temperature increase (the temperature is shown in Table 3), although the precipitation will appreciably increase, the evapotranspiration will generally increase to a larger extent than the precipitation, resulting in a decrease in runoff. It is unlikely that floods exceeding the maximum historical scale will occur, but it is possible that extreme droughts may occur.

From Fig. 5 we can see that annual runoff has the biggest fluctuation, followed by annual precipitation, and annual evapotranspiration has the smallest fluctuation. In the years with more precipitation than the historical average, such as 2021, 2024 and 2043, the evapotranspiration will increase by a smaller extent than the precipitation, resulting in the runoff increasing by a larger extent than the precipitation. In the years without big fluctuation compared with the historical average, such as 2026, 2027 and 2031, the runoff will decrease to a larger extent than the precipitation compared with the historical average. In the years with big fluctuations compared with the historical average, such as 2032, 2039 and 2046, the evapotranspiration will not obviously fluctuate, and the runoff will decrease to a large extent. Therefore, in the future, in the years with more precipitation compared with the historical average, the flood scale may be larger; whereas in the remaining years, the drought may be more severe.

Average monthly change rule

The average temperature, average precipitation, average evapotranspiration and average runoff for 12 months both under historical conditions and the three future scenarios are listed in Tables 3–6.

Table 2 Annual average characteristics values of the main hydrological elements under historical conditions as well as three scenarios in the future (unit : mm).

Scenario	Pre_{ave}	Pre_{max}	Pre_{min}	Eva_{ave}	Eva_{max}	Eva_{min}	$Runoff_{ave}$	$Runoff_{max}$	$Runoff_{min}$
History	565.94	780.15	407.21	499.77	601.69	427.53	25.00	67.08	4.84
SRES-A1B	569.27	792.79	419.42	510.9	617.72	440.57	21.17	55.88	2.82
SRES-A2	569.90	713.40	407.18	510.82	592.89	429.53	21.65	65.98	4.76
SRES-B1	580.14	747.16	414.87	516.62	609.95	443.74	23.63	60.04	6.34

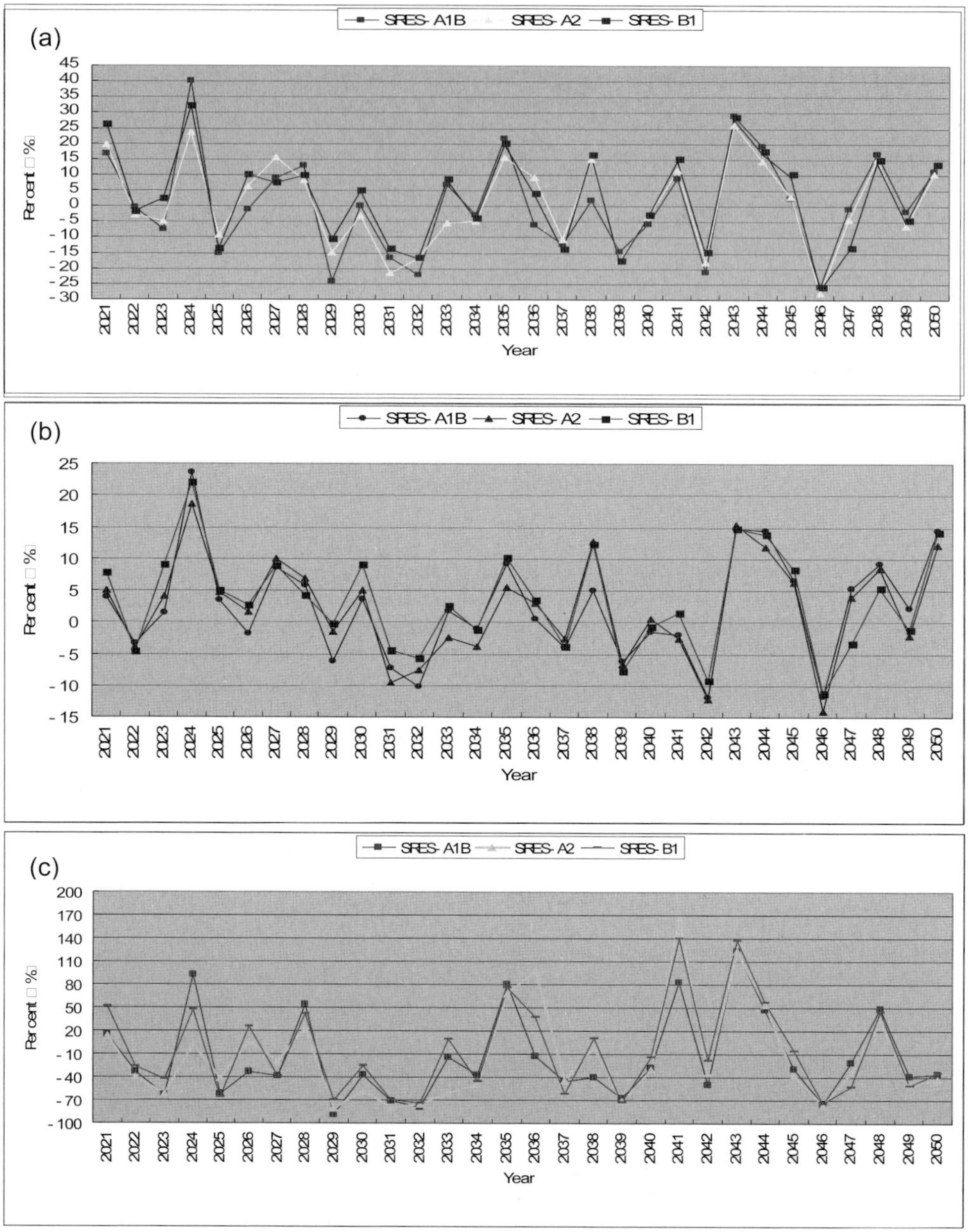

Fig. 5 Changes of average annual values under 3 future scenarios compared with those under historical conditions: (a) annual precipitation, (b) annual evapotranspiration, and (c) annual runoff.

Table 3 Average monthly temperature under different conditions (°C).

Scenario	Jan.	Feb.	Mar.	Apr.	May	Jun.	Jul.	Aug.	Sep.	Oct.	Nov.	Dec.
History	–4.6	–1.6	4.2	10.7	15.7	19.9	21.9	20.6	15.5	9.7	2.9	–2.8
SRES–A1B	–2.6	–0.4	5.6	12.1	17.3	21.4	23.2	22.4	16.9	11.4	4.4	–1.3
SRES–A2	–2.9	–0.4	5.4	12.0	17.2	21.3	23.1	22.1	16.7	11.1	4.2	–1.5
SRES–B1	–3.1	–0.8	5.3	11.7	16.9	21.1	23.0	22.0	16.5	10.9	4.0	–1.8

Table 4 Average monthly precipitation under different conditions (mm).

Scenario	Jan.	Feb.	Mar.	Apr.	May	Jun.	Jul.	Aug.	Sep.	Oct.	Nov.	Dec.
History	4.7	8.1	20.1	39.5	55.0	60.7	108.6	103.0	96.6	47.6	18.1	4.0
SRES-A1B	5.1	8.9	21.8	42.3	58.8	60.8	109.4	100.4	94.3	45.0	18.0	4.6
SRES-A2	5.0	9.3	22.0	43.0	55.7	62.4	108.1	103.5	91.8	45.8	18.9	4.4
SRES-B1	5.0	8.9	21.4	44.7	57.8	61.7	109.6	101.4	97.7	48.1	19.4	4.4

Table 5 Average monthly evapotranspiration under different conditions (mm).

Scenario	Jan.	Feb.	Mar.	Apr.	May	Jun.	Jul.	Aug.	Sep.	Oct.	Nov.	Dec.
History	1.8	6.8	30.8	44.2	58.9	66.9	91.5	82.8	57.9	37.6	16.7	4.0
SRES-A1B	3.2	8.5	29.1	42.8	60.3	68.3	93.1	84.3	59.3	37.7	18.7	5.5
SRES-A2	2.9	8.8	29.7	43.2	60.0	68.4	92.9	84.2	58.8	38.0	18.6	5.5
SRES-B1	3.1	8.5	30.5	44.3	61.0	68.9	93.1	84.6	59.6	38.6	19.2	5.4

Table 6 Average monthly runoff under different conditions (mm).

Scenario	Jan.	Feb.	Mar.	Apr.	May	Jun.	Jul.	Aug.	Sep.	Oct.	Nov.	Dec.
History	0.1	0.16	0.33	0.51	0.84	0.87	3.59	5.04	8.5	4.46	0.49	0.11
SRES-A1B	0.1	0.15	0.3	0.54	1.1	0.9	3.36	4.09	6.57	3.54	0.42	0.11
SRES-A2	0.1	0.16	0.31	0.56	0.78	0.82	3.46	5.0	6.79	3.1	0.48	0.12
SRES-B1	0.1	0.15	0.31	0.62	0.99	0.85	3.77	4.41	7.61	4.16	0.54	0.12

It can be seen from Table 3 that, in the future 30 years, the temperatures in all 12 months will increase by more than 1°C. From Table 4 we can see that, during drought seasons, compared with the historical average, the monthly precipitation will increase, and the biggest increase may be 15%; whereas during flood seasons, the monthly precipitation will decrease and the biggest decrease may be 5%. From Table 5 we can see that, compared with the historical average, in March and April the monthly evapotranspiration will appreciably decrease and the biggest decrease may be 5%, but the monthly evapotranspiration will increase in all the remaining months and the biggest increase may be 81%. From Table 6 we can see that, compared with the historical average, during the flood seasons (June–October), the monthly runoff will decrease and the biggest decrease may be 30%, whereas in April and May, the monthly runoff will increase and the biggest increase may be 30%. In the remaining months, the monthly runoff will not obviously fluctuate.

Therefore, from the view of average monthly change, annual evapotranspiration has the biggest fluctuation, followed by annual runoff, and annual precipitation has the smallest fluctuation. Compared with the historical average (1961–1990), in the future 30 years (2021–2050), evapotranspiration in every month will increase; the precipitation and runoff will appreciably decrease in flood seasons, while in the remaining months, although the precipitation will appreciably increase, the runoff will decrease.

CONCLUSION

In this article the climate change impacts on water resources are studied by using the multi-model average dataset and the distributed hydrological model WEP-L, and some preliminary research results are obtained, such as qualitative analyses on the change trends of main hydrological elements including precipitation, runoff and evapotranspiration under different scenarios in 2021–2050. But at the same time, there are some problems and technical difficulties to which the next step study forwards: (1) the data scale of GCMs and hydrological models is different, which is of great significance and it is difficult to reasonably solve the problem of data scale conversion and couple GCMs with distributed hydrological models; (2) it could be more convincing coupling GCMs with hydrological models aiming to qualitatively analyse the change trends of hydrological components, while how to quantitatively analyse the impacts or how to improve the simulation accuracy to an acceptable extent is still a challenge in future study.

Acknowledgements This study received financial support from the NSFC Creative Research Group Fund Project of China (50721006), the National 11th Five-year Scientific Support Project of China (2006BAB06B06) and the NSFC International Cooperation Fund Project (50721140161). We also acknowledge the modelling groups for providing their data for analysis, the Program for Climate Model Diagnosis and Intercomparison (PCMDI) and the World Climate Research Programme's (WCRP'S) Coupled Model Intercomparison Project for collecting and archiving the model output, organizing the model data analysis activity. The data has been collected, analysed, and are provided by the National Climate Center.

REFERENCES

Arnell, N. W. (1999) Climate change and global water resources. *Global Environmental Change* **9**, 31–49.

Benioff, R., Guill, S. & Lee, J. (1996) *Vulnerability and Adaptation Assessments* (Version 1. 1, An International Handbook). Environmental Science and Technology Library, Kluwer Academic Publishers, Dordrecht, The Netherlands.

Cao, L. Q., Yu, J. H. & Ge, C. X. (2004) Variation of atmospheric water vapor with climate and its impact on water resources of North China. *J. Hohai University (Natural Sciences)* **32**(5), 504–507.

Jia, Y. W., Wang, H., Ni, G. H., Yang, D. W., Wang, J. H. & Qin, D. Y. (2005) *Theory and Practices of Distributed Watershed Hydrological Models.* China Water Resources and Hydropower Publishing, Beijing, China.

Liu, C. Z. (1997) Potential Impact of climate change on hydrology and water resources in China. *Adv. Water Sci.* **8**(3), 220–225.

Matondo, J. I., Peter, G. & Msibi, K. M. (2004) Evaluation of the impact of climate change on hydrology and water resources in Swaziland: Part I. *Physics and Chemistry of the Earth* **29**, 1181–1191.

Mimikou, M. A., Baltas, E., Varanou, E. *et al.* (2000) Regional impacts of climate change on water resources quantity and quality indicators. *J. Hydrol.* **234**, 95–109.

Wan, G. Q., Wang, Y. Z. & Shang, C. K. (2000) Impact of climate change on water resources in the Yellow River. *Yellow River* **22**(9), 40–41.

Zhou, Z. H. Jia, Y. W. & Wang, H. (2006) Station-based temporal and spatial interpolation in large basins. *Hydrol.* **26**(1), 35–43.

Improvement of the distributed hydrological AFFDEF model for runoff simulation

XIANGHU LI[1,2], LILIANG REN[2] & GUIZUO WANG[2]

1 *State Key Laboratory of Lake Science and Environment, Nanjing Institute of Geography and Limnology, Chinese Academy of Sciences, Nanjing 210008, China*
lxh8010@yahoo.com.cn

2 *State Key Laboratory of Hydrology Water Resources and Hydraulic Engineering, Hohai University, Nanjing 210098, China*

Abstract The distributed hydrological model AFFDEF represents physical processes with little detail, but still gives a spatial distributed representation of the watershed. In order to improve the model scheme more closely to the practical processes and integrate more remote sensing data, an exponential equation replaces the original linear function to divide surface and sub-surface flow, moreover, the leaf area index (LAI) based canopy interception model and the Shuttleworth-Wallace two source evapotranspiration model, are incorporated into the model structure. The improved model was tested in the Huangnizhuang watershed from 1 January 1982 to 31 December 1987. The results show that the Nash efficiency in calibration period was 0.75–0.845, the relative runoff depth error was –4.94 to 8.28% and the correlation coefficient was 0.872–0.925. In the validation period the Nash efficiency was 0.826 and 0.792, the relative runoff depth error was 4.13% and –6.92%, and the correlation coefficient was 0.913 and 0.903. The simulated stream flow hydrographs, except for flood peak periods, are in good agreement with the observed ones. Every evaluation criteria was better than their counterparts in the original model. In a word, the modification can be validated to be efficient for the improvement of the original model's performances.

Key words distributed hydrological model; LAI; soil storage capacity; two source evapotranspiration model°

INTRODUCTION

Distributed hydrological models have become the main tool to study hydrology natural law and solve the practical hydrology questions, especially the ones that are physically-based, which can fulfill the necessity of describing basin heterogeneity, assessing the impact of natural and human induced changes and providing detailed descriptions of the hydrological processes in a watershed to satisfy various needs in spatial modelling (Abbott *et al.*, 1996). In the last two decades, a large number of distributed hydrological models have been developed and applied. Examples of such models include the SHE model, the IHDM model, the CASC2D model, the SWAT model, etc. However, owing to their complex structure, a huge number of parameters, and large data requirements, the application of these modelling techniques is not without limitations (Beven *et al.*, 2001). A major limitation in hydrology is the lack of availability of adequate data to quantitatively describe a hydrological process accurately (Pandey *et al.*, 2008). Remote sensing (RS) techniques, which inherently have the ability to provide spatial and temporal information of the land surface, may be the only viable way to obtain the data needed for distributed process models (Engman *et al.*, 1996; Ritchie *et al.*, 1996). The application of RS techniques in hydrological studies and water resources management has progressed in the past decades (Kite *et al.*, 1996). Numerous studies described the use of RS in hydrological modelling (Sekhar *et al.*, 2002; Chowdary *et al.*, 2004). In all these studies, the potential benefits of RS in hydrological and water quality modelling have been clearly demonstrated.

The main objective of this study is therefore to improve and test an alternative distributed hydrological model, the AFFDEF model (Moretti *et al.*, 2007), which represents physical processes with far less detail, but still gives a spatial distributed representation of the watershed. The original version of AFFDEF model is a spatially distributed (grid based), continuous simulation hydrological model. The main characteristic of the model is that long simulation runs can be performed in limited computational times. In order to improve the model scheme more closely to the practical processes and integrate it with more remote sensing data, an exponential equation replaces the original linear function to divide surface and sub-surface flow, moreover, leaf area index (LAI)-based canopy interception model and the Shuttleworth-Wallace two source evapotranspiration model (S-W model) are incorporated into the model structure. The S-W model is developed only using parameter values from the literature, neither experimental measurement nor calibration is introduced, and all input data are publicly available, so that it can be applied to the data-poor or ungauged basins, particularly to the large basins (Zhou *et al.*, 2006). The improved model is tested and compared with the original version in the Huangnizhuang watershed, which lies in the Shiguan River basin in Anhui province, China.

MODEL DESCRIPTION AND IMPROVEMENT

According to the concept of AFFDEF, the watershed is discretized into a number of square grids considering the heterogeneity of the basin's topography. It accounts for interception of vegetation by means of a local "interception reservoir" at every grid. The capacity of the interception reservoir is calculated in direct proportion to the local soil storage capacity, which is computed depending on land use and soil type, according to the Curve Number method (Chow *et al.*, 1988). The rainfall that reaches the ground is split into surface runoff and infiltrated water through the ratio of water content to the capacity of the infiltration reservoir (Moretti *et al.*, 2007). The evapotranspiration in the AFFDEF model is expressed as a linear function of temperature. The surface and sub-surface flow are transferred to the basin outlet by using the Muskingum-Cunge model with variable parameters, which are determined on the basis of the "matched diffusivity" concept (Orlandini *et al.*, 1996). Distinction between hill slope rill and the channel network is based on the concept of "constant critical support area" (Montgomery *et al.*, 1993), and the sub-surface river network is assumed to coincide with the surface one (Brath *et al.*, 2004). Most of the parameters can be estimated from the digital elevation model (DEM), digital soil and land cover maps, and the remainder of the parameters can be automatically optimized by the SCE-UA algorithm. The AFFDEF model has been used in a wide range of applications with good results (Brath *et al.*, 2004; Montanari *et al.*, 2004; Moretti *et al.*, 2007).

Although the basic mathematical equations are kept unchanged, several improvements have been made to the AFFDEF model in the rainfall split equation and evapotranspiration module.

Potential evapotranspiration

Shuttleworth & Wallace (1985) extended the Penman-Monteith method (P-M model) to sparse vegetation to consider two coupled sources in a resistance network: transpiration from vegetation and evaporation from substrate soil. Some researchers applied it successfully in numerous studies and gained good results. As such, the S-W model was incorporated in the model to replace the original evapotranspiration module.

The total evapotranspiration in the S-W model is expressed as (Zhou *et al.*, 2006):

$$\lambda ET = C_c ET_c + C_s ET_s \tag{1}$$

$$ET_c = \frac{\Delta(R_n - G) + \left[(24\times 3600)\rho c_p(e_s - e_a) - \Delta r_a^c (R_n^s - G)\right]/(r_a^a + r_a^c)}{\Delta + \gamma\left[1 + r_s^c/(r_a^a + r_a^c)\right]} \tag{2}$$

$$ET_s = \frac{\Delta(R_n - G) + \left[(24\times 3600)\rho c_p(e_s - e_a) - \Delta r_a^s (R_n - R_n^s)\right]/(r_a^a + r_a^s)}{\Delta + \gamma\left[1 + r_s^s/(r_a^a + r_a^c)\right]} \tag{3}$$

$$C_c = \frac{1}{1 + (R_c R_a)/[R_s(R_c + R_a)]} \tag{4}$$

$$C_s = \frac{1}{1 + (R_s R_a)/[R_c(R_s + R_a)]} \tag{5}$$

$$R_a = (\Delta + \gamma) r_a^a, \quad R_c = (\Delta + \gamma) r_a^c + \gamma r_s^c, \quad R_s = (\Delta + \gamma) r_a^s + \gamma r_s^s \tag{6}$$

where ET is the evapotranspiration (mm/day); λ is the latent heat of water vaporization (MJ/kg); ET_c and ET_s are equivalent to transpiration and evaporation by applying the P-M equation to closed canopy and bare substrate, respectively; C_c and C_s are weighting coefficients; R_n and R_n^S are the net radiation at the reference height and substrate soil surface (MJ/m^2·day); G is the substrate soil heat flux (MJ/m^2·day); e_s and e_a are the saturation and actual vapour pressures (kPa); Δ is the slope of saturation vapour pressure curve (kPa/°C); ρ is the mean air density at constant pressure (kg/m^3); c_p is the specific heat of moist air at constant pressure (MJ/kg·°C); γ is the psychrometric constant (kPa/°C); r_s^c and r_a^c are the bulk stomatal and boundary layer resistances of the canopy (s/m); r_a^s and r_a^a are the aerodynamic resistances between soil and canopy and between canopy and reference height (s/m); and r_s^s is the surface resistance of soil (s/m).

Interception evaporation

The interception storage capacity, I_{max} (mm), is related to the vegetation LAI (Vazquez & Feyen, 2004) as:

$$I_{\max} = C_{\text{int}} \cdot LAI \tag{7}$$

where C_{int} is the interception coefficient, and 0.2 mm is usually used (Mo *et al.*, 2004).

The interception evaporation, $ET_i(t)$, is calculated as:

$$ET_i(t) = \begin{cases} \min[I_{\max}, PET_i(t)] & P(t) > D_{\text{int}}(t) \\ \min[I(t-1) + P(t), PET_i(t)] & P(t) < D_{\text{int}}(t) \end{cases} \tag{8}$$

$$D_{\text{int}}(t) = I_{\max} - I(t-1) \tag{9}$$

where $PET_i(t)$ is potential interception evaporation (mm); $P(t)$ is rainfall (mm); $D_{int}(t)$ is interception deficit (mm); $I(t–1)$ is storage state of interception (mm).

Soil water evapotranspiration

In the root zone, if the soil moisture content is larger than or equal to the water storage capacity, then actual evapotranspiration, $ET(t)$, occurs according to the potential evapotranspiration, otherwise calculated through the relationship:

$$ET(t) = \frac{1 - \exp\{-\alpha_{ET}[F(t)/(H \cdot S)]\}}{1 - \exp(-\alpha_{ET})} \cdot PET(t) \tag{10}$$

where $PET(t)$ is potential evapotranspiration (mm); α_{ET} is an empirical constant (dimensionless).

Rainfall split equation

The rate of the rainfall depth which contributes to the local surface runoff is computed in accordance with the improved relationship:

$$\frac{P_s[t,(i,j)]}{P_n[t,(i,j)]} = \left\{\frac{F[t,(i,j)]}{H \cdot S(i,j)}\right\}^B \tag{11}$$

where $P_s[t, (i, j)]$ is the local surface runoff (mm); $P_n[t, (i, j)]$ is rainfall that reaches the ground (mm); $F[t, (i, j)]$ is the water content of the infiltration reservoir (mm); $H \cdot S(i, j)$ is the capacity of the infiltration reservoir (mm), given by the parameter H (dimensionless) multiplied by the soil storage capacity $S(i, j)$ (mm); B is the parameter (dimensionless).

The soil storage capacity $S(i, j)$ is expressed as:

$$S(i,j) = 254\left(\frac{100}{CN(i,j)} - 1\right) \tag{12}$$

where $CN(i, j)$ is the Curve Number at the given cell location and relates to the gradient, land cover, soil type, antecedent moisture condition, etc.

And, the infiltrated water $I[t, (i, j)]$ is calculated as:

$$I[t,(i,j)] = P_n[t,(i,j)] - P_s[t,(i,j)] = P_n[t,(i,j)]\left[1 - \left\{\frac{F[t,(i,j)]}{H \cdot S(i,j)}\right\}^B\right] \tag{13}$$

At the same time, the outflow from the infiltration reservoir can be expressed as (Moretti *et al.*, 2007):

$$W[t,(i,j)] = \frac{F[t,(i,j)]}{H_S} \tag{14}$$

where $W[t, (i, j)]$ is the outflow from the infiltration reservoir (m^3/s); H_S is the parameter reflect outflow from the infiltration reservoir (h).

So, the mass balance equation for the infiltration reservoir can be written as:

$$\frac{dF[t,(i,j)]}{dt} = -\frac{F[t,(i,j)]}{H_S} - ET[t,(i,j)] + P_n[t,(i,j)]\left[1 - \left\{\frac{F[t,(i,j)]}{H \cdot S(i,j)}\right\}^B\right] \tag{15}$$

Sub-surface kinematic velocity

The sub-surface kinematic velocity c_k is given by Darcy's law (Moretti *et al.*, 2007):

$$c_k = K_{sat} \cdot i_s \tag{16}$$

where i_s is the slope of the sub-surface river network, which is assumed to be equal to the surface river network; K_{sat} is the saturated conductivity of the soil (m/day) and can be calculated as:

$$K_{sat} = K_{sand} \cdot U_{sand} + K_{silt} \cdot U_{silt} + K_{clay} \cdot U_{clay} \tag{17}$$

where K_{sand}, K_{silt} and K_{clay} are the saturated conductivity of the sand, silt and clay (m/day); U_{sand}, U_{silt} and U_{clay} are the percent content of sand, silt and clay in the soil (%).

STUDY AREA AND DATA COLLECTION

Study area

The model was tested in the Huangnizhuang watershed (31.1–31.7°N and 115.25–115.72°E), a drainage area of 805 km^2, which lies in Shiguan River basin, the first-order southern tributary of the Huaihe River in Anhui Province, China. There are four rainfall stations and one discharge station in the watershed. The mean annual temperature of this watershed is 11–16°C and mean annual precipitation is 1077 mm. However, the spatial and temporal distribution is uneven, as precipitation in the rainy season accounts for almost 50–80% of annual rainfall.

The Huangnizhuang watershed contains a range of topographical settings, including mountainous and hilly areas (maximum elevation of 1495 m) where streamflow velocities are high, as well as low alluvial plains where the drainage network is well developed. As such, the Shiguan River basin was selected as the intensified observation field site for the HUBEX (Huaihe River Basin Experiment) project as a component of GAME (the GEWEX Asian Monsoon Experiment), a part of the Global Energy and Water cycle EXperiment (GEWEX).

Data collection

The digital elevation model (DEM) is one of the most important data sets required for the S-W and AFFDEF models. In this study, the DEM is derived from the GLOBE (Global Land One-kilometer Base Elevation) data set with a horizontal resolution of 30 arc seconds. The land cover data is downloaded from Global Land Cover Characteristics Data Base Version 2, using the IGBP classification scheme with 1 km spatial resolution. There are seven land cover categories in the Huangnizhuang watershed, as Fig. 1 shows. The largest area is the woody savannas, which accounts for 39.77%, followed by deciduous broadleaf forest accounting for 26.36%. The cropland/natural vegetation mosaic and closed shrub lands account for 10.7% and 9.12%, respectively, with other categories comprising less than 6%. Table 1 shows some parameters values for various land cover type. The Food and Agriculture Organization (FAO) soil map is also used in this study. The Huangnizhuang watershed is composed of two categories of soil texture, which are represented as the content percentage of sand, silt and clay, as Table 2 shows. Using the USDA soil triangle, the categories are transferred into USDA soil classifications (Fig. 2). The meteorological data such as mean daily temperature, mean diurnal temperature range, cloud cover and vapour pressure are derived from the IPCC Data Distribution Centre at 0.5° grids, and the CRU CL 2.0 dataset provides the wind speed at 10 min grids.

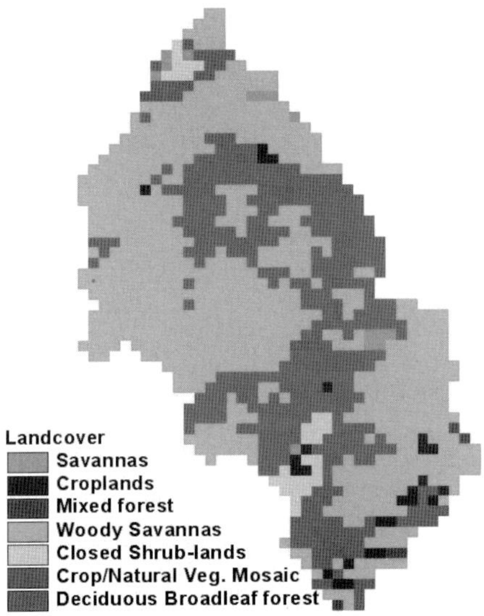

Fig. 1 Land cover map of study area.

Fig. 2 Soil type map of study area.

Table 1 Main parameters values for various land cover type (Zhou *et al.*, 2006).

IGBP land cover type	n	α_m	h_c	w_{max}	Z_{0g}
Savannas	0.065	0.25	0.1~0.8	0.01	0.02
Croplands	0.045	0.20	0.0~0.8	0.01	0.005
Mixed forest	0.146	0.17	20.0~20.0	0.04	0.02
Woody savannas	0.097	0.20	0.8~0.8	0.01	0.02
Closed shrub lands	0.097	0.20	1.5~1.5	0.01	0.02
Cropland/natural vegetation mosaic	0.045	0.20	0.1~0.8	0.01	0.01
Deciduous broadleaf forest	0.146	0.18	25.0~25.0	0.08	0.02

n, Manning roughness; α_m, maximum albedo; h_c, vegetation height (m); w_{max}, maximum vegetation leaf width (m); Z_{0g}, roughness length of substrate ground (m).

Table 2 Soil grain proportion and main parameter values for each soil type.

FAO soil type	Soil grain proportion (%) Sand	Silt	Clay	USDA soil type	Field capacity (cm^3/cm^3)	Wilting point (cm^3/cm^3)
3085	38.05	22.27	39.69	Clay loam	0.318	0.075
3963	45.02	29.73	25.25	Loam	0.27	0.027

SIMULATION RESULTS

Simulation was carried out from 1 January 1982 to 31 December 1987. The first 4 years data were used for calibration and the remaining 2 years data were used for validation. Some of the model parameters were estimated from DEM, digital soil and land cover maps and the remainder were automatically optimized by the SCE-UA algorithm. The Nash-Sutcliffe efficiency (*Nash*) was used as the objective function for the optimization algorithm, moreover, the relative runoff depth error (*BIAS*) and the Pearson correlation coefficient (*R*) were used for evaluation criteria during the calibration and validation periods. The optimized parameter values are listed in Table 3.

Table 3 Calibrated model parameter values for daily streamflow simulation.

Parameter	Value	Parameter	Value	Parameter	Value
A_0	9.0	w_v	8100	α_{ET}	–6.924
H	0.786	w_r	45	K_{sand}	12.5
B	0.869	n_{sr}^0	0.045	K_{silt}	8.5
H_S	210	n_{sr}^1	0.33	K_{clay}	2.5

A_0, critical area (km^2); H, parameter for the infiltration reservoir capacity; B, parameter in the rainfall split exponential equation; H_s, bottom discharge parameter for the infiltration reservoir (h); W_v, channel width/height ratio for the hill slope; W_r, channel width/height ratio; n^i_{sr}, minimum and maximum roughness for the channel; α_{ET}, parameter for evaporation; K_{sand}, K_{silt} and K_{clay}, the saturated conductivity of the sand, silt and clay (m/day).

Table 4 shows the evaluation criteria of model performance. It is seen that the improved AFFDEF model gained a good performance. The Nash efficiency in calibration period was 0.75–0.845, the relative runoff depth error was –4.94 to 8.28% and the correlation coefficient was 0.872–0.925, the average of whole calibration period were 0.821, 0.07% and 0.907, respectively; in the validation period, the Nash efficiency was 0.826, 0.792, and 4.13%, –6.92% for the relative runoff depth error, and the correlation coefficient was 0.913 and 0.903. Figures 3 and 4 show the hydrograph of observed and simulated daily streamflow at the Huangnizhuang station in calibration period and validation period, respectively. It can also be seen that the simulated streamflow hydrographs, except for the flood peak periods, are in good agreement with the observed ones.

Table 4 Comparison of the performance of improved and original AFFDEF model.

Year		Improved AFFDEF			Original AFFDEF		
		Nash	*BIAS* (%)	*R*	*Nash*	*BIAS* (%)	*R*
Calibration period	1982	0.801	–1.48	0.897	0.814	2.66	0.903
	1983	0.845	0.01	0.924	0.805	–4.23	0.907
	1984	0.815	8.28	0.925	0.807	0.37	0.901
	1985	0.750	–4.94	0.872	0.733	–5.06	0.858
	Average	0.821	0.07	0.907	0.804	–1.79	0.898
Validation period	1986	0.826	4.13	0.913	0.747	–5.16	0.884
	1987	0.792	–6.92	0.903	0.779	–8.96	0.893

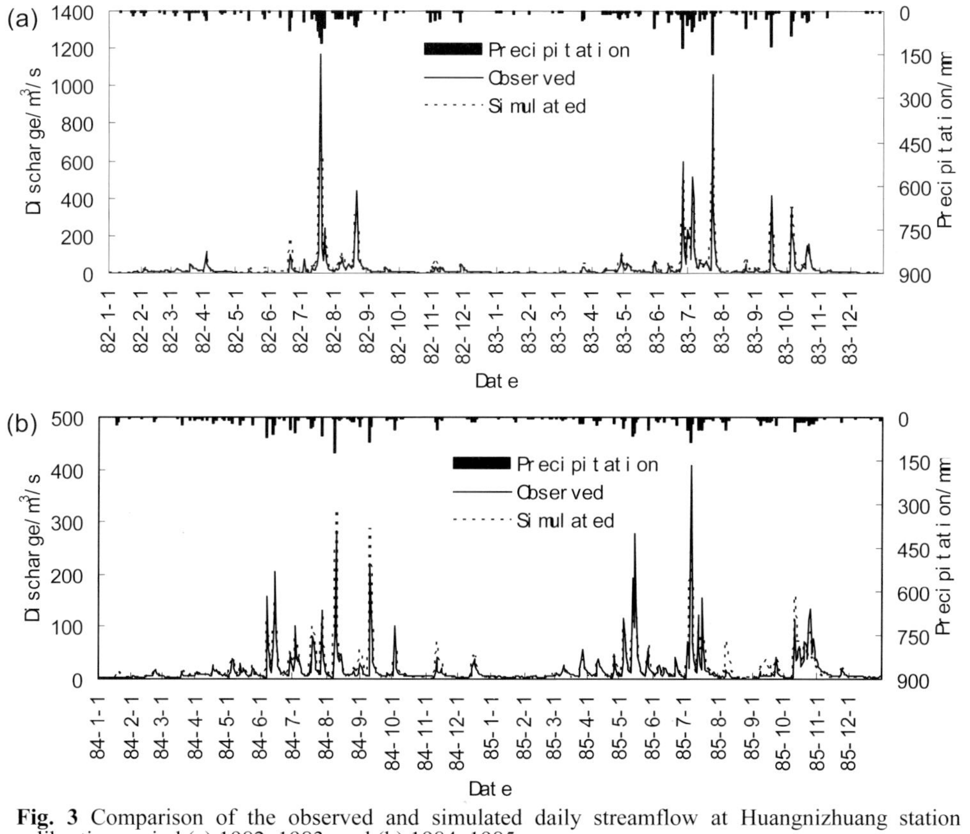

Fig. 3 Comparison of the observed and simulated daily streamflow at Huangnizhuang station in calibration period (a) 1982–1983, and (b) 1984–1985.

In addition, in order to justify the improvement of the model's performance by modifying the rainfall split exponential equation and incorporating the LAI-based canopy interception model and Shuttleworth-Wallace two source evapotranspiration model, simulation results from both improved and original AFFDEF model are compared (Table 4). It can also be seen that in the calibration period, the Nash efficiency and correlation coefficient of improved AFFDEF were almost the same as the ones of the original model, but the relative runoff depth error were much better (except 1984); in validation period, both the Nash efficiency, the relative runoff depth error and the correlation coefficient were better than their counterparts. In summary, the improved AFFDEF model has a better performance than the original model; the modification done in this study can be validated to be efficient.

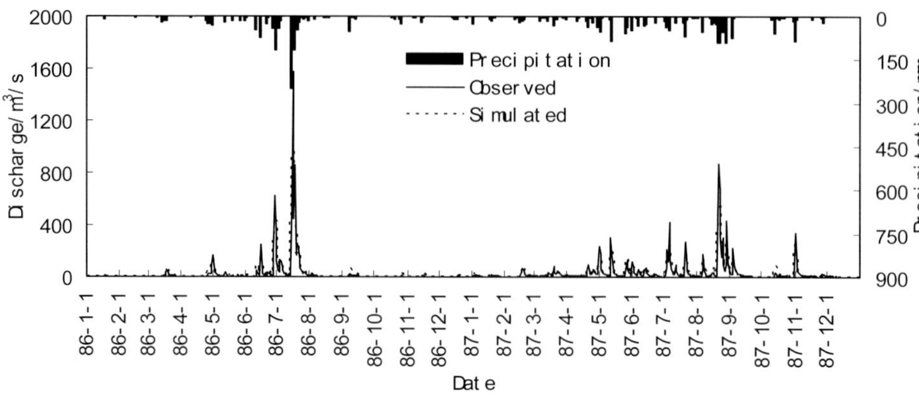

Fig. 4 Comparison of the observed and simulated daily streamflow at the Huangnizhuang station in validation period (1986–1987).

CONCLUSIONS

The AFFDEF model represent physical processes with far less detail than other models, but still gives a spatial distributed representation of the watershed, and long simulation can be performed in limited computational times. This paper made a modification to the rainfall split equation and incorporated the LAI-based canopy interception model and Shuttleworth-Wallace two source evapotranspiration model. The improved model was tested in the Huangnizhuang watershed and gave a good performance. The Nash efficiency in the calibration period was 0.75–0.845, the relative runoff depth error was –4.94–8.28% and the correlation coefficient was 0.872–0.925; in the validation period, the Nash efficiency were 0.826 and 0.792, the relative runoff depth error was 4.13% and –6.92% and the correlation coefficient was 0.913 and 0.903. The simulated streamflow hydrographs, except for flood peak periods, match well with the observed ones. Compared with the original model's performances, not only in calibration period but also in validation period, almost every evaluation criteria of model performance was better than their counterparts.

As a whole, the improved AFFDEF model scheme is closer to the practical processes and incorporates more remote sensing data without adding parameters, and can be validated to be efficient for the improvement of the original model's performances. The modifications made in this study were successful.

Acknowledgements This research was supported by the 111 Project under Grant B08048, Ministry of Education and State Administration of Foreign Experts Affairs, China and the Program for Changjiang Scholars and Innovative Research Team in University under Grant no. IRT0717, Ministry of Education, China.

REFERENCES

Abbott, M. B. & Refsgaard, J. C. (1996) *Distributed Hydrological Modelling*. Kluwer Academic Publishers, The Netherlands.

Beven, K. J. (2001) *Rainfall–runoff Modelling: A Primer*. Wiley, Chichester, UK.

Brath, A., Montanari, A. & Toth, E. (2004) Analysis of the effects of different scenarios of historical data availability on the calibration of a spatially-distributed hydrological model. *J. Hydrol.* **291**, 272–288.

Chow, V. T., Maidment, D. R. & Mays, L. W. (1988) *Applied Hydrology*. McGraw-Hill, Singapore.

Chowdary, V. M., Yatindranath, Y., Kar, S. *et al.* (2004) Modelling of non-point source pollution in watershed using remote sensing and GIS. *J. Indian Soc. Remote Sensing* **32**(1), 59–73.

Engman, E. T. (1996) Remote sensing applications to hydrology: future impact. *Hydrol. Sci. J.* **41**, 637–647.

Kite, G. W. & Pietroniro, A. (1996) Remote sensing applications in hydrological modelling. *Hydrol. Sci. J.* **41**(1), 563–591.

Mo, X., Liu, S., Lin, Z. *et al.* (2004) Simulating temporal and spatial variation of evapotranspiration over the Lushi basin. *J. Hydrol.* **285**, 125–142.

Montanari, A. & Brath, A. (2004) A stochastic approach for assessing the uncertainty of rainfallerunoff simulations. *Water Resour. Res.* **40**(1), doi:10.1029/2003WR002540.

Montgomery, D. R. & Foufoula-Georgiou, E. (1993) Channel network source representation using digital elevation models. *Water Resour. Res.* **29**(12), 3925–3934.

Moretti G. & Montanari A. (2007) AFFDEF: A spatially distributed grid based rainfall–runoff model for continuous time simulations of river discharge. *Environ. Model. Software* **22**, 823–836.

Orlandini S. & Rosso R. (1996) Diffusion wave modelling of distributed catchment dynamics. *J. Hydrol. Engng ASCE.* **1**(3), 103–113.

Pandey, A., Chowdary, V. M., Mal, B. C. *et al.*, (2008) Runoff and sediment yield modelling from a small agricultural watershed in India using the WEPP model. *J. Hydrol.* **348**, 305–319.

Ritchie, J. C. & Rango, A. (1996) Remote sensing applications to hydrology: introduction. *Hydrol. Sci. J.* **41**, 429–431.

Sekhar, K. R. & Rao, B. V. (2002) Evaluation of sediment yield by using remote sensing and GIS: a case study from the Phulang Vagu watershed. *Int. J. Remote Sensing* **23**(20), 4449–4509.

Vazquez, R. F. & Feyen, J. (2004) Effect of potential evapotranspiration estimates on effective parameters and performance of the MIKE SHE-code applied to a medium-size catchment. *J. Hydrol.* **270**, 309–327.

Zhou, M. C., Ishidaira, H., Hapuarachchi, H. P. *et al.* (2006) Estimating potential evapotranspiration using Shuttleworth-Wallace model and NOAA-AVHRR NDVI data to feed a distributed hydrological model over the Mekong River basin. *J. Hydrol.* **327**, 151–173.

Predicting inflows to barrier lakes formed in the Wenchuan earthquake using the WEP model and TRMM-based rain data

YANGWEN JIA[1], ZHIXUAN ZHAO[1,2], DENGHUA YAN[1], HUI PENG[1], CHUNFENG HAO[1] & HUI GAO[1]

1 *Water Resources Department, China Institute of Water Resources & Hydropower Research (IWHR), Beijing 100038, China*
jiayw@iwhr.com; ywjia@yahoo.com

2 *Civil Engineering College, Tianjin University. Tianjin 300072, China*

Abstract Inflow prediction to barrier lakes formed in earthquakes is important to flood disaster warning and prevention, but it is difficult because barrier lakes are usually in mountainous regions and lack gauge data. This study tries to predict inflows to the Tangjiashan barrier lake in the Upper Fujiang River of Mianyang municipality formed during the Wenchuan earthquake on 12 May 2008 by combining the WEP distributed hydrological Model and the TRMM-based precipitation estimates. At first, the DEM data with a scale of 1:250 000 are utilized to generate the digital river system in the study area (12032.25 km^2), and related land cover, soil and historical hydro-meteorological data are collected. Then, five-year daily data of four meteorological stations (one inside and three neighbouring the study area) and four flow gauge stations are applied to validate the WEP model. Lastly, the TRMM three-hourly data from 1 January to 1 July is downloaded to run the model for the prediction of inflow to the barrier lake and lake storage. The application results show an acceptable accuracy, illustrating that the combination of distributed hydrological model and spatial observation and analysis technology are useful for flood disaster warning purposes in the poorly gauged basin.

Key words barrier lake; inflow prediction; WEP model; TRMM; earthquake

INTRODUCTION

Barrier lakes in mountainous areas often result from earthquakes, of which high-risk barrier lakes bring great threats to downstream people's lives and possessions (Nie *et al*., 2004). How to exactly forecast the runoff into barrier lakes is very important for the prevention of floods and the preparation of correlative evacuation. But it is quite a difficult work because barrier lakes are usually in mountainous regions and lack gauge data. Meanwhile, traditional hydrological statistical regressive or lumped hydrological models are less precise because they have difficulty in revealing the impacts of global climate and land surface changes, lead to much uncertainty in runoff forecasting and go against the preparation of prevention and mitigation measures for disasters. If wrong decisions are made based on these inaccurate predictions, there will be great losses. A physically-based distributed hydrological model like WEP (Water and Energy transfer Processes) (Jia *et al*., 2006), may provide profound analysis for the hydrology cycle processes in areas that lack data. Making full use of modern techniques for data observation and data processing, and understanding limited observed data may be the way to help solve the data problem. The physical mechanisms in distributed hydrological models may ensure their precision to some extent, but the rainfall input is still one of the inputs which greatly influence the results, just like the traditional lumped or semi-lumped models. Advantages of distributed hydrological models will be weakened because of this. It may produce even worse results than traditional models if they are without accurate meteorological and rainfall input data (Xu *et al*., 2007). Generally speaking, the observed rainfall data from hydrological gauges are of sufficient precision, but there are only a few hydrological gauges in the mountain areas; observed data from these gauges usually can not offer a satisfactory representative distribution of the rainfall. Also, point observed rainfall data of hydrological gauges can not provide the instantaneous spatial distribution of the precipitation, and the direction of the rainstorm centre also can not be tracked. There are great uncertainties when the point rainfall data are extended to the polygon rainfall data, which are restricted by the methodologies of data assimilation (Arnaud, 2002). Weather radar can catch temporal trends and the spatial distribution of the rainfall in a certain area with its high resolving power, and solve the problems of inadequate representativeness of rain data and distortion after data assimilation to some extent (Zhang *et al.*, 2007). However, the observation precision of precipitation from radar, restricted by current technology, is expected to be improved (Collier, 1996). To satisfy the demand of practice, long hydrological series (1956–2005) are firstly simulated by the distributed hydrological WEP model in this paper, and the model is calibrated and validated with the monthly and daily river flow data of four hydrological gauges from 1976 to 1980. Lastly, the rainfall data from TRMM radar combined with the observed data from hydrological gauges are taken as the data input of the model, the inflows into the Tangjiashan barrier lake are simulated and forecasted to see the applicability of the methodology for the simulation and forecast in the data scarcity areas, and to improve the model to deal with floods in the future.

STUDY BACKGROUND

The Tangjiashan barrier lake in Beichuan county is the biggest formed in the Wenchuan earthquake, which was 3 km away from Beichuan county. Minimum elevation of the lake weir crest is 750.2 m, and the corresponding weir height is 82.8 m. The upstream basin area of the barrier lake is 3550 km^2, with the Jiangjunshi hydrological gauge in its downstream controlling an area of 4162 km^2. The storage below the weir crest is about 3.2×10^8 m^3. Also, the lives and possessions of more than 1 million people downstream would be threatened if the dam breaks.

The Tangjiashan barrier lake is located on Tongkou River, upstream of Fujiang River, which is the biggest branch on the right bank of Fujiang River. Tongkou River originates from Yuleishan Mountain in the southeast of Minshan Mountain in Songpan County, and flows via Baiyang of Songfan, Sier of Pingwu, Xiaoba of Beichuan, Qushan and other counties. It also goes through Tongkou and Qinglian of Jiangyou, and joins Fujiang River from the right bank. Tongkou River is about 145 km long, the area of its basin is 4160 km^2, and the average slope of the river bed is 9‰. There are many mountains and woods in the upstream, and river channels are meandering. After going into Beichuan, it divides into many branches. The middle and downstream of the Tongkou River basin are located at the rainstorm area of Lutou Mountain and Longmen Mountain, which have plenty of water and steep river channels, and there are large numbers of flash floods during the flood season. The Fujiang River is narrow and long, originates from the north of Xuebaoding of the Min Mountain in the Songpan County of Sichuan Province, and flows from the northwest to the southeast. It flows from mountain areas of the northeast into hilly areas, including Pingwu, Jiangyou, Mianyang, Santai, Shehong, Suining, Tongnan and Diaoyu Towns of Hechuan County, and then it joins the Jianglingjiang River. It is about 670 km long and its basin area is 35 892 km^2. The inflow of the river is mostly from rainfall, and there is also some snow replenishment in the mountain areas upstream. The floods in the Fujiang River basin mostly result from rainstorms in summer from June to September, and big rainstorms usually happen in July–September, sometimes in June. The study area in this paper was selected as the upstream of Mianyang City along Fujiang River basin, and its area is 12 032.25 km^2.

MODEL STRUCTURE

The spatial calculation units of the WEP model are regular square grid cells. The vertical structure of the calculation unit includes an interception layer of vegetation and city buildings, ground surface depression layer, unsaturated soil layer, transition layer, shallow groundwater layer and deep groundwater layer, from top to bottom. State variables are interception by vegetation, storage in ground depression, soil moisture content, temperature of ground surface, storage in transition layer, groundwater table and streamwater stage, etc. (Jia *et al.*, 2005). The "Mosaic" method is applied to take unevenness of land use in the calculation cell into consideration by classifying the land uses into bare land and vegetation area, waters and impermeable area in the cell for calculation of water and heat fluxes through the land surface. The bare land and vegetation area is further subdivided into bare land, high vegetation and short vegetation, and the impermeable area into urban land, urban buildings and rocks. In addition, in order to reflect change in surface soil moisture content with depth and describe evaporation from soil and water absorption by root systems of high vegetation and short vegetation, the unsaturated soil layer of permeable areas is further divided into three layers, i.e. top, middle and bottom soil layers. Runoff routing in rivers and on overland slopes are conducted based on the elevation, gradient and the Manning roughness (harmonious average of various types of land uses) by applying a one-dimensional (1-D) kinematic wave approach from upstream to downstream. Numerical simulation is carried out for groundwater flows in mountain and plain areas, separately considering exchange of groundwater with surface water, soil moisture and streamflow (Jia *et al.*, 2001).

MODEL CALIBRATION AND VERIFICATION

Input data

The river system, main hydrologic gauges and meteorologic gauges of the basin in which the Tangjiasha barrier lake is located are shown in Fig. 1. Geographic data of geology and physiognomy, soil and vegetation, hydrology and weather and land use are extensively collected and stored in a GIS system.

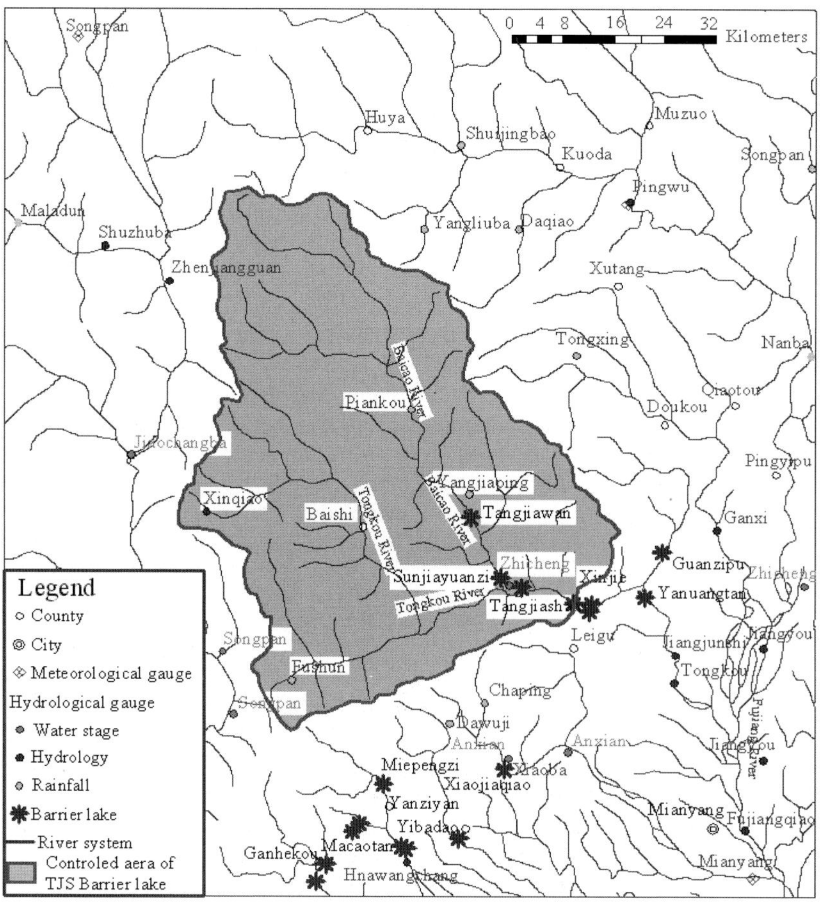

Fig. 1 Distribution of hydrological gauges neighbouring the Tangjiashan barrier lake.

The DEM data of the study area are obtained from national 1:250 000 basic geographic information data. Based on the basic geographic information of elevation, slope, slope direction and runoff direction, sub-basins are divided and a digital river system is generated. Considering the above information, flow accumulations are obtained. The reaches' natural attributes, the locations of the barrier lake and hydrological gauges are considered to divide the reaches, and the reach parameters (bottom width, top width, maximum depth, the Manning roughness, etc.) are generalized and interpolated in each sub-section based on the large section measurement and planning information of hydrology (water stage) stations. Then, the stream networks are constructed. The Fujiang River basin within the study area is divided into 7 sub-basins, 13 rivers, and 134 river sections in total. Because of the river data limitations, river runoff routings are performed only in the main channel of Fujiang River and its upstream tributaries of Huoxihe River, Baicaohe River, Anchang River and Pingtong River, and other small rivers are approximately calculated by overland runoff routing. Daily weather data were collected from 1956 to 2005: the weather data at four weather stations of Pingwu, Ganxi, Jiangjunshi and Fujiangqiao are used in the model; weather observation items includes rainfall, temperature, wind speed, sunshine and relative humidity, etc.; precipitation, temperature, relative humidity of calculation units are corrected according to the correlation with elevation based on the Thiessen polygon interpolation. Since this area belongs to a poorly gauged area, according to the available data situation, monthly runoff data of the four stations of Ping Wu, Ganxi, Jiangjunshi, and Fujiang-qiao are selected to verify the model at first, and the period of 1976–1980 was selected as the model verification period. In order to prevent the deficiencies of initial states of hydrological variables, the period of 1970–1975 is selected as the model warming-up period.

The land-use source information that the model used is the TM digital image data of 1985, 1995 and 2000. Based on interpretation of the National Land Remote Sensing Detailed Investigation and Classification System and grids as the basic units, the original 32 land-use types are re-classified into six land use types of waters, high vegetation (including forests and shrubs), low vegetation (including grass and crops), bare land, buildings and impermeable areas, and the percentage of each type is calculated. In the monthly calculation of 1956–2005 series, percentage of each land type of each year in each grid unit is achieved from yearly linear interpolation in accordance with the above-mentioned data of three periods. In the model input, soil parameters, soil types are generalized into three soil types of sand, silt and clay. Thickness of the soil aquifer is set to 2 m in the area.

Parameter estimation

The main parameters of the model include soil parameters, groundwater aquifer parameters, vegetation parameters, roughness of slope and river channel, infiltration coefficient of river bed etc. Most parameters in the model do not need to be calibrated, but for several important parameters such as saturated hydraulic conductivity of soil, groundwater aquifer parameters, infiltration coefficient, and thickness of river bed, as well as the Manning roughness, are needed to be modified when comparing the simulated river flow with observed data in the calibration period.

Table 1 Soil moisture characteristic parameters (Haverkamp *et al.*, 1977; Mualem, 1978; Jia & Tamai, 1997).

Parameters	Sand	Silt	Clay	Parameters	Sand	Silt	Clay
θs	0.400	0.455	0.473	ks (m/s)	2.5E-4	7.0E-5	3.5E-5
θr	0.077	0.115	0.137	Havercamp parameter α	1.75E10	6451	6.576E6
θm	0.020	0.053	0.124	Havercamp parameter β	16.95	5.56	9.00
θf	0.174	0.318	0.363	Mualem parameter n	3.37	3.97	4.23

θs: saturated soil moisture rate; θr: residual soil moisture rate; θm: hygroscopic soil moisture rate; θ_f: field capacity; ks: saturated hydraulic conductivity.

Soil On the basis of the soil moisture movement experimental research results in the Fujiang River basin and adjacent regions, soil moisture characteristics parameters are estimated (Table 1). At the same time, according to the soil composition and soil moisture rate, the change of soil thermodynamic properties parameters with soil humidity are also considered through the establishment of empirical formula.

Groundwater aquifer Groundwater aquifer parameters in the Fujiang River basin are deduced using geological exploration data and groundwater simulation results. The saturated soil layer parameters in mountains are established, combining runoff simulation results, and the conductivity coefficient is assumed as 65.5 m / m × soil layer thickness (m), and specific yield is assumed as 0.05.

Vegetation Vegetation parameters change with seasons. Main parameters of four vegetation types include vegetation covering rate (veg), leaf area index (LAI), vegetation height (hc), root depth (lr) and the minimum stomatal resistance (rsmin). The monthly change of vegetation parameters is considered in the model.

Runoff routing parameters The Manning roughness used in river runoff routing is assumed according to the Hydrological Yearbook, while the Manning roughness used in overland runoff convergence routing is assumed according to the land-use type in the grid cell. After model debugging, the Manning roughness of woodland, grassland, farmland, bare land and urban surface, water body is set as 0.3, 0.1, 0.05, 0.02 and 0.01, separately. The exchange amount between river and groundwater is closely related to riverbed material properties. In this study, riverbed material infiltration coefficient divided by riverbed material thickness is established as the model debugging parameter. The final value is 4.0E-04/s.

Snow melting parameters Snow melting coefficient and melting critical temperature are also model debugging parameters. After model adjustment, snow melting coefficient is 1 mm/°C/day for woodland, 2 mm/°C/day for grassland, 2 mm/°C/day for bare land, 5 mm/°C/day for urban land use and 1 mm/°C/day for glacier separately; melting critical temperature is 0°C; rain-snow critical temperature is 1°C.

Model calibration and validation

In the model validation, the Fujiang River basin is divided into 48 130 units by 500 m × 500 m square grid cells, and a 50-year simulation of from 1956 to 2005 is done with a time step of 1 day. The model is calibrated using monthly and daily runoff processes observed at Pingwu, Ganxi, Jiangjunshi and Fujiangqiao hydrological gauges from 1976 to 1980. The main calibration parameters including soil saturated hydraulic conductivity coefficient, groundwater aquifer conductivity coefficient, specific yield, riverbed material infiltration coefficient, and the Manning roughness. Calibration rules include: (a) average annual runoff relative error during the simulation period should be as small as possible; (b) the Nash-Sutcliffe efficiency should be as large as possible; (c) correlation coefficient of simulated flow and observed flow should be as large as possible.

The equation of Nash-Sutcliffe efficiency coefficient is:

$$\eta = 1 - \frac{\sum_{i=1}^{n}(Q_{sim} - Q_{obs})^2}{\sum_{i=1}^{n}(Q_{sim} - \overline{Q}_{obs})^2} \tag{1}$$

where η is the Nash-Sutcliffe efficiency coefficient, Q_{sim} is the simulated streamflow, Q_{obs} is the observed streamflow and $\overline{Q}_{obs}$ is the annual average observed streamflow.

Model calibration and the parameter debug method is the same as for other physically-based distributed hydrological models using the method of trial and error, that is, firstly determine the initial value according to physical properties, experimental data and reference data, then adjust the parameters according to the above three rules. Specifically, the soil saturated hydraulic conductivity coefficient is firstly adjusted according to soil types since it is sensitive to the total runoff amount and the peak runoff; then aquifer conductivity coefficient, specific yield and riverbed infiltration coefficient are adjusted according to soil types since they mainly affect total annual runoff and runoff in flow regressive periods; and lastly the Manning roughness of overland flow is adjusted according to different land use since it mainly affects the shape of flood peak in runoff hydrographs. In addition, the snow melting coefficient and melting critical temperature have a great influence on summer runoff and rising time, the conductivity coefficient of frozen soil has a great influence on dry season runoff in winter, so they are all adjusted according to the runoff hydrograph. As the barrier lake inflow is influenced by the initial conditions of hydrological state variables like soil moisture, the calibrated snow melting coefficient and melting critical temperature are used to update the state variables during the model warming-up period (i.e. 1 January to 11 May 2008).

Model validations at four hydrological gauge stations are shown in Figs 2–5. Simulated results are not very good at some river sections because of limited data. Absolute and relative errors are usually bigger in dry seasons than wet seasons. But the result at the Jiangjunshi gauge which controls the Tangjiashan earthquake lake is quite good generally speaking, especially for the period from May to August (though a little overestimation but good for risk management), as we can see in Fig. 4. The simulation effect assessment is shown in Table 2, demonstrating that the relative error and the Nash-Sutcliffe efficiency indices are acceptable in the poorly gauged basin.

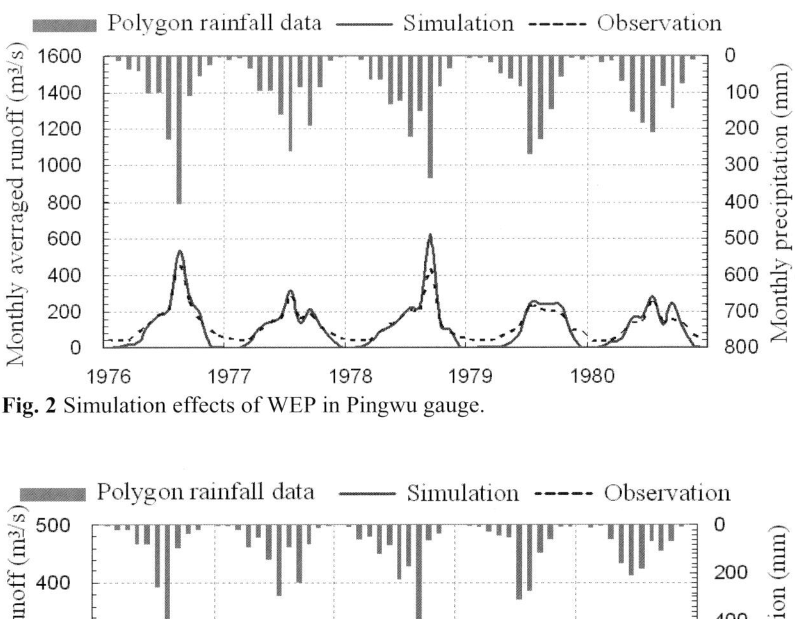

Fig. 2 Simulation effects of WEP in Pingwu gauge.

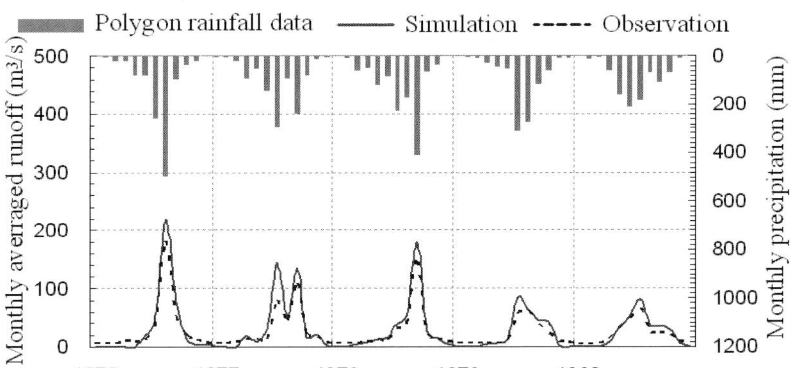

Fig. 3 Simulation effects of WEP in Ganxi gauge.

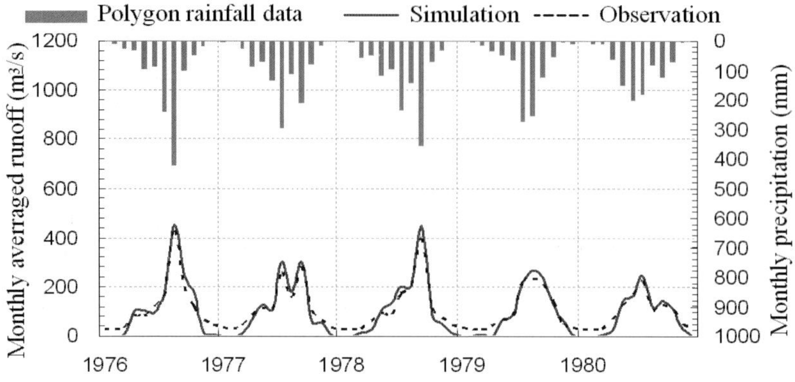

Fig. 4 Simulation effects of WEP in Jiangjunshi gauge.

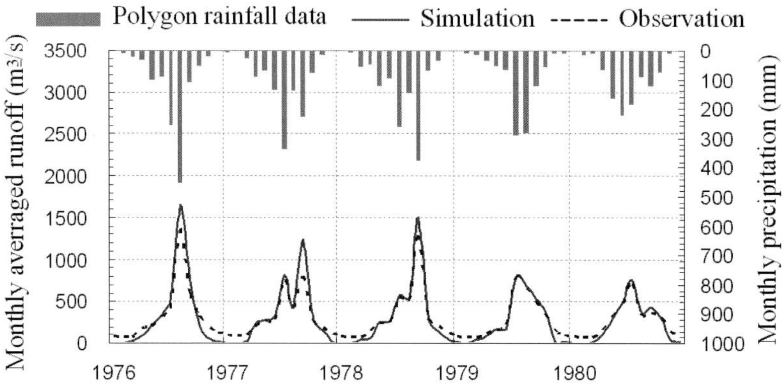

Fig. 5 Simulation effects of WEP in Fujiangqiao gauge.

Table 2 Simulation effect assessment of WEP at four hydrological gauges of the Fujiang River basin.

Hydrological gauges	Ann. Ave. R. Error	Nash-Sutcliffe efficiency	R. Error of May–Aug	Nash-Sutcliffe efficiency of May–Aug
Pingwu	6.28%	0.64	3.79%	0.75
Ganxi	12.38%	0.60	28.76%	/
Jiangjunshi	5.37%	0.67	4.12%	0.71
Fujiangqiao	2.95%	0.68	9.00%	0.62

TRMM RADAR DATA

Rainfall is a very important model input and has a direct effect on the accuracy of the prediction. The observation data of precipitation in meteorologic gauges is relatively accurate, but it is difficult to describe the spatial distribution of the precipitation, and almost impossible to catch the centre of the precipitation. Radar rainfall measurement has remarkable advantages in describing the characteristic of the spatial distribution of the precipitation, which is also very important for predicting, and it helps to diminish errors and improve the accuracy of the simulation and prediction. For the sake of combining the advantages of these two methods and avoiding the disadvantages, the Tropical Rainfall Measurement Mission (TRMM) three-hourly data from 1 January to 1 July was first downloaded (http://trmm.gsfc. nasa.gov/data dir/data. html). Then the radar rainfall data was calibrated and modified based on historical daily observation data at Songpan meteorological station. Lastly, the modified results of the radar rainfall data were used to run the model for the prediction. The predicting period was selected as 12 May–6 June 2008.

For safety, the missing data and initial hydrological status variables are adopted as follows: the warming-up period is chosen from 1 January to 11 May 2008. Parameters such as air temperature, wind speed, hours of sunlight, relative humidity, etc. are assumed by referring to historical values of 1990–2000 in Song-pan meteorological station. Initial soil moisture and groundwater level on 1 January are assumed as the maximum values during 1990–2000 in the Song-pan meteorological station. The simulation and prediction is set into three different schemes according to three types of rainfall inputs: (1) data from meteorological station without DEM modification; (2) data from meteorological station with DEM modification; (3) data from TRMM radar with meteorological station observed data modification. The accumulated inflow simulation and prediction results of these three different scenarios are shown in Fig. 6.

The total accumulated lake storage predicted based on the scheme (3) during the forming period (12–19 May) of the barrier lakes is 169.4 million cubic meters; i.e. 7.4 million cubic meters more than the observed data. During the real time monitoring period (29 May–9 July), the relative error of scheme (3) is 4.4% when compared to the observed data, thus meeting the needs for diminishing the disaster; whilst the relative errors of the other two schemes are as big as 50%. The results illustrate that the combination of distributed hydrological model, TRMM radar data and meteorological station observed data are useful for the flood disaster warning purpose in the poorly gauged basin.

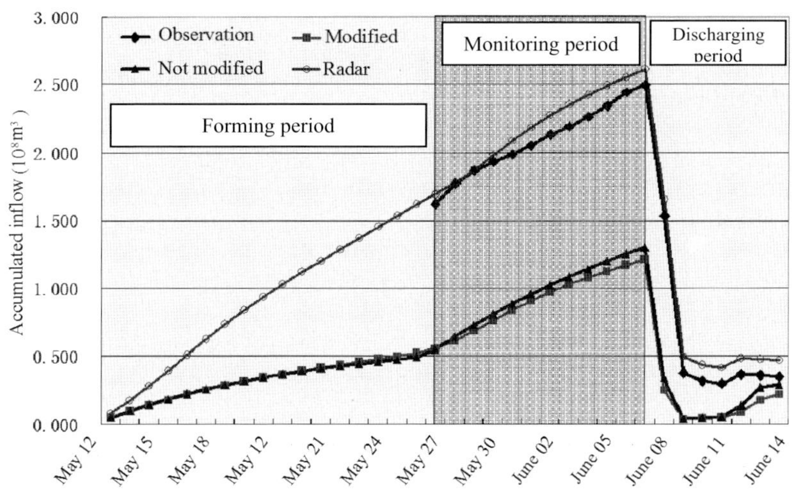

Fig. 6 Simulation effects of the total accumulated lake storage with 3 different rainfall inputs.

SUMMARY

The WEP distributed hydrological model is suitable for flow simulation and prediction in poorly gauged basins, though a further improvement of the model is expected to raise the accuracy of the simulation and prediction. Limited local gauge data are valuable for prediction, but not enough, and simulated results are not very good at some river sections because of the limited data. Radar rainfall data describe the characteristic of the spatial distribution of the precipitation, so the methodology of combining the meteorological station observed data and the TRMM radar rainfall data can help to improve the prediction results. However, there are still some errors in the predicted inflows and lake storage compared with the observations, which may be mainly restricted by the precisions of the satellites and correlative observation equipments except for the structure and the parameters of the model itself. Therefore, the uncertainty problems need to be studied further.

Acknowledgements This study is funded by the NSFC projects of China (50721006, 50779074), the National 973 Program of China (2006CB403404), and the National Science and Technology Supporting Project of China (2006BAB04A08).

REFERENCES

Arnaud, P., Bouvier, C. & Cisneros, L. *et al.* (2002) Influence of rainfall spatial variability on flood prediction. *J. Hydrol.* **260**, 216–230.

Collier, C. G. (1996) *Applications of Weather RADAR System*. M. Praxis Publishing Ltd, UK.

Haverkamp, R. Vauclin, M., Touma, J. *et al.* (1977) Comparison of numerical simulation models for one-dimensional infiltration. *J. Soil Sci. Soc. Am.* **41**, 285–293.

Jia, Y. W. & Tamai, N. (1997) Modeling infiltration into a multi-layered soil during an unsteady rain. *J. Ann. Hydraul. Engng JSCE* **41**, 31–36.

Jia, Y. W., Ni, G. H. & Kawahara, Y. *et al.* (2001) Development of WEP model and its application to an urban watershed. *J. Hydrol. Processes* **15**, 2175–2194.

Jia, Y. W., Wang, H. & Ni, G. H. *et al.* (2005) *Theory and Practices of Distributed Watershed Hydrological Models*. China Water Resources and Hydropower Publisher, Beijing, China.

Jia, Y. W. Wang, H. & Yan, D. H. (2006) Distributed modeling of hydrological cycle system in Heihe River basin (I): Model development and verification. *Chinese Journal Hydraul. Engng* **37**, 534–542.

Mualem, Y. (1978) Hydraulic conductivity of unsaturated porous media: generalized macroscopic approach. *J. Water Resour. Res.* **14**, 325–334.

Nie, G. Z., Gao, J. G. & Deng, Y. (2004) Preliminary study on earthquake-induced dammed Lake. *J. Quatern. Sci.* **24**, 293–300.

Xu, J. J. Yang, D. W. & Cai, Z. H. G. (2007) Distributed hydrological model combining with radar rainfall data for regional flood forecast of the Three Gorges. *J. Yangtze River Scientific Research Institute* **24**, 42–48.

Zhang, L. P., Li, Lu. & Ye, A. Z. H. *et al.* (2007). Contrast research on precision of estimating regional precipitation with weather radar and rain gauge. *Engng J. Wuhan University* **40**, 1–5.

A modified generalized likelihood uncertainty estimation method by using Copula function

KAIRONG LIN, XIAOHONG CHEN, QIANG ZHANG & ZHIHE CHEN
School of Geographical Science and Planning, Sun Yat-sen University, no. 135 Xingangxi Road, Guangzhou 510275, China
linkr@mail.sysu.edu.cn

Abstract In this study, we propose a modified Generalized Likelihood Uncertainty Estimation method (GLUE) by using Copula function (CGLUE) with the aim of describing the correlation of parameters. Six scenarios are adopted by taking the sampling size as 10 000, 50 000, and 100 000, with threshold values of 0.5, 0.6, and 0.7. With these six scenarios, we compared the CGLUE and the GLUE methods in terms of strength and weakness of these two techniques in the evaluation of the parameter uncertainty in hydrological modelling results. The results show that, when compared to the GLUE method, the CGLUE method performs better in reflecting the correlation of the parameters by replacing the two one-variable probability distributions with a joint probability distribution. Therefore, the uncertainty of parameter within the hydrological model can be fully evaluated with more behavioral parameter sets.

Key words Copula functions; Generalized Likelihood Uncertainty Estimation; hydrological model; uncertainty evaluation

INTRODUCTION

Hydrological models have been accepted as effective tools in the description of dynamic relations between hydrological processes, meteorological behaviours, land use and land cover, and also the changes of vegetation coverage within a watershed, providing theoretical and practical support for river management (Wagener & Gupta, 2005). However, the hydrological system is complicated and is the integrated consequence of more than one factor such as climate changes, land use and land cover, vegetation and soil conditions, etc. The complexity of the hydrological system poses great challenges for hydrological modelling practices. Besides, due to our limited knowledge of hydrological system and simulation tools, we usually face many problems when we propose, design and use hydrological models, and some of them are far from being settled. Ye & Xia (2002) pointed out that the complexity and uncertainty of hydrological models are the two most important aspects in hydrological science research. Therefore, the uncertainty of hydrological models has been drawing increasing concern from hydrologists.

The complexity of the model structure and uncertainty of modelling results come to be increasingly serious and intolerable with the development of hydrological theory and wide practical application of hydrological models in the water resource planning and management, and requires urgent settlement. Research focused on the efficiency and precision of the optimization algorithm for model parameters and the traditional parameter identification can not satisfy the requirements of the fast development of hydrological theories and hydrological modelling practices (Beck, 1987). The model structure is usually validated by statistical techniques after the identification of parameters. Therefore, uncertainty analysis of parameters plays the key role in hydrological modelling practices.

Generalized Likelihood Uncertainty Estimation (GLUE) proposed by Beven & Binley (1992), as a kind of parameter uncertainty estimation method, is widely used today in many complex and nonlinear models (Freer *et al.*, 1996; Beven & Freer, 2001). However, this method is based on the assumption that the parameters of the model are independent. Actually, complexity and correlation of the distribution of parameter space is the major source of the parameter uncertainty. In this case, by using Copula functions (CGLUE in this study), we try to address uncertainty of parameters of hydrological model by clarifying the correlation of the model's parameters.

COPULA

The definition of Copula

The Copula technique, after it was proposed in 1959, has been an important method aimed at describing the multivariate joint distribution (Nelson, 1999). The Gumbel-Hougaard Copula is another kind of Copula hydrologists are familiar with, which is taken as the correct technique to describe hydrological extreme variables (Salvadori & De Michlele, 2004). This paper uses the Gumbel-Hougaard Copula to describe the correlative structure of model parameters, and can be expressed as:

$$C(u,v) = \exp\left\{-[(-\ln u)^{\theta} + (-\ln v)^{\theta}]^{1/\theta}\right\}, \quad \theta \geq 1 \tag{1}$$

where θ is the parameter of Copula. Larger θ value implies stronger correlation. $\theta = 1$ denotes independency of variables.

The parameter estimation

The parameter of Copula, θ, is related to the Kendall's rank correlation coefficient τ as:

$$\theta = 1/(1-\tau) \tag{2}$$

UNCERTAINTY ESTIMATION METHOD

In this study, we propose the CGLUE method based on the GLUE technique. The computation procedure can be explained as follows:

(a) Calculating the correlative parameters, the Kendall's τ, and the parameter of Copula, i.e. θ.
(b) Generating the independent model parameter sets with the Monte Carlo method by some specified distributions. In addition, correlative parameter sets are generated directly by Copula function.
(c) The model is run with aforementioned generated parameter sets. An appropriate likelihood function is chosen and then the likelihood values of each parameter sets are obtained.
(d) A threshold value of likelihood function is decided and the behavioural parameter sets whose likelihood values are greater than the thresholds are chosen. Then, the prediction limits are obtained according to the likelihood weights.
(e) Updating the model likelihood weights based on new calibration data.

CASE STUDY

Catchment data

In this study, Hanzhong catchment is selected as the case study. The Hanzhong catchment drainage is 9329 km^2 and is located at the upper Hanjiang River, a branch of the Yangtze River in China. The Hanzhong catchment is characterized by hills and river valleys. There are 50 rain stations, 3 evaporation stations and 7 hydrological stations (Fig. 1). The digital drainage network of Hanzhong catchment is generated automatically by using TOPAZ software (Martz & Garbrecht, 1992). The hydrometeological data cover from 1980 to 1990. The grid size of DEM of the river catchment is 100 m × 100 m.

The hydrological model of study

TOPMODEL (Beven & Kirkby, 1979) is a topographically-based hydrological model aimed at reproducing the hydrological behaviours of watersheds in a semi-distributed way, in particular the dynamics of surface and subsurface contributing areas. Figure 2 shows the distribution curve of the soil-topographic index calculated by the multiple flow directions method.

Fig. 1 River network map of Hanzhong catchment.

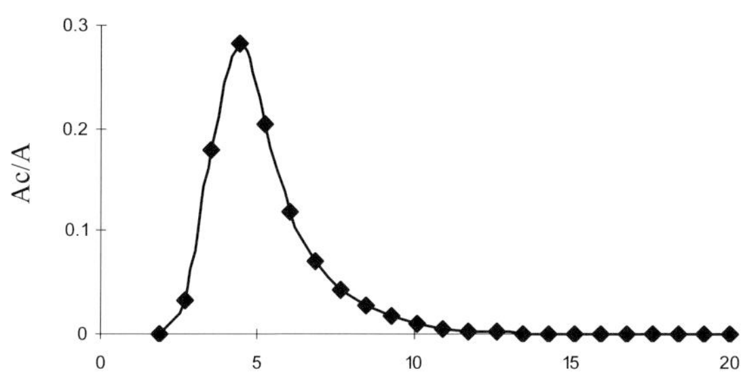

Fig. 2 Distribution curve of topographic index.

The application of the new method

This study selects four primary parameters of TOPMODEL which are displayed in Table 1, where S_{zm} is the maximal moisture deficit; T_0 is the transmissivity of the soil at saturation; T_d is the time constant; $SR_{\max}$ is maximum allowable root storage deficit. It assumes that the parameters follow the uniform distribution, and the ranges of the parameters are listed in Table 1.

The results indicate that the distribution of complexity and correlation of parameter space is one of the main sources of the parameter uncertainty. According to the hypotheses in TOPMODEL, we can assume that correlations may occur between S_{zm} and T_0. At first, we use the traditional

Table1 Ranges of parameters used in TOPMODEL model.

Parameter		Minimum value	Maximum value	Mean value
S_{zm}	(m)	0.01	1	0.5
T_0	(m^2/h)	0.01	3	1.5
T_d	(h)	1	100	50
SR_{max}	(m)	0.01	0.5	0.25

GLUE method to obtain all the parameter sets higher than the likelihood function threshold. In this study, we choose the threshold efficiency of 0.7 (Nash-Sutcliffe efficiency criterion), and then we analyse the correlation of the parameters. We find that correlations do occur between S_{zm} and T_0 (Fig. 3). The corresponding Kendall τ value is 0.55.

The Copula parameter, θ, is 2.22 by equation (2). The Monte Carlo method is used to sample the model parameters within the value ranges of the presupposed parameter distribution. One of the correlative parameters is directly generated by Copula. Simulation is based on the model parameter samples available, and the Nash-Sutcliffe efficiency criterion is accepted as the likelihood function. The threshold value is set to 0.7 and all the parameter sets whose likelihood values are greater than the threshold are adopted. Figure 4 shows the uncertainty zone of the modelled forecast according to the likelihood weights. The upper and the lower bound show 95% limits.

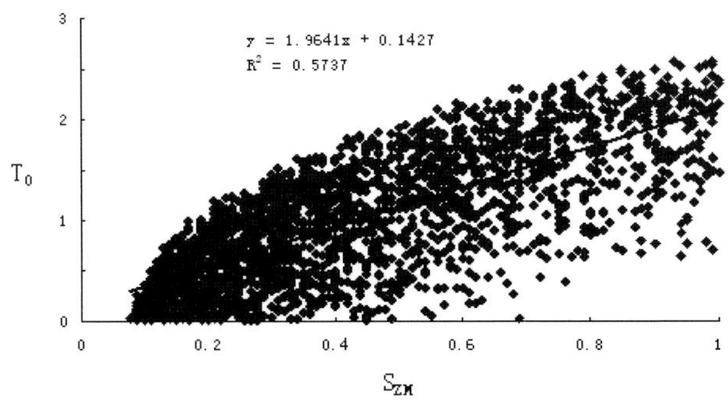

Fig. 3 Scatter plot of the parameter S_{zm} and T_0.

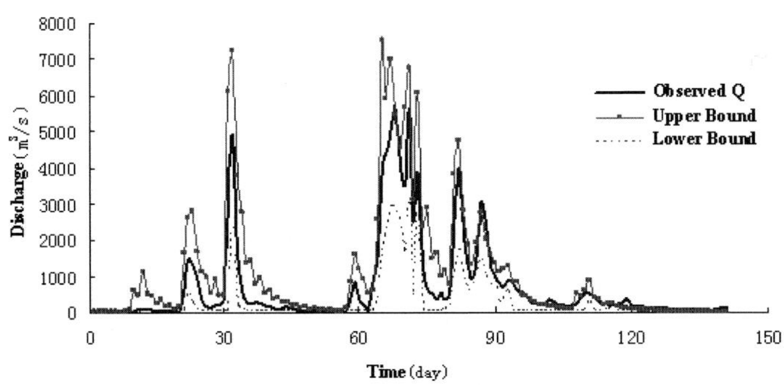

Fig. 4 Comparison of prediction limits for 14 June–2 November 1981.

ANALYSIS OF RESULTS

This study accepts six scenarios by taking the sampling size of 10 000, 50 000, and 100 000 with threshold values of 0.5, 0.6, and 0.7. The objective is to thoroughly investigate the uncertainty simulation. The GLUE method and the CGLUE method proposed in this study are used to the uncertainty simulation respectively. Figure 5 shows the correlation of bound discharge between the GLUE and the CGLUE methods within 95% prediction limits. Figure 5(a) and (b) show the correlation of discharge within the upper and the lower bounds when the sample size of 10 000 and the threshold value of 0.7 are taken. Figure 5(c) and (d) illustrate the correlation of the discharge within the upper and the lower bounds when the sampling size of 100 000 and the threshold value of 0.7 are chosen. The difference of the results is not too obvious between the proposed method or the traditional method or the above six different scenarios. The lower-bound discharge of the proposed method in this paper is less than the discharge by the traditional method, and it is particularly the case when the discharge is greater in the case that the simulation time is 10 000 (Fig. 5(b)), implying larger prediction limits by the proposed method than by the traditional method. However, the bound discharge by two methods is generally in good agreement with increasing sampling times.

Little difference is observed between GLUE and CGLUE in terms of prediction limits. However, remarkable differences can be found between the above-mentioned two methods with respect to the uncertainty estimation of the model parameters. The total number of behavioural parameter sets of the above six scenarios are listed in Table 2, which shows that the number of behavioral parameter sets decreases with the threshold and increases with the sampling times. The sampling time needs to be increased in order to evaluate the uncertainty of the parameters in a more comprehensive way. However, a 10 000- or a 100 000- or even a 1 000 000-parameter sample is necessary due to the large amount of parameter sets in the multi-parameter model structures. Thus, Monte Carlo computation procedure consumes massive computing resources. If the samples are not adequate, the sampling procedure which assumes independence of the parameters may not involve all the behavioural parameter combinations, which is expected to underestimate the uncertainty. Table 2 also shows that the method proposed in this study considers the correlation in the right way when the model parameters are in good correlation, and this method can obtain more behavioural parameter sets in the same sample times so that the

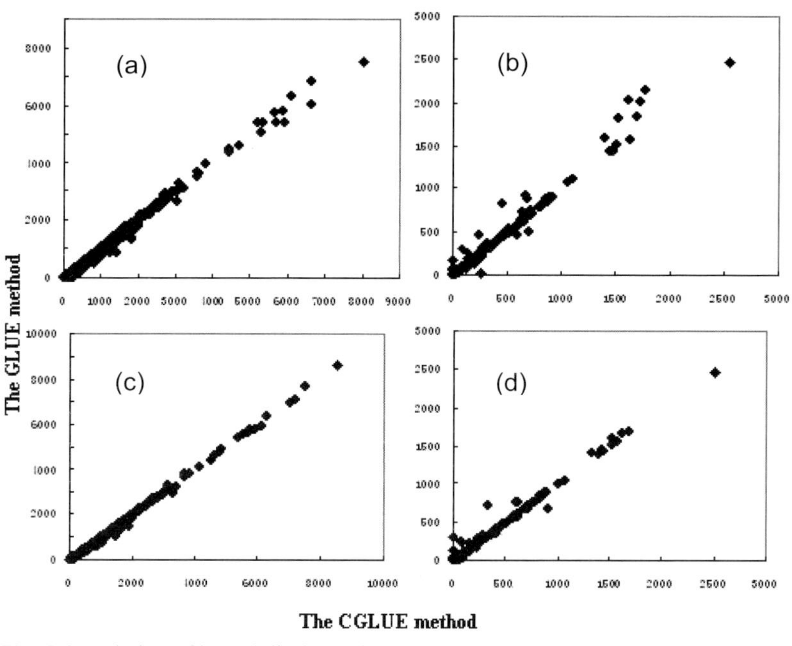

Fig. 5 Correlation of bound discharge between the GLUE and the CGLUE methods.

Table 2 List of the total number of the behavioral parameter set.

Threshold	10 000		50 000		100 000	
	Traditional method	Proposed method	Traditional method	Proposed method	Traditional method	Proposed method
0.5	918	1393	4497	6995	9043	14018
0.6	537	803	2584	3945	5215	7865
0.7	278	420	1384	2055	2796	4125

uncertainty of parameters could be estimated thoroughly. The analytical method for uncertainty in this study offsets the deficiency of the traditional methods and achieves better effects.

Besides, the result shows that the upper and lower bounds of the simulated discharge processes cannot include the observed discharge changes, and there are always several observed discharges outside the 95% prediction limits (Fig. 4). This may indicate that TOPMODEL cannot simulate the discharge processes in great detail. Many studies (Beven & Free, 2001; Huang & Xie, 2007) indicated that it may be the result of other factors triggering uncertainty.

CONCLUSION

Hydrological models are based on determinate hydrology regularity, including physics and statistical rules (Lin *et al.*, 2007), and research on uncertainty could be helpless without considering determinateness. The recent uncertainty assessment methods assume that the parameters of the model are mutually independent. However, in fact it is not true due to complexity and correlation of the distribution of the parameter, which is seen as the source of the uncertainty of the parameters. This study uses Copula to describe the correlation of the model's parameters and proposes an uncertainty estimate method, namely CGLUE. When the parameters of the model are correlative, this method considers the correlation of parameters in the right way by using joint distributions instead of single parameter distributions. This method can both avoid the uncertainty induced by underestimating the model parameters in the traditional GLUE method, and improve the searching efficiency of the Monte Carlo sampling technique. Meanwhile, there are a great number of factors which can introduce the uncertainty into the hydrological simulation and forecast in complicated environment systems. It should be noted here that the problems of uncertainty are far more than what is discussed in this study. Thus, further work is still necessary.

Acknowledgements This study was financially supported by the National Natural Science Foundation of China (Grant no. 50809078; 50839005; 40701015), the National Key Technology Support Program (no. 2006BAB14B07), the Open Foundation of the State Key Laboratory of Water Resources, Hydropower Engineering Science (Wuhan University) (2008B043), and the Scientific Research Foundation for Young Scholars of Sun Yat-sen University (3171912).

REFERENCES

Beck, M. B. (1987) A review of the analysis of uncertainty. *Water Resour. Res.* **23**(8), 1393–1442.

Beven, K. J. & Kirkby, M. J. (1979) A physically-based variable contributing area model of basin hydrology. *Hydrol. Sci. Bull.* **24**(1), 43–69.

Beven, K. J. & Binley, A. (1992) The future of distributed models, model calibration and uncertainty prediction. *Hydrol. Processes* **6**, 279–298.

Beven, K. J. & Freer, J. (2001) Equifinality, data assimilation, and uncertainty estimation in mechanistic modeling of complex environmental systems using the GLUE methodology. *J. Hydrol.* **249**, 11–29.

Freer, J. & Beven, K. J. & Ambroise, B. (1996) Bayesian estimation of uncertainty in runoff prediction and the value of data, an application of the GLUE approach. *Water Resour. Res.* **32**(7), 2161–2173.

Huang, G. & Xie, H. (2007) Uncertainty Analyses of Watershed HydrologicalModel Based on GLUE Method. *J. South China University of Technology (Natural Science Edition)* **35**(3), 137–142.

Lin, K, Guo, S. Zhang, W. & Liu, P. (2007) A new baseflow separation method based on analytical solutions of the Horton infiltration capacity curve. *Hydrol. Processes* **21**(13), 1719–1736.

Martz, L. W & Garbrecht, J. (1992) Numerical definition of drainage network and subcatchment areas from digital elevation models. *Computers Geosciences* **18**(6), 747–761.

Nelson, R. B. (1999) *An Introduction to Copulas*. Springer, New York, USA.

Salvadori, G. & De Michlele, C. (2004) Frequency analysis via Copulas, theoretical aspects and applications to hydrological events. *Water Resour. Res.* **40**(12), 1–17.

Wagener, T. & Gupta, H. V. (2005) Model identification for hydrological forecasting under uncertainty. *Stochast. Environ. Res. Risk Assessment* **19**, 378–387.

Ye, S. & Xia, J. (2002) Century's retrospect and looking into the future of hydrological science. *Chinese Adv. Water Sci.* **13**(1), 93–104.

Distributed hydrologic modelling in Huai River basin with a high density of dams and floodgates

YONGYONG ZHANG[1]**, JUN XIA**[1,2]**, ZHONGGEN WANG**[1] **& XUESONG ZHANG**[3]

1 *Key Laboratory of Water Cycle and Related land Surface Processes, CAS, Beijing, 100101, China*
zhangyy003@163.com

2 *State Key Laboratory of Water Resources and Hydropower Engineering Science, Wuhan University, Wuhan 430072, China*

3 *Joint Global Change Research Institute, Pacific Northwest National Laboratory, College Park, Maryland 20740, USA*

Abstract In Huai River basin, China, the construction of dams and floodgates has substantially changed the basin's natural hydrological cycle. In this study, the dams and floodgates discharge simulation module of the Soil and Water Assessment Tool (SWAT) was extended to incorporate the dams and floodgates' dispatch rules. This upgraded version of SWAT model was applied to simulate the monthly flow at 23 dams and floodgates and 4 hydrological stations for 1991–2000. The evaluation coefficients (i.e. relative volume error, correlation coefficient, and efficiency coefficient) are used to elevate the modelling performance. In the calibration period, the volume errors at 18 monitoring stations are within ±0.15. The average correlation coefficient and efficiency coefficient are 0.75 and 0.41, respectively. In the verification period, there are 13 stations with volume error within ±0.15. The average correlation coefficient and efficiency coefficient are 0.82 and 0.53, respectively. These results show that the modified model can be used as the hydrological modelling tool to provide technical support for the integrated management of Huai River basin.

Key words water projects; hydrological simulation; SWAT; Huaihe River basin

INTRODUCTION

Water projects construction is critical for water resources management, ecology and environment protection. According to the statistics of the International Commission on Large Dams (ICOLD; http://www.icold-cigb.org/), the number of completed dams (more than 15 m or larger than 3 million m^3 in storage capacity) was 49 248 until 1998. Approximately two thirds of these dams are in developing countries. Dams and floodgates can play a vital role in flood control, water supply, farmland irrigation, hydropower generation, aquaculture, recreation and tourism. This is helpful for rapid regional economic development over a certain time. However, excessive dam and floodgate construction disturbs the natural river flow regime, resulting in environmental degradation and reducing biodiversity (Berkamp *et al*., 2000; Suo, 2005).

Huai River basin (HRB) with a drainage area of 270 000 km^2 is located in eastern China (Fig. 1). The total population in HRB is 165 million and the average population density is approx. 611 per km^2, which is 5 times the national average. Although annual mean precipitation and runoff of HRB are 888 mm and 83.5 billion m^3, respectively, water resources per capita and unit area is less than one-fifth of the national average. In order to control floods and relieve droughts, >5700 reservoirs and 5000 floodgates have been constructed. On average, each branch has nearly 10 dams or floodgates. The total storage capacity of dams and floodgates in HRB is 303 billion m^3 which accounts for 51% of the annual runoff. Along with the excessive dam and floodgate constructions, the natural hydrological cycle regime in HRB has been interrupted dramatically. How to simulate hydrological process of the dam-crowded area and analyse the impact of dams and floodgates on the riverine flow regime are urgent issues in HRB's water resources management.

The major objective of this paper was to develop a distributed hydrological model of HRB incorporating dams and floodgates' dispatch rules – Soil and Water Assessment Tool (SWAT). The upgraded SWAT model has been applied to simulate the monthly discharge of the main dams and floodgates in the upper and middle reaches of HRB. The following parts of this paper are organized as follows: the next section provides a brief description of SWAT model and our improvement. The modified SWAT model is then introduced and parameters calibration and validation is discussed. Finally, a summary is provided.

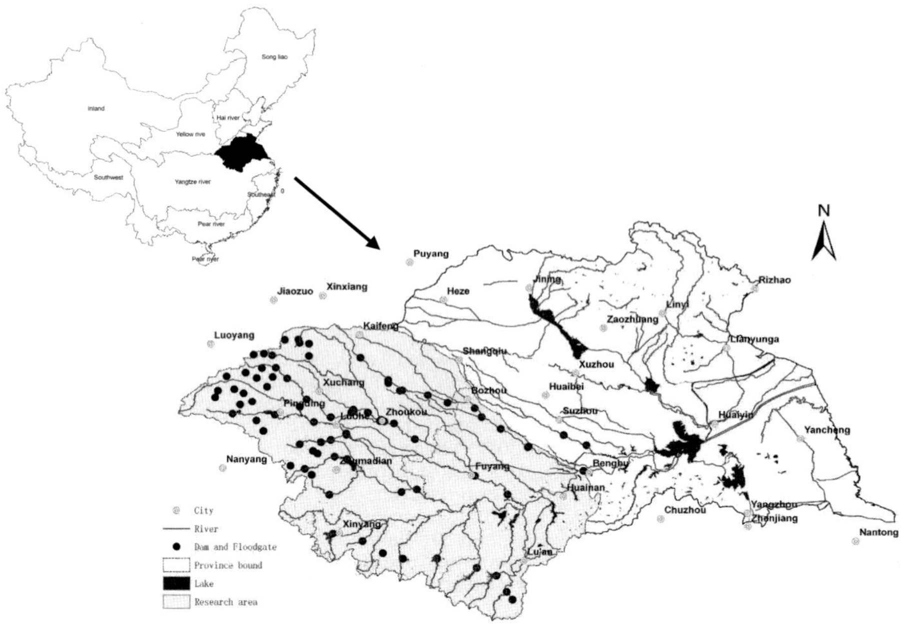

Fig.1 Location of Huai River Basin in China.

MODEL SELECTION AND IMPROVEMENT

SWAT is a continuous-time, long-term, distributed-parameter model (Arnold *et al.*, 1998) and has been applied worldwide for hydrological and water quality modelling, e.g. it has been incorporated into the US Environmental Protection Agency (USEPA) Better Assessment Science Integrating Point and Nonpoint Sources (BASINS) software package (http://www.epa.gov/waterscience/basins/) and been applied by the United States Department of Agriculture for the Conservation Effects Assessment Project (USDA-CEAP) (http://www.nrcs.usda.gov/TECHNICAL/NRI/ceap/). In China, SWAT has been applied widely, e.g. Hei River basin, Hai River basin, Hanjiang basin, Yellow River basin, etc. (Liu *et al.*, 2003, 2006; Wang *et al.*, 2003; Chen *et al.*, 2004; Zhang *et al.*, 2005). It is widely demonstrated and agreed that SWAT is well established because of its flexibility and suitability for hydrologic simulation in complex basins. Moreover, SWAT includes the reservoir simulation function and it can be used to analyse the impact of reservoirs on the hydrological cycle.

In the SWAT model, the water balance for a reservoir is a function of V_{stored} (water stored volume in the reservoir at the beginning of a day), V_{flowin} (water volume entering the reservoir during the day), V_{pcp} (precipitation volume falling into the reservoir), V_{evap} (water removed volume by evaporation), V_{seep} (water lost volume by seepage):

$$V = V_{stored} + V_{flowin} - V_{flowout} + V_{pcp} - V_{evap} - V_{seep} \tag{1}$$

where V is water storage of the reservoir at the end of a day. The three variables (V_{pcp}, V_{evap}, and V_{seep}) have the relationship with the water surface area (SA) of reservoir, and the area of water surface varies with the change in the water storage volume in the reservoir.

$$\begin{cases} V_{pcp} = 10 \cdot R_{day} \cdot SA \\ V_{evap} = 10 \cdot \eta \cdot E_0 \cdot SA \\ V_{seep} = 240 \cdot K_{sat} \cdot SA \\ SA = \beta_{sa} \cdot V^{\exp sa} \end{cases} \tag{2}$$

where η is an evaporation coefficient; E_0 is the potential evapotranspiration for a given day; K_{sat} is the effective saturated hydraulic conductivity of the reservoir bottom; β_{sa} is a coefficient and *expsa* is an exponent.

For the calculation of $V_{flowout}$, there are four methods in SWAT, i.e. measured daily outflow, measured monthly outflow, average annual release rate for uncontrolled reservoir, and controlled outflow with target release (Neitsch *et al.*, 2002). But these four methods do not consider the regulation of water projects and are not suitable for the complicated hydrological simulation in the dam-crowded basins where the data is absent. Therefore, SWAT model should be modified before applying it to Huai River basin.

In China, the dispatch rule of dams and floodgates is that water discharge of a dam or floodgate ($V_{flowout}$) is determined by the water level or water storage volume. By analysing water discharge, water storage and water level of 23 dams and floodgates in HRB, we developed several equations to describe the relationship between water discharge of dam and floodgate and the corresponding water storage and water level. It was found that there are strong correlations between water discharge and water storage or water level. For example, in Baiguishan reservoir:

$$\begin{cases} V_{flowout} = 388 \cdot V^2 - 378.92 \cdot V + 37.351 & R^2 = 0.9958 \\ V = 0.0085 \cdot H^2 - 2.5973 \cdot H + 198.29 & R^2 = 0.9999 \end{cases} \quad (3)$$

In Bengbu floodgate:

$$\begin{cases} V_{flowout} = -31.45 \cdot V^2 + 888.95 \cdot V + 1180.8 & R^2 = 0.9535 \\ V = 12.676 \cdot H^2 + 521.97 \cdot H - 9697.9 & R^2 = 0.9604 \end{cases} \quad (4)$$

where V and H are water storage at the end of a day ($10^8 m^3$) and water level of the reservoir (m), respectively. R^2 is the coefficient of determination (Fig. 2). These newly developed relationship equations for the dams and floodgates in HRB were incorporated into the reservoir routing module of SWAT. These equations substantially improved the prediction of reservoir discharge in HRB.

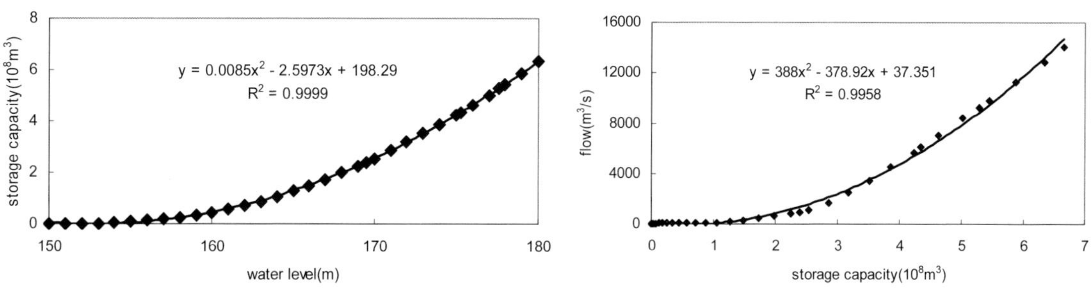

Fig. 2 Illustration of the relationship between storage capacity with water level and the relationship of flow with storage capacity of Baiguishan reservoir.

MODEL DEVELOPMENT

Setting up SWAT model

The extended SWAT model was set up using the AVSWAT2000 interface. Bengbu floodgate in the middle of Huai River was selected as the outlet of the whole basin. 129 sub basins and 468 hydrological response units (HRU) were divided, and the SWAT model considering the regulation of water projects was developed (see Fig. 3).

The codes in many files (e.g. res.f, readres.f, allocate_parms.f) of SWAT were modified according to equations (1)–(4) in order to incorporate the dams and floodgates' dispatch rules to simulate their outflow.

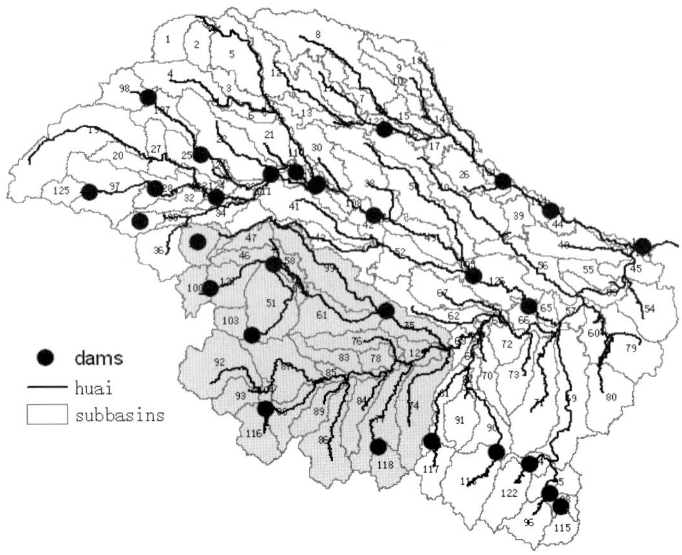

Fig. 3 The digitized streams, the position of dams and floodgates and the sub-basin division.

Calibration and validation

Ten years (1991–2000) daily precipitation, temperature, wind speed, solar radiation and relative humidity from 199 weather stations, and monthly flow data from 23 dams and floodgates (14 reservoirs and 9 floodgates) and 4 hydrological stations (Wangjiaba, Changtaiguan, Xixian and Lutaizi) in Shaying River, Hongru River, Guo River and Huai River were selected to calibrate and validate the distributed modelling system. The first 8 years (1991–1998) data were used for parameter calibration and the last two years (1999–2000) data were used for validation.

Six sensitive parameters (CN2, ESCO, EPCO, SOL_AW, SOL_K and SOL_Z) in the HRB SWAT model have been adjusted to match the simulated and observed flow data at multiple monitoring stations (Romanowicz, 2005). The agreement between simulated and observed data was evaluated using several evaluation indices, which are the relative volume error (*rVol*), correlation coefficient (*R*), and Nash-Sutcliffe efficiency coefficient (NSEC). The definitions of the different evaluation coefficients are provided in Table 1.

Table 1 The evaluation indices.

Definition	Reference	Comments
$rVol=\frac{\Sigma\left(Q_{ob,i}-Q_{si,i}\right)}{\Sigma Q_{ob,i}}$	Relative volume error	The optimal statistical value is close to 0
$R=\frac{\sum(Q_{ob,i}-\overline{Q}_{ob})\cdot(Q_{si,i}-\overline{Q}_{si})}{\sqrt{\sum(Q_{ob,i}-\overline{Q}_{ob})^2\cdot\sum(Q_{si,i}-\overline{Q}_{si})^2}}$	Correlation coefficient	The optimal statistical value is close to 1
$NSEC=1-\frac{\sum(Q_{ob,i}-Q_{si,i})^2}{\sum(Q_{ob,i}-\overline{Q}_{ob})^2}$	Nash–Sutcliffe (Nash & Sutcliffe, 1970)	The optimal statistical value occurs when the value does reach 1

Note: Q_{ob} Observed value; $\overline{Q}_{ob}$ mean observed value; Q_{si} simulated value; $\overline{Q}_{si}$ mean simulated value.

RESULTS AND DISCUSSION

The selected 14 reservoirs located in the headstream of Huai River basin, all of which have large storage capacity for flood control and water supply. Comparing the runoff simulation with and without considering the reservoir storage, 50% (7/14) of *rVol* are within ±0.15 in the calibration period. The average *R* increased from 0.59 to 0.72 with the minimum value increased from 0.36 to 0.77 (Baisha reservoir). The average NSEC increased from 0.19 to 0.23 with the minimum value of –0.41 (Gushitan reservoir) and increased a lot after improving the model. In the validation period, the average *R* changed from 0.57 to 0.75 with the minimum value increased from –0.16 to 0.55 (Baisha reservoir). The average NSEC increased from 0.13 to 0.35 and the minimum value rose from –0.78 to 0.18 (Boshang reservoir). The results are shown in Fig. 4 and Table 2.

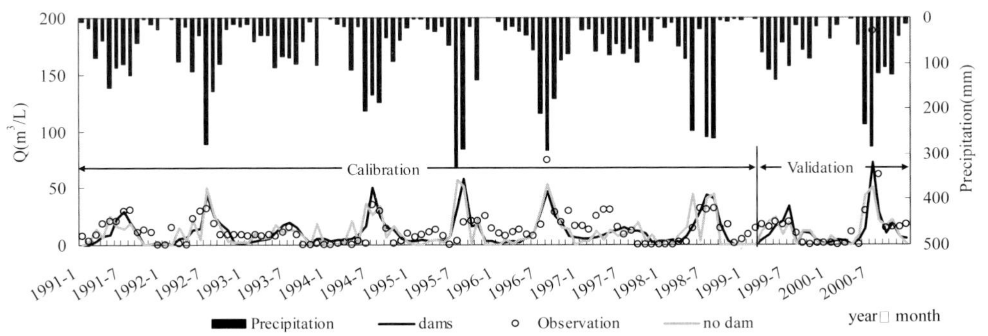

Fig. 4 the simulated and observed runoff at Zhaopingtai reservoir.

Table 2 The modelling performance of 14 reservoirs in Huai River basin.

Dams and floodgates	Calibration period (1991–1998)						Validation period (1999–2000)					
	With dams			Without dams			With dams			Without dams		
	rVol	r	NSEC	rVol	r	NSEC	rVol	r	NSEC	rVol	r	NSEC
Baiguishan	–0.18	0.59	0.07	–0.19	0.47	–0.32	0.26	0.95	0.67	0.25	0.86	0.54
Baisha	0.36	0.77	0.57	0.57	0.36	0.01	–1.02	0.55	–1.16	0.61	–0.16	–0.53
Banqiao	–0.23	0.68	0.26	–0.15	0.69	0.32	0.26	0.88	0.55	0.3	0.88	0.54
Boshan	0.05	0.79	0.61	–0.03	0.65	0.41	–0.39	0.89	0.18	–0.47	0.62	–0.79
Gushitan	–0.71	0.54	0	–0.79	0.53	–0.41	0.42	0.82	0.45	0.31	0.7	0.4
Meishan	–0.01	0.7	0.49	–0.02	0.55	0.3	–0.01	0.81	0.65	–0.06	0.17	–0.37
Mozitan	0.01	0.81	0.65	0.01	0.68	0.46	–0.34	0.78	0.53	–0.39	0.57	–0.03
Nanwan	0.05	0.8	0.64	–	–	–	0.15	0.82	0.64	–	–	–
Nianyushan	0.01	0.73	0.53	0.02	0.69	0.47	0.16	0.37	0.11	0.13	0.5	0.24
Xianghongdian	0.08	0.74	0.55	0.08	0.52	0.27	0.27	0.87	0.62	0.1	0.42	0.14
Zhaopingtai	0.09	0.64	0.25	0.09	0.47	–0.26	0.42	0.89	0.52	0.4	0.72	0.34
Suyahu	–0.26	0.67	0.43	–0.26	0.67	0.43	0.01	0.76	0.47	–0.01	0.79	0.56
Fuziling	0.07	0.89	0.63	0.07	0.82	0.54	0.06	0.82	0.63	–0.27	0.77	0.54
Pohe	–1.21	0.68	–2.5	–	–	–	0.35	0.34	0	–	–	–
Average	–	**0.72**	**0.23**	–	**0.59**	**0.19**	–	**0.75**	**0.35**	–	**0.57**	**0.13**
Bengbu	–0.27	0.74	0.52	–0.29	0.74	0.52	0.39	0.71	0.36	0.36	0.59	0.23
Fuyang	–0.03	0.84	0.69	0.03	0.74	0.54	–0.19	0.83	0.67	–0.03	0.97	0.9
Huaxing	0.07	0.89	0.63	–	–	–	0.06	0.82	0.63	–	–	–
Huaidian	–0.13	0.82	0.67	–0.06	0.78	0.58	–0.05	0.69	0.48	0.14	0.95	0.72
Huangqiao	0.17	0.71	0.42	0	0.67	0.19	–0.18	0.8	0.63	–0.49	0.83	0.61
Mawan	–0.19	0.79	0.6	–0.2	0.74	0.5	0.05	0.96	0.79	–0.2	0.74	0.5
Mengcheng	–0.13	0.73	0.54	–0.15	0.71	0.5	0.11	0.92	0.83	0.09	0.9	0.81
Zhoukou	0	0.97	0.94	0.07	0.79	0.61	0.02	1	0.97	0.21	0.95	0.69
Xuanwu	–0.26	–0.03	–0.46	–0.43	0.02	–1.85	0.18	0.96	0.58	0.1	0.84	0.66
Average	–	**0.72**	**0.51**	–	**0.65**	**0.2**	–	**0.85**	**0.66**	–	**0.85**	**0.64**
Wangjiaba	–0.03	0.94	0.86	–0.15	0.9	0.8	0	0.99	0.97	–0.05	0.92	0.85
Changtaiguan	–0.03	0.89	0.78	–0.03	0.89	0.78	0.03	0.92	0.81	0.03	0.92	0.81
Xixian	–0.03	0.91	0.8	0.37	0.91	0.61	–0.1	0.95	0.88	0.37	0.91	0.61
Lutaizi	–0.06	0.93	0.82	–0.19	0.89	0.77	–0.14	0.93	0.84	–0.24	0.86	0.7
Average	–	**0.88**	**0.75**	–	**0.85**	**0.63**	–	**0.93**	**0.83**	–	**0.89**	**0.72**

The research floodgates located in the middle and lower reach, whose purpose are flood control have less effect on water storage. Comparing runoff simulation with and without the storage of floodgates, *rVol* of the five floodgates were within ±0.15 in the calibration period, and the average *R* increased from 0.65 to 0.72. The average NSEC increased from 0.20 to 0.51 with the minimum value changed from –1.85 to –0.46 (Xuanwu gate). In the validation period, the average *R* has no remarkable change, but the minimum value increased from 0.23 to 0.36 (Bengbu gate). Compared to the modelling performance of the reservoirs, the floodgate results have been greatly improved, and the main reason is that the relatively small storage capacity has less effect on runoff. The results are shown in Fig. 5 and Table 2.

For the four hydrological stations, *rVol* of all cross-sections increased obviously after considering the storage of dams, and all of them were less than ±0.15. In the calibration period, the average *R* increased from 0.85 to 0.88 and the average NSEC increased from 0.61 to 0.75 and the minimum value increased from 0.61 to 0.78 (Xixian station). In the validation period, the average *R* increased from 0.89 to 0.93. The average NSEC increased from 0.72 to 0.83 with the minimum value increased from 0.61 to 0.78 (Xixian station). Comparing the simulation the performance of reservoirs and floodgates, runoff process of hydrological stations has not disturbed by human activities, thus the results are the best. The results are shown in Fig. 6 and Table 2.

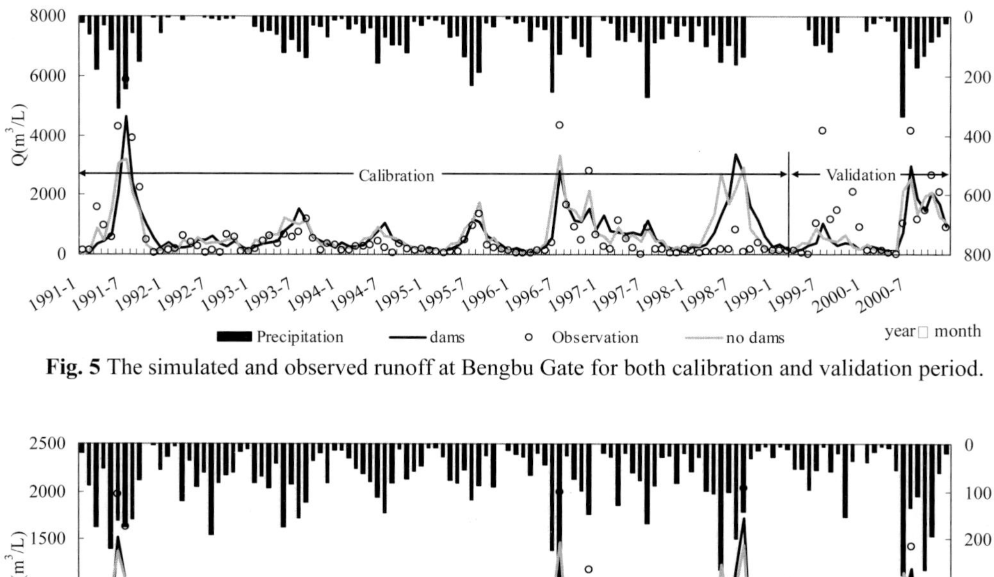

Fig. 5 The simulated and observed runoff at Bengbu Gate for both calibration and validation period.

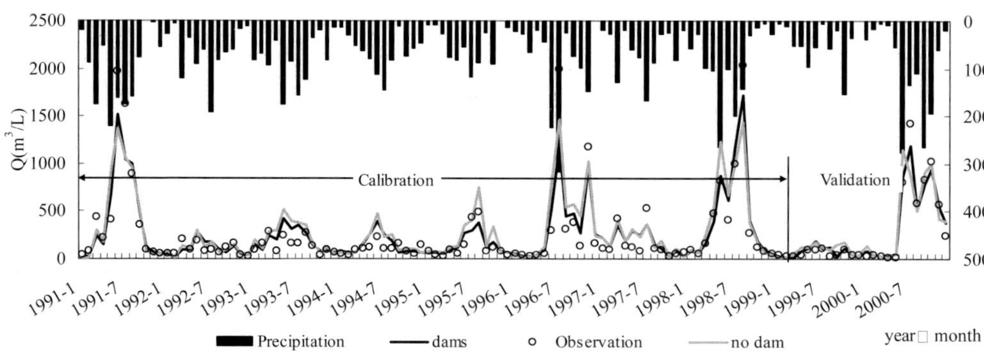

Fig. 6 the simulated and observed runoff at Wangjiaba station for both calibration and validation period.

As shown in Table 3, in summary, for the calibration period, 67% (18) of all the 27 monitoring stations have *rVol* within ±0.15. The average *R* is 0.75, and the average NSEC is 0.41. For the validation period, 48% (13) of all the stations have *rVol* within ±0.15. The average *R* is 0.82, and the average NSEC is 0.53. In comparison to the results without considering the dams'

Table 3 The modelling performance contrasting the upgraded SWAT with the original model.

Period	Average correlation coefficient (*R*)			Average efficiency coefficient (NSEC)		
	With dams	Without dams	Increase amplitude	With dams	Without dams	Increase amplitude
Calibration period	0.75	0.66	0.14	0.41	0.28	0.46
Validation period	0.82	0.72	0.14	0.53	0.40	0.33

storage, the average *R* and the average NSEC increased 14% and 46%, respectively, in the calibration period, and increased 14% and 33%, respectively, for the validation period. This indicates that the modelling performance of HRB SWAT has been elevated after improvement, especially for NSEC. The modified SWAT model can better simulate the flow processes of HRB.

CONCLUSIONS

After continuous construction of water projects for many years, Huai River basin has many self-contained water project systems and become the area of highest density of dams and floodgates in China. These intensive human activities have disturbed the natural water cycle. Analysing the water cycle process and evaluating the impact of dams and floodgates on the flow regime are significant tasks for the integrated management of HRB. In this study, we modified SWAT model using the dispatch rules derived for the dams and floodgates in HRB. The calibration and validation indicate that this upgraded SWAT model can predict monthly flow volume well at multiple monitoring sites within HRB. This newly developed distributed modelling system is expected to be used to assess the dams and floodgates management practices on the water cycle, and provide technical support for the integrated management of HRB.

Acknowledgements This study was supported by Key Project of International Cooperation of the Natural Science Foundation of China (no. 40721140020) and the 44th postdoctoral sciences foundation of China "Research on the mechanism of paroxysmal water pollution in Huai River Basin". Thanks also to Dr Liu Yongbo in the Department of Geography in Guelph University for his help.

REFERENCES

Arnold, J. G., Srinivasan, R., Muttiah, R. S. & Williams, J. R. (1998) Large area hydrologic modeling and assessment part I: model development. *J. Am. Water Resour. Assoc.* **34**(1), 73–89.

Berkamp, G., McCartney, M. *et al.* (2000) *Dams, Ecosystem Functions and Environmental Restoration.* Thematic Review II.1 prepared as an input to the World Commission on Dams, Cape Town, South Africa.

Chen, J. F. & Chen, X. W. (2004) Water balance of the SWAT model and its application in the Suomo basin. *Acta Scientiarum Naturalium Universitatis Pekinensis* **40**(2), 265–270.

Liu, C. M., Li, D. F. & Tian, D. (2003) An application study of DEM based distributed hydrological model on macro scale watershed. *Progr. Geogr.* **22**(5), 437–445 (in Chinese).

Liu, C. M., Zheng, H. X. & Wang, Z. G. (2006) *Distributed Simulation of Watershed Hydrological Cycle.* Yellow River Conservancy Press, Zheng zhou, China (In Chinese).

Liu, Y. B., Yang, W. H. & Wang, X. X. (2008) Development of a SWAT extension module to simulate riparian wetland hydrologic processes ata watershed scale. *Hydrol. Processes* **22**, 2901–2915.

Nash, J. E. & Sutcliffe, J. E. (1970) River flow forecasting through conceptual models. Part I. A discussion of principles. *J. Hydrol.* **10**(3), 282–290.

Neitsch, S. L., Arnold, J. G., Kiniry, J. R., King, K. W. & Williams, J. R. (2002) *Soil and Water Assessment Tool (SWAT) theoretical documentation.* Blackland Research Center, Texas Agricultural Experiment Station, Temple, Texas, USA, BRC Report 02-05.

Romanowicz, A. A., Vanclooster, M. *et al.* (2005) Sensitivity of the SWAT model to the soil and land use data parametrisation: a case study in the Thyle catchment, Belgium. *Ecol. Modelling* **187**, 27–39.

Suo, L. S. (2005) Dams and ecology. *Rural Hydropower & Electrigicatian in China* **8**, 3–5 (in Chinese).

Wang, Z. G., Liu, C. M. & Huang, Y. B. (2003) The theory of SWAT model and its application in Heihe Basin. *Progr. Geogr.* **22**(1), 79–86 (in Chinese).

Zhang, D., Zhang, W. C. & Xu, Q. Z. (2005) Comparison and modification of evapotranspiration estimation method in Upper Hanjiang Basin. *Resources Sci.* **27**(1), 97–103 (in Chinese).

Impacts of climate variation and land-use change on river discharge in the upstream of the Miyun Reservoir

BING GAO & DAWEN YANG

State Key Laboratory of Hydro-Science and Engineering & Department of Hydraulic Engineering, Tsinghua University, Beijing 100084, China

gb03@mails.tsinghua.edu.cn

Abstract Both climate variation and change in land-use and cover caused by human activities influence river discharge. It is necessary to distinguish the impacts of climate variation and land-use and cover change in the analysis of the variation of river discharge and water resources. The study area is the upstream of the Miyun Reservoir, which consists of the Chao and the Bai rivers, in northern China. In the present study, a distributed hydrological model, GBHM, is used to simulate the natural runoff without artificial water control and intake. Corresponding to different land-use types in two different periods, the 1980s and 1990s, land surface parameters are used in the model for characterizing land-use and cover changes. We use the fixing-changing method to run the model, firstly the climate variation factor is fixed, and the land-use is changed; then the land-use factor is fixed and the climate is changed. The results of model simulation show the quantitative effects of climate variation and land-use change on river discharge. It is found that without climate variation the land-use/cover change reduces the river discharge in the Chao River, but in the Bai River basin the impact of land-use change is the opposite. The spatial distribution of the evaporation is highly influenced by land-use/cover change, and the climate variation contributes the majority of the river flow change from the 1980s to the 1990s.

Key words river discharge; impact of land-use and cover change; distributed hydrological model; the Miyun Reservoir

INTRODUCTION

River discharge, which is closely related to reliable water resources, is given much attention for water resources management. Climate variation and land-use/cover change (LUCC) affect the runoff generation. Therefore, analysing the impacts of climate variation and LUCC upon the hydrological cycle will help us to evaluate the sustainability of regional water resources. Recently, lots of research works focused on this issue (Samaniego *et al.*, 2006; Guo *et al.*, 2008). Barlage *et al.* (2002) used the modified BATS model to quantify the impacts of LUCC and future climate change on runoff in a Great Lakes watershed. Seguis *et al.* (2004) used a physically-based distributed hydrological model to study the impacts of climate change and land-clearing on runoff in a small Sahelian catchment and found that the effect of land-use change was larger than the effect of climate change. A number of studies had also been carried out in the upstream of Miyun Reservoir. Sun *et al.* (2007, 2008) studied the impacts of changes in land-use/cover on annual streamflow of the Chao River basin using a lumped conceptual model IHACRES. Wang *et al.* (2006, 2009) used a distributed conceptual model DTVGM to identify the impact of climate variation and LUCC by a fix-changing method in the upstream of Miyun Reservoir.

The main objective of this study is to analyse the influences of climate variations and LUCC upon river discharge over the upstream of Miyun Reservoir. A physically-based distributed hydrological model GBHM is employed to analyse these impacts in this region. Compared with the conceptual models, in this study, high spatial resolution land-use and vegetation cover data in two periods of 1980s and 1990s are directly used as the inputs of the model. The parameters which relate to the land-use/cover have physical meanings and the mechanism of LUCC on water cycle is shown by the model simulation. The spatial heterogeneity of LUCC, evaporation, precipitation and runoff are analysed in this study.

STUDY AREA AND DATA PREPARATION

Study area

The study area is the upstream catchment of the Miyun Reservoir, which belongs to the Hai River basin in northern China. There are two tributaries in this catchment: the Chao River and the Bai

River, and their confluence is at the Miyun Reservoir. The Chao River basin, with an area of 5340 km^2, is controlled by Xiahui station, and the Bai River basin, with an area of 8506 km^2, is controlled by Zhangjiafen station. The Miyun Reservoir is the main source for water supply to Beijing City. Since the 1980s, the Reservoir has been supplying about 0.9 billion m^3 water to Beijing city every year.

Data used in the study

The data used in the study includes the DEM, soil type, climate data, land-use, NDVI data, and observed river discharge data. Table 1 summarizes the details of data used in the study. The DEM of the basin is shown in Fig. 1. In the study, 6 hydrological gauges on the river (see Fig. 2) are selected to analyse the change of river discharge. The observed discharge of the Xiahui gauge and the Zhangjiafen gauge cannot represent the natural runoff because of upstream reservoir regulation and water intake. The annual natural runoff data of the Xiahui and Zhangjiafen station is obtained from the Water Resources Assessment Report of China, which is the sum of the observed discharge and the artificial regulation and consumption.

Table 1 Description of data used in the study.

Data type	Data contents	Data source
DEM	Approximately 90 m resolution digital elevation data of the whole China	Global topography database
Land-use	Land cover type, totally 26 types in China, reclassified into 9 types in this study, 100 m resolution	The Resources and Environment Data Center of Chinese Academy of Sciences
Soil type	Totally 237 types in China, there are mainly 5 types in this study catchment	1:1 000 000 Digital Soil Map of the World and Derived Soil Properties (FAO, 2003)
NDVI	Monthly NDVI data, 8 km resolution , 1982 onwards	NOAA
Climate data	9 stations, daily record of precipitation/temperature/humidity /windspeed /sunshine time, from 1956 to 2005	China Meteorological Administration
Precipitation	70 rainfall gauges, daily record of rainfall or snowfall from 1980 to 2005	The Bureau of Hydrology, Ministry of Water Resources of China
Observed discharge	6 hydrological station, observed daily river discharge from 1980 to 2005	The Bureau of Hydrology, Ministry of Water Resources of China

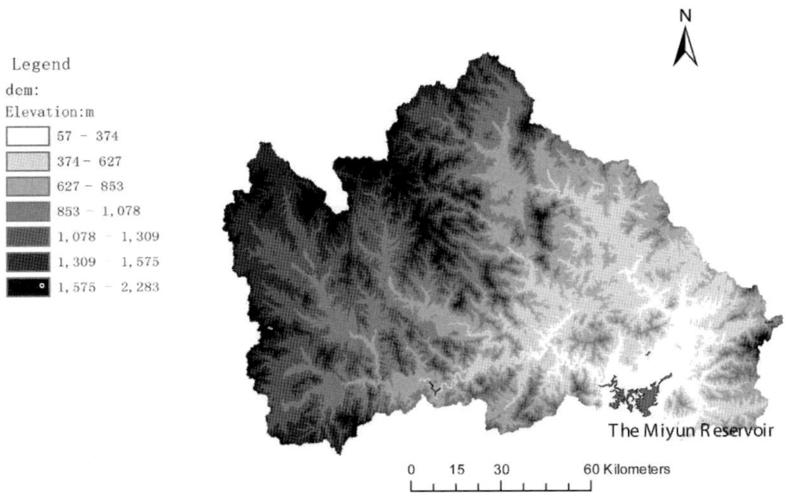

Fig. 1 DEM of the study area.

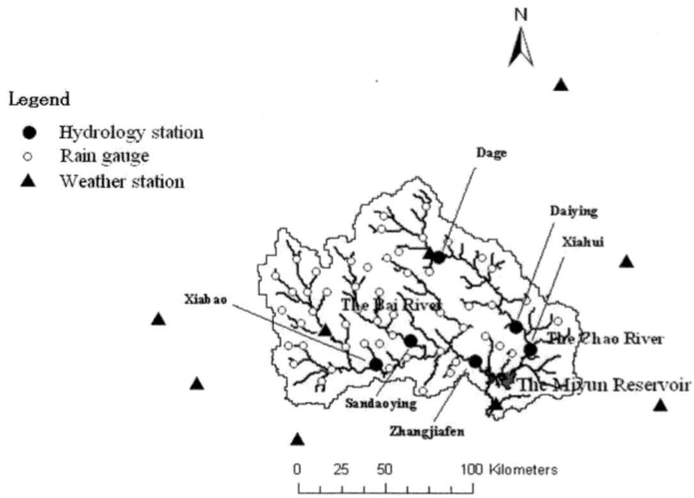

Fig. 2 Locations of weather stations, hydrology stations and raingauges in the study area.

Model description

The present study employs the GBHM model (Yang *et al.*, 1998, 2000) for estimating the natural runoff in the upstream catchment of the Miyun Reservoir, in which only natural hydrological processes are included (there is no consideration of irrigation or reservoir control).

A total of 81 sub-catchments have been identified in the basin. The model uses a 1-km grid which is simplified into a number of hillslopes to describe the topography. The hillslopes located in a 1-km grid are grouped according to the land cover types. The hydrological processes considered in the hillslope are snowmelt, canopy interception, evapotranspiration, infiltration, surface flow, subsurface flow and the exchange between the groundwater and the river (Yang *et al.*, 2002).

The required gridded hydrological input is interpolated from the gauge data (Yang *et al.*, 2000, 2004), taking into consideration the distance and direction of the gauges to the target grid. The degree of canopy interception is described using a linear function of the leaf-area index (LAI), which is estimated from the monthly normalized difference vegetation index (NDVI) data according to different land-use/cover types. The actual evapotranspiration is calculated from the potential evaporation by considering seasonal variation of LAI, root distribution and soil moisture availability. The vertical water flow in the topsoil is represented by Richards equation and solved by an implicit numerical solution scheme. The surface runoff flows through the hillslope into the stream via kinematic wave. The groundwater aquifer is treated as an individual storage corresponding to each grid. The exchange between the groundwater and the river water is considered as steady flow and is calculated by Darcy's law (Yang *et al.*, 2002, 2004). The runoff generated from the grid is the lateral inflow into the river at the same flow interval. Flow routing in the river network is solved using the kinematic wave approach.

Model calibration and validation

The observed discharge at the Xiahui gauge and the Zhangjiafen gauge cannot be used to calibrate the model because of reservoir operation and water taken for irrigation in the upstream of this gauge. Two sub-catchments were selected for the calibration: the upstream of the Dage gauge in the Chao River with a drainage area 1850 km^2 and the upstream of the Sandaoying gauge in the Bai River basin with a drainage area of 1600 km^2 (see Fig. 2). There is little human impact in the two regions, and the model parameters calibrated in the two sub-catchments can represent the natural hydrological characteristics in the whole study area. Two periods of 1981–1990 and 1991–2000 are selected for model calibration and validation, respectively. Land-use data of the late 1980s and NDVI data from 1982 to 1990 are selected to represent the land-use in the calibration period, and land-use data of the late 1990s and NDVI data from 1991 to 2000 are selected to represent land-use in the validation period. The Nash coefficient (*R*2) of the monthly runoff is calculated to evaluate the model performance. The results are showed in Table 2 and the comparison of the simulated and observed discharge is shown in Figs 3 and 4. It can be seen that the hydrological model has sufficient accuracy for simulating the river discharge and can be used for analysing the impacts of climate fluctuation and LUCC.

Table 2 Model performance during different periods.

	The Dage gauge		The Sandaoying gauge	
	1981–1985	1986–1990	1981–1985	1986–1990
*R*2	0.88	0.60	0.83	0.77
	1991–1995	1996–2000	1991–1995	1996–2000
*R*2	0.60	0.71	0.95	0.81

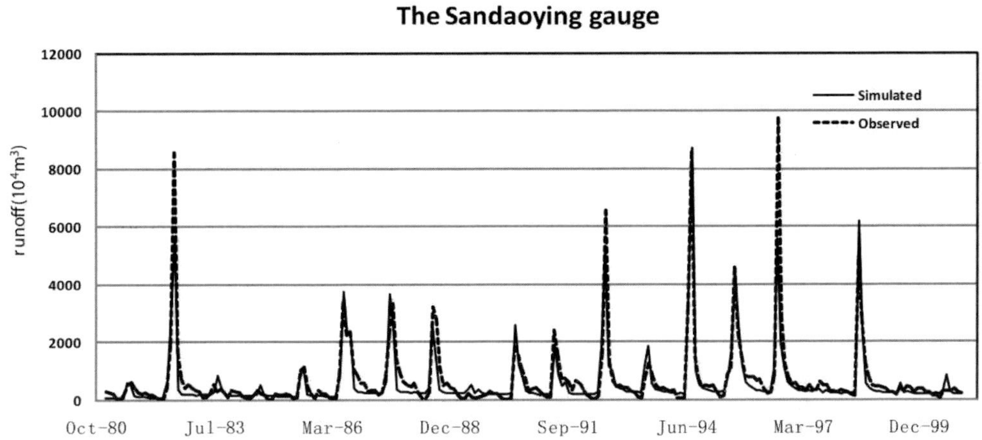

Fig. 3 Comparison of the observed and simulated monthly runoff at the Sandaoying station.

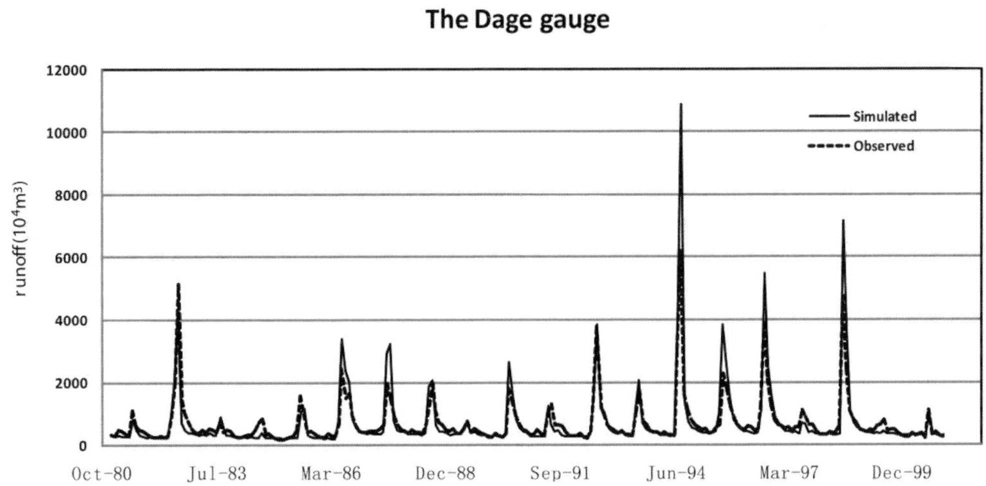

Fig. 4 Comparison of the observed and simulated monthly runoff at the Dage station.

ANALYSING THE IMPACTS OF LUCC AND CLIMATE VARIATION

Table 3 shows the comparison of the land-use of the study basin in the 1980s and 1990s. In the Chao River, compared with the 1980s, the area of grassland decreased by 11.4% and farmland changed little, while the area of forest and shrub increased by 11.5% in the 1990s. In the Bai River basin, compared with the 1980s, the area of grassland decreased by 12.6% and farmland decreased by 7.3%, and the area of forest changed little, the area of shrub increased by 19.7% in the 1990s.

Figure 5 shows the spatial distribution of LUCC in the two periods. In the Chao River, the largest change of land-use and cover from the 1980s to the 1990s was located downstream, where the grassland was changed into forestland and shrub. In the Bai River, the largest change occurred in two regions: in the upstream basin, part of the grassland and crop land was changed into forest, and in the downstream basin part of forest land was changed into shrub. Though the area of forestland has only changed 1%, its spatial distribution has been greatly changed.

Table 3 LUCC in the upstream of the Miyun Reservoir.

Land-use classification and percentage	The Chao River			The Bai River		
	1980s (1)	1990s (2)	(2)–(1)	1980s (3)	1990s (4)	(4)–(3)
Forest	31.3	36.7	5.4	35.0	36.0	1.0
Shrub	18.8	24.9	6.1	11.5	31.2	19.7
Grassland	27.0	15.6	–11.4	29.0	16.4	–12.6
Crop land	21.0	20.9	–0.1	22.9	15.6	–7.3
Other	1.9	1.9	0	1.6	0.8	–0.8

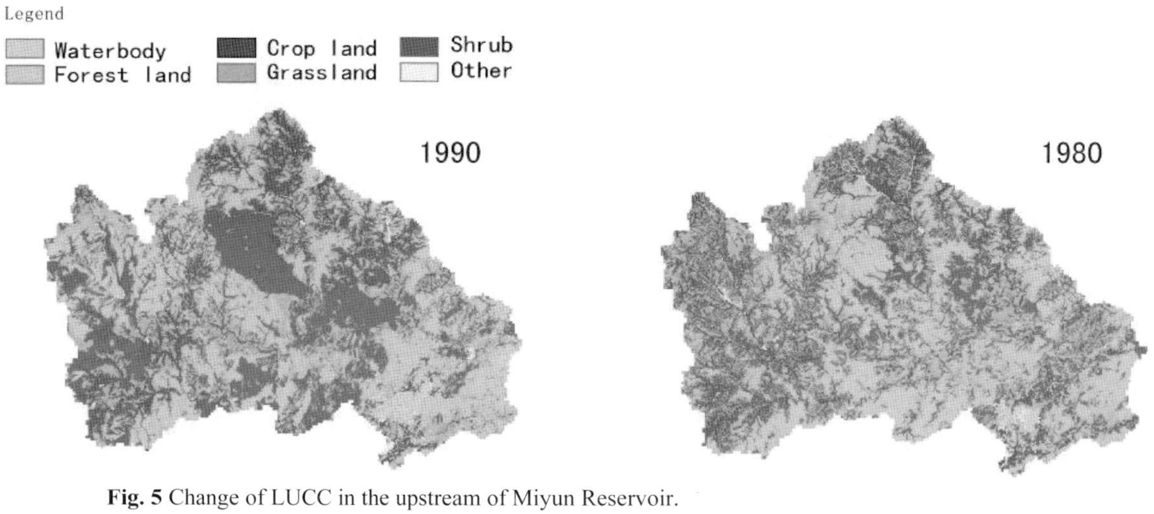

Fig. 5 Change of LUCC in the upstream of Miyun Reservoir.

There are two factors considered in this study: the climate factor and LUCC factor. Two periods of 1981–1990 and 1991–1998 are selected for comparison. The last two years of the 1990s are not included because the annual precipitation is quite a lot lower than in the first 8 years. The runoff of the two periods is simulated under the two land-use and cover scenarios. We run the distributed hydrological model through a fixing-changing method. A combination of the land-use/cover scenario and climate scenario was designed for hydrological simulation. As showed in Table 4, L1 means the land-use/cover data in period 1 (1981–1990), and C1 means the climate data in period 1. L2 means the land-use/cover data in period 2 (1991–1998), and C2 means the climate data in period 2.

Figure 6 shows the spatial distribution of mean annual evaporation over the period 1981–1990 under two land-use and cover scenarios, demonstrating that evaporation is greatly influenced by the LUCC.

Comparing Figs 5 and 6, it is known that the evaporation from the grassland and cropland is lower than the forest land and shrub, and the evaporation of shrub is lower than the forest under the same climate conditions.

Table 4 Water balance under different land-use and cover scenarios.

Scenario	Precipitation (mm)	Evaporation (mm)	Simulated runoff (mm)	Natural runoff (mm)	Potential evaporation (mm)
The Chao River					
C1, L1	498.7	439.0	48.9	46.0	1089.7
C1, L2	498.7	443.8	44.7	46.0	1089.7
C2, L1	513.9	439.9	71.2	72.2	1067.2
C2, L2	513.9	445.8	66.3	72.2	1067.2
The Bai River	Precipitation (mm)	Evaporation (mm)	Simulated runoff (mm)	Natural Runoff (mm)	Potential evaporation (mm)
C1, L1	447.7	390.7	48.9	46.0	1088.0
C1, L2	447.7	389.9	51.0	46.0	1088.0
C2, L1	455.3	389.4	69.9	68.4	1066.1
C2, L2	455.3	388.0	72.5	68.4	1066.1

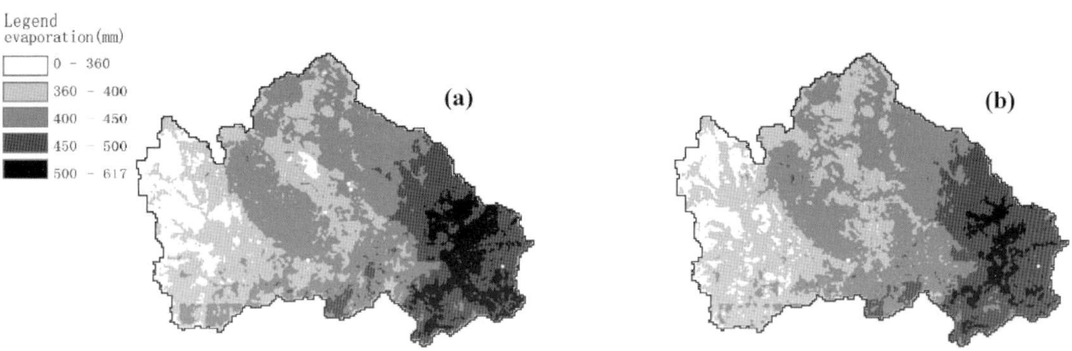

Fig. 6 Spatial distribution of mean annual evapotranspiration over the period of 1981–1990 under two land-use and cover scenarios: (a) 1990s, (b) 1980s.

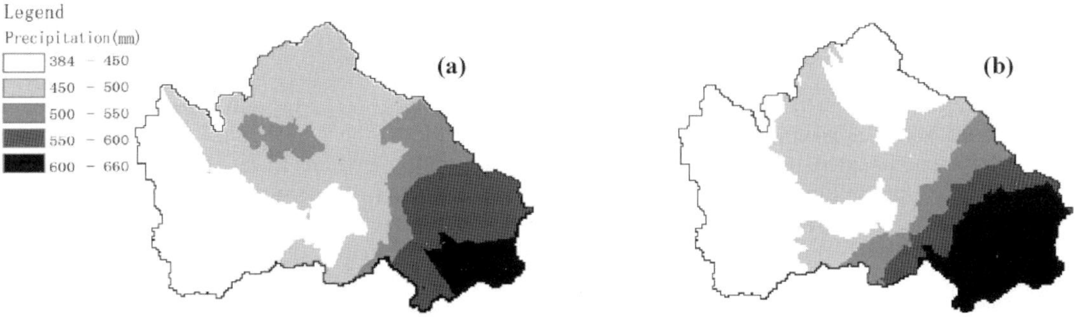

Fig. 7 Annual precipitation over the two periods: (a) 1991–1998, (b) 1981–1990.

From Table 4 we can see that under the same climate scenario, in the Chao River basin, compared with the 1980s, LUCC of 1990s, reduces the runoff, and in the Bai River basin, LUCC of 1990s, increases the runoff. These impacts of LUCC on runoff can be explained by the change of the spatial distribution of the land-use. From the *NDVI* data and the land-use map, the annual vegetation coverage of main land-use types were calculated by using the method proposed by Gutman & Ignatov (1998):

$$M = (NDVI - NDVI_{min})/(NDVI_{max} - NDVI_{min}) \tag{1}$$

where M is the vegetation coverage. $NDVI_{max}$ and $NDVI_{min}$ are the *NDVI* signals from bare soil and dense green vegetation, respectively. In our study, we prescribe the $NDVI_{max} = 0.05$ and $NDVI_{min} = 0.80$. As shown in Table 5, it is found that the vegetation coverage of forest and shrub is larger than cropland and grassland. Forest and shrub take more canopy interception than crop land and grassland. Increasing of forest and shrub will reduce the surface runoff and increase the canopy evaporation. The vegetation coverage of forest land is larger than shrub (see Table 5) and the transpiration of forest is larger than shrub. In the Chao River basin, the areas of forest and shrub increase, the evaporation increases and reduces the runoff. In the upstream of the Bai River basin, the grassland turns into forest, which reduces the runoff, but in the downstream of the Bai River, the forest turns into shrub, which increases runoff. The precipitation is higher in the downstream than in the upstream and contributes more runoff. Therefore, it is shown that the land-use change increases the river runoff in the Bai River basin.

Table 5 Vegetation coverage (in percentage) of each land-use type.

	The Chao River				The Bai River			
	Forest	cropland	grassland	shrub	forest	cropland	grassland	shrub
1980s	0.447	0.430	0.430	0.434	0.456	0.394	0.407	0.429
1990s	0.451	0.433	0.439	0.441	0.457	0.398	0.409	0.432

If the land-use and cover does not change, the climate variation will increase the runoff in both sub-catchments because the precipitation increases in period 2 compared with period 1 (see Fig. 7). The spatial distribution of precipitation also changed; the areas with precipitation of between 450 mm and 500 mm increases.

The contribution rate of two factors (LUCC and climate variation) on the runoff change between the two periods can be computed. The differences of runoff simulated under scenario ($C1$,$L1$) and runoff simulated under scenario ($C1$,$L2$) shows the contribution of LUCC. The contribution rate of LUCC is calculated as:

$$CR = \frac{R(C1,L2) - R(C1,L1)}{R(C2,L2) - R(C1,L1)} \tag{2}$$

Similarly, the contribution rate of climate variation is calculated as:

$$CR = \frac{R(C2,L1) - R(C1,L1)}{R(C2,L2) - R(C1,L1)} \tag{3}$$

$R(C1,L1)$ means the runoff simulated under one scenario ($C1$,$L1$), and other items in the equations above are similar. Based on the results in Table 4, the contribution rates of the two factors are calculated. For the Chao River, the contribution rate of LUCC is –24%, and the contribution rate of climate variation is 128%. For the Bai River, the contribution rate of LUCC is 9%, and the contribution rate of climate variation is 89%. It is known that in the Chao and Bai River basin, the change of runoff is mainly caused by the climate variation. In the Chao River basin, the contribution rate of LUCC is negative and this means that the LUCC reduced river discharge if climate does not change.

CONCLUSIONS

A distributed hydrological model is developed for the upstream of the Miyun Reservoir to analyse the impacts of climate variation and land-use/cover change. Through a fix-changing method, the contributions of LUCC and climate variation are identified between two periods: 1991–1998 and 1981–1990. From the simulated results, we can conclude that in the Chao River basin, the LUCC between period 1 and period 2 may reduce the runoff without climate variation, and in the Bai River basin, the LUCC may increase the runoff slightly without climate variation. The runoff change in the study area between the two periods is mainly caused by the climate variation. The spatial distribution of evaporation is highly influenced by LUCC. Under the same climate scenario, the increasing areas of forest and shrub increase the evaporation and reduce the river discharge, but the LUCC from forest to shrub increases the river discharge. From this study, it is known that the fix-changing method is an efficient way to identify the impacts of climate variation and LUCC change. However, the LUCC can also affect the regional climate, and the combination impacts of climate variation and LUCC are nonlinear. Better understanding of the complex interaction between the hydrological processes and climate is needed for better estimating impact of climate variation and land-use change on water resources.

Acknowledgements This study was partially supported by the key research program of the National Science & Technology Pillar Program in the Eleventh Five-year Plan Period (no. 2006BAB14B02) and the research project of the Ministry of Water Resources (no. 200801012).

REFERENCES

Barlage, M. J., Richards, P. L., Sousounis, P. J. & Brenner, A. J. (2002) Impacts of climate change and land-use change on runoff from a Great Lakes watershed. *J. Great Lakes Res.* **28**, 568–582.

FAO (2003) Digital soil map of the world and derived soil properties. *Land and Water Digital Media Series Rev.* **1**.

Gao, Y. C., Yao, Z. J., Liu, B. Q. & Lv, A. F. (2002) Evolution trend of Miyun Reservoir inflow and its motivating factors analysis. *Progr. Geogr.* **21**(6), 546–553 (in Chinese).

Guo Hua, Hu Qi & Jiang Tong (2008) Annual and seasonal streamflow responses to climate and land-cover changes in the Poyang Lake basin, China. *J. Hydrol.* **355**, 106–122.

Gutman, G. & Ignatov, A. (1998) The derivation of the green vegetation fraction from NOAA/AVHRR data for use in numerical weather prediction models. *Int. J. Remote Sens.* **19**(8), 1533–1543.

New, M. & Hulme, M. & Jones, P. (2000) Representing twentieth-century space-time climate variability. Part II: Development of a 1961-96 monthly grids of terrestrial surface climate. *J. Climate* **13**, 2217–2238.

Ott, B. & Uhlenbrook, S. (2004) Quantifying the impact of land-use changes at the event and seasonal time scale using a process-oriented catchment model. *Hydrol. Earth System Sci.* **8**, 62–78.

Samaniego, Luis & Bardossy, Andras (2006) Simulation of the impacts of land-use/cover and climatic changes on the runoff characteristics at the mesoscale. *Ecol. Modelling* **96**, 45–61.

Seguis, L., Cappelaere, B., Milesi, G., Peugeot, C., Massuel, S. & Favreau, G. (2004) Simulated impacts of climate change and land-clearing on runoff from a small Sahelian catchment. *Hydrol. Processes* **18**, 3401–3413.

Sun Ning, Li Xiubin, Ran Shenghong & Li Zijun. (2007) The changes of rainfall–runoff relationship and the impacts of human activities in the Upper Chao watershed. *Progr. Geogr.* **26**(5), 41–47 (in Chinese).

Sun, N., Li, X. B, Li, Z J., Yu, X. X. (2008) Simulation of impacts of changes in land-use and cover on annual stream flow in the upper reach of Chaohe River Basin. *J. Beijing Forestry Univ.* **30**(Supp. 2) (in Chinese).

Terpstra, J. & van Mazijk, A. (2001) Computer aided evaluation of planning scenarios to assess the impact of land-use changes on water balance. *Phys. Chem. Earth (B)* **26**(7–8), 523–527.

Wang Gang-sheng, Xia Jun, Wan Dong-hui & Ye Ai-zhong (2006) A distributed monthly water balance model for identifying hydrological response to climate change and human activities. *J. Nat. Resour.* **21**(1), 86–91 (in Chinese).

Wang, G. S., Xia, J. & Chen, J. (2009) Quantification of effects of climate variations and human activities on runoff by a monthly water balance model: a case study of the Chaobai River basin in northern China. *Water Resour. Res.* 45 (W00A11), doi: 10.1029/2007WR006768.

Yang, D., Herath, S. & Musiake, K. (1998) Development of a geomorphology-based hydrological model for large catchments. *Ann. J. Hydraul. Engng, JSCE* 42, 169–174.

Yang, D., Herath, S. & Musiake K.. (2000) Comparison of different distributed hydrological models for characterization of catchment spatial variability. *Hydrol. Processes* **14**, 403–416.

Yang, D., Herath, S. & Musiake K.. (2002) Hillslope-based hydrological model using catchment area and width functions. *Hydrol. Sci. J.* **47**(1), 49–65.

Yang, D., Li, C., Hu, H., Lei, Z., Yang, S., Kusuda, T., Koike, T. & Musiake, K. (2004) Analysis of water resources variability in the Yellow River of China during the last half century using historical data. *Water Resour. Res.* 40(W06502), doi:10.1029/2003WR002763.

A comparative study of simulation results between IHMS-VSWSC and hybrid runoff model in Dianzi Watershed

LILIANG REN[1],GUIZUO WANG[1], FANG LU[2] & TIANFANG FANG[3]

1 *State Key Laboratory of Hydrology-Water Resources and Hydraulic Engineering, Hohai University, Nanjing 210098, China*
njrll9999@126.com

2 *YRCC Shandong Bureau, Shandong Province, China*

3 *Department of Geography, Texas State University, San Marcos, Texas 78666, USA*

Abstract For the purpose of hydrological process simulation in arid and semi-arid areas, the authors developed the Integrated Hydrological Modelling System based on Various Spatial Water Storage Capacity (IHMS-VSWSC), based on the spatial distributed model of water storage capacity. IHMS-VSWSC included a canopy interception module, snow accumulation and melt module, evapotranspiration module and runoff generation and routing module. To verify the simulation result of IHMS-VSWSC, both IHMS-VSWSC and a hybrid runoff model were applied to the Dianzi watershed to simulate the daily runoff process in the period 1973–1979. After simulations, the authors found that the annual average Nash-Sutcliffe coefficient of IHMS-VSWSC reached 0.80; and the annual average Nash-Sutcliffe coefficient of the hybrid runoff model was only 0.67. Simulated results indicate that the annual Nash-Sutcliffe coefficient is acceptable, and this integrated hydrological modelling system is reasonable for arid and semi-arid area daily runoff simulation.

Key words IHMS-VSWSC; hybrid runoff model; various spatial water storage capacity; Dianzi watershed

INTRODUCTION

The study of hydrological processes simulation in arid and semi-arid areas is a priority focus in the field of hydrology recently. Zhao (1984) developed the Shaanbei model to simulate the runoff process in arid regions. Hu *et al.* (2004) brought forward the runoff–evaporation hydrological model for arid plain oasis. Hu (1993), Bao (1995), Wang (1998), Liang & Xie (2001) and Zheng *et al.* (2002) applied the hybrid runoff model in Loess Plateau arid and semi-arid area. Jia *et al.* (2006) developed the distributed model of hydrological cycle system in the Heihe River watershed which acquired good results. Liu *et al.* (2006) developed the TsingHua Integrated Hydrological Modeling System-for Small Watershed (THIHMS-SW) and applied it in Chabagou watershed. The studies mentioned above mainly covered the northwest inland part of China and the Yellow River Basin in the Loess Plateau. There is less research covering the northeast cold arid region of China. The authors took this aspect as a breakthrough point, developing the Integrated Hydrological Modeling System Based on Various Spatial Water Storage Capacity (IHMS-VSWSC) on the basis of the spatial distributed model of water storage capacity. Furthermore, the available hybrid runoff model was applied in the study area as a contrast to verify the applicability of IHMS-VSWSC.

INTRODUCTION OF MODELS

IHMS-VSWSC

The Integrated Hydrological Modeling System based on Various Spatial Water Storage Capacity (IHMS-VSWSC) was built on the base of this spatial distributed model, saturation excess runoff mechanism and multi-sourced data (digital elevation model data, land cover data, soil type data, etc.). Its simulation processes included canopy interception, snow accumulation and melt, evapotranspiration, runoff yield and concentration.

Spatial distributed model of watershed water storage capacity

The spatial distributed model of watershed water storage capacity was developed based on a tension water capacity curve and watershed storage (or storage deficit) of the Xin'anjiang model (Zhao, 1984) and TOPMODEL (Beven, 1979, 1992). According to the spatial change pattern of vegetation root system depth and different kinds of soil water parameters, the authors described the spatial distribution of watershed water storage capacity and then built the spatial distributed model of watershed water storage capacity. The authors brought out two concepts – grid water storage capacity and grid maximum water storage capacity. The grid water storage capacity of different points in a specified watershed can be calculated by grid vegetation root system depth and field moisture capacity, while maximum water storage capacity has a function relationship with grid water storage capacity. Vegetation root system is a complicated system which includes its depth's uneven spatial distribution in the watershed and the soil layer's constraint to its depth. To solve this problem, on the basis of the acquisition of the data of different kinds of vegetation root system depth, the authors brought out spatial distributed model of watershed water storage capacity and geomorphologic concepts of up-critical slope gradient, down-critical slope gradient and critical soil depth to describe the spatial change of vegetation root system depth.

The critical soil depth is a function of typical vegetation root system, which is shown in equation (1).

$$L_1 = \kappa \times Zrc \tag{1}$$

In equation (1), L_1 indicates the critical soil depth, and the unit is *mm*; κ indicates critical soil depth parameter; *Zrc*, indicates typical vegetation root, system depth, and the unit is mm.

The spatial change pattern of vegetation root system is shown in equation (2).

$$Zr = \begin{cases} L_1 & \theta \geq \theta_1 \\ \dfrac{(\theta - \theta_1)}{(\theta_1 - \theta_2)} \times (Zrc - L_1) + L_1 & \theta_2 < \theta < \theta_1 \\ Zrc & \theta \leq \theta_2 \end{cases} \tag{2}$$

In equation (2), *Zr* indicates vegetation root system depth, and the unit is mm; θ indicates slope gradient, and the unit is (°); θ_1 indicates up-critical slope gradient, and the unit is (°); θ_2 indicates down-critical slope gradient, and the unit is (°).

According to the spatial change pattern of vegetation root system depth and different kinds of soil water parameters, the authors described the spatial distribution of watershed water storage capacity and then built the spatial distributed model of watershed water storage capacity.

Grid water storage capacity is the product of vegetation root system depth and field water capacity as shown in equation (3).

$$Sc = \phi \times Zr \tag{3}$$

In equation (3), *Sc* indicates grid water storage capacity, and the unit is mm; ϕ indicates field water capacity; and *Zr* indicates grid vegetation root depth, and the unit is mm.

Grid maximum water storage capacity is a function of grid water storage capacity which is bigger than grid water storage capacity and replies on grid maximum soil water storage capacity, as shown in equation (4).

$$S = \beta \times Sc \tag{4}$$

In equation (4) *S* indicates grid maximum water storage capacity, and the unit is mm; β indicates a parameter.

Grid saturation excess runoff model

The grid saturation excess runoff module could be built on the basis of the spatial distributed model of water storage capacity and saturation excess runoff mechanism. In the saturation excess runoff module, net rainfall flows into soil directly, and then increases the soil water content; vegetation transpiration and soil water evaporation depletes soil water and then decreases the soil water content.

In Fig. 1, P_g indicates the precipitation volume (mm); *P* indicates the net rainfall volume (mm); *EI* indicates canopy interception evaporation (mm); *I* canopy interception volume; I_{max} indicates maximum interception capacity.

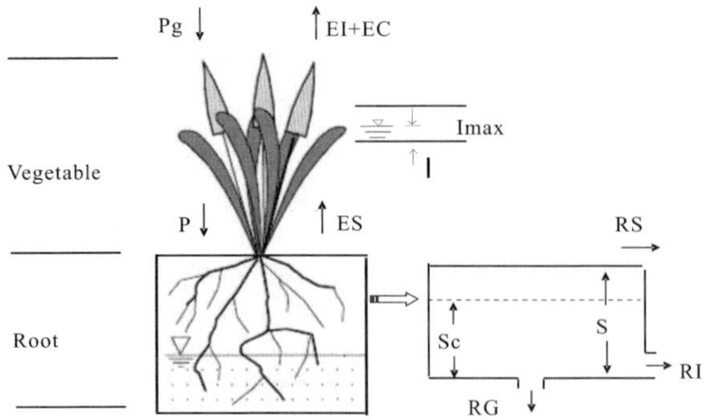

Fig. 1 Schematic diagram of the grid saturation excess runoff calculation module.

When the difference between the sum of excess rain and soil water content and the sum of vegetation transpiration and soil evaporation is still bigger than the grid maximum water storage capacity, surface runoff will occur, and the soil free water storage volume equals to the difference between grid maximum water storage capacity and grid water content.

$$\left.\begin{aligned} RS &= P + S0 - EC - ES - S \\ SS &= S - Sc \end{aligned}\right\} \quad P + S0 - EC - ES > S \tag{5}$$

In equation (5), RS indicates surface runoff (mm); P indicates excess rain volume (mm); $S0$ indicates soil water content (mm); SS indicates soil free water storage volume (mm).

When the difference between the sum of excess rain and soil water content and the sum of vegetation transpiration and soil evaporation is smaller than the grid maximum water storage capacity and bigger than grid water storage capacity, surface runoff will not occur, only interflow and groundwater runoff will occur; soil free water storage volume equals to the difference between the sum of excess rain and soil water content and the sum of vegetation transpiration and soil evaporation and grid water storage capacity.

$$\left.\begin{aligned} RS &= 0 \\ SS &= P + S0 - EC - ES - Sc \end{aligned}\right\} \quad Sc < P + S0 - EC - ES \le S \tag{6}$$

When the difference between the sum of excess rain and soil water content and the sum of vegetation transpiration and soil evaporation is smaller than or equals to the grid water storage capacity, no runoff will occur, and soil free water storage volume equals to 0.

$$\left.\begin{aligned} RS &= 0 \\ SS &= 0 \end{aligned}\right\} \quad P + S0 - EC - ES \le Sc \tag{7}$$

The volumes of interflow and groundwater runoff were calculated through the partition of soil free water storage volume with the outflow coefficient of interflow and outflow coefficient of groundwater.

$$RI = KI \times SS \tag{8}$$

$$RG = KG \times SS \tag{9}$$

In equations (8) and (9), RI indicates the volume of interflow (mm); RG indicates the volume of groundwater runoff (mm); KI indicates outflow coefficient of interflow, unitless; KG indicates outflow coefficient of groundwater, unitless.

The temperature-index approach was applied (Takeuchi *et al.*, 1999; Ao *et al.*, 2006) to simulate the processes of snow accumulation and snow melt. The authors applied the following hypothesis to build the canopy interception module: the authors assumed that when rain drops to the canopy of vegetation and the canopy interception volume reaches its maximum interception capacity, the remaining rain will pass through the vegetation and reach the ground (Wigmosta *et al.*, 1994). The double source evapotranspiration model brought by Mo *et al.* (2004) was applied to compute the actual evapotranspiration of vegetation, interception and soil directly. Surface runoff flows into the neighbouring lower reaches; interflow and subsurface runoff apply linear reservoir accordant junction; a watercourse accordant junction applies to the section Muskingum Method.

Hybrid runoff model

In the hybrid runoff model (Hu, 1993), denoted HYB, the saturation excess runoff mechanism is mixed with the infiltration excess runoff mechanism by combining spatial distribution curves of soil tension water storage capacity and infiltration quantity.

The basic concept of HYB is illustrated in Fig. 2. The watershed is spatially divided by the distribution curve of soil tension water storage capacity into two parts, α and 1–α. The rainfall over the subarea, α, becomes runoff which includes surface runoff, interflow, and groundwater runoff. Over the other subarea, 1–α, runoff cannot be produced until the repletion of soil tension water storage to the soil field capacity according to the principle embodied in XAJ. In the process of soil moisture content increasing up to the field capacity, surface runoff may possibly be produced when rainfall intensity exceeds infiltration rate over a partial area where soil moisture does not reach its field capacity.

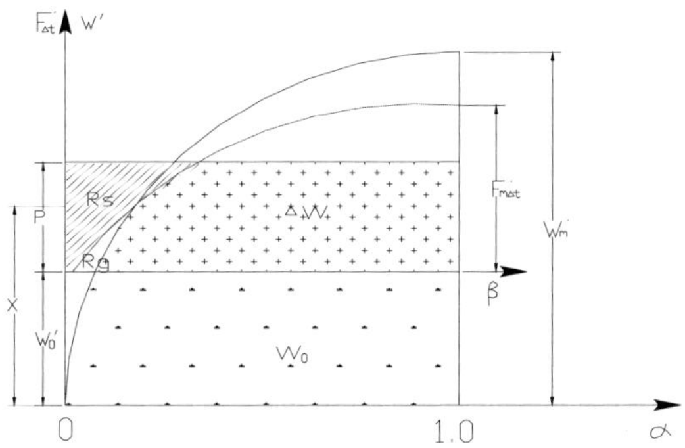

Fig. 2 Sketch of hybrid runoff model.

STUDY AREA AND DATA COLLECTION

Study area general situation

Dianzi Hydrologic station is located in the Laoha River source area, China. Its geographical position is 41°25′ North latitude, 118°50′ East longitude. The watershed area is located in the south part of the watershed and has an area of 1643 km^2. There are 10 precipitation stations and 3 evaporation stations entirely in the watershed region. Local forest cover rate is high, temperature is low, and annual precipitation is about 520 mm.

Data collection

Global Land One-kilometre Base Elevation (GLOBE) Digital Elevation Model (DEM) data from the National Geophysical Data Center (NGDC) was applied to describe the topographical distribution of Dianzi watershed, which has a resolution of 30*s*.

The University of Maryland's 1-km global land cover data was applied to determine local land cover condition. There are 14 land cover types in the UMD classification criteria, and the data's resolution is 1 km.

The Food and Agriculture Organization (FAO) soil map (FAO, 1995) was applied as soil type data in this study, which had a resolution of 1 km. The soil types were converted into United States

Department of Agriculture (USDA) classification types through a USDA soil triangular diagram. Relative meteorological data was provided by the China Meteorological Bureau.

APPLICATION

Parameter calibration

The authors took runoff gross error and Nash-Sutcliffe coefficient as the object function of parameter calibration, 1973–1977 as model calibration period, 1978–1979 as model verification period. Parameters of IHMS-VSWSC and hybrid runoff model are shown in Tables 1 and 2.

Table 1 IHMS-VSWSC parameters.

Parameter	α	β	κ	θ_1	θ_2	*KI*	*KG*	*CI*	*CG*	*KE* (h)	*XE*
Value	5.94	1.32	0.06	8.26	7.29	0.37	0.27	0.84	0.9991	22.99	0.38

α, drying function parameters; *CI* and *CG*: recession constant of interflow and recession constant of groundwater, respectively; *KE*, storage constant (h); *XE*, flow weight factor.

Table 2 Hybrid runoff model parameters.

Parameter	*kc*	*WUM* (mm)	*WLM* (mm)	*WDM* (mm)	*b*	*bx*	f_0 (mm)	*Fc* (mm)	*K*	*kg*	*imp* (%)	*x*
Value	0.67	20	80	50	0.48	0.53	18	11	0.88	0.85	4.2	0.18

kc, evapotranspiration coefficient; *WUM*, upper layer tension water capacity; *WLM*, lower layer tension water capacity; *WDM*, deep layer tension water capacity; *b*, tension water storage capacity curve index; *bx*, infiltration ability attenuation coefficient; f_0, the average maximum infiltration capacity of watershed; *Fc*, the average steady infiltration capacity of watershed; *kg*, outflow coefficient of groundwater; *K*, infiltration capacity attenuation coefficient; *imp*, impermeable area ratio; *x*, the calculation coefficient with Muskingum method.

Simulation results analysis

Simulation results are compared in Table 3. The Nash-Sutcliffe coefficient reflects the model's simulation accuracy towards the flood peak, whereas absolute error and relative error reflect the model's simulation accuracy towards flood amount. In the daily runoff process simulation in Dianzi watershed, the annual average Nash-Sutcliffe coefficient of IHMS-VSWSC reached 0.80, while the annual average Nash-Sutcliffe coefficient of the hybrid runoff model only reached 0.67. Generally speaking, the above mentioned simulation results were acceptable in cold arid hydrological process simulation study, and the simulation runoff process was similar to the observed runoff process. In the parameter calibration period, the annual average Nash-Sutcliffe coefficient of IHMS-VSWSC was higher than the hybrid runoff model by 11.2%; in the verification period, the annual average Nash-Sutcliffe coefficient of IHMS-VSWSC was higher than the hybrid runoff model by 19%. As a rule, a good simulation result reflects a model's acceptability, its regional applicability, and its parameter calibration correctness. Results showed that both IHMS-VSWSC and the hybrid runoff model were adequate for the simulation study in the study area, and IHMS-VSWSC had a better simulation outcome than the hybrid runoff model. From the aspect of runoff simulation flow error control, IHMS-VSWSC obviously surpassed the hybrid runoff model. But in the process of simulating runoff in 1976, both IHMS-VSWSC and the hybrid runoff model had relatively large water quantity errors, which had a direct relationship with the rare precipitation and small runoff amount in 1976. In summary, both models' simulation accuracies for a low flow year are not good and need to be enhanced in future study.

Table 3 Simulation results contrast.

Year		IHMS–VSWSC			Hybrid runoff model		
		Nash	absolute error /mm	relative error /%	Nash	absolute error /mm	relative error /%
Calibration period	1973	0.80	–0.88	–0.74	0.72	–29.41	–24.67
	1974	0.75	0.18	0.17	0.67	–6.47	–6.25
	1975	0.79	3.47	2.86	0.79	21.76	17.96
	1976	0.73	20.33	42.75	0.39	34.06	71.60
	1977	0.86	–2.92	–2.95	0.80	30.68	31.04
Verification period	1978	0.86	19.83	13.57	0.67	72.84	49.84
	1979	0.86	0.74	0.44	0.67	73.58	43.90

Nash : Nash-Sutcliffe coefficient

The comparison between simulated flow of the hybrid runoff model towards the observed flow and simulated flow of IHMS-VSWSC towards the observed flow are shown in Figs 3 and 4. From these figures, the authors found that IHMS-VSWSC was always more precise than the hybrid runoff model, except in very few periods, which indicates that the distributed hydrological model is very useful and spatial distributed model of water storage capacity is applicable.

IHMS-VSWSC built a direct relationship between rainfall–runoff and grid geomorphic unit, and it is easier to predict water watershed response distribution than the distributed probabilistic humidity model. The water storage capacity spatial distribution of Dianzi watershed is shown in Fig. 5. In IHMS-VSWSC, factors such as watershed topography, vegetation type, soil type and soil water parameters and their influences on the spatial distribution of watershed water storage capacity were taken into account. Therefore Fig. 5 reasonably presents the spatial distribution of water storage in Dianzi watershed.

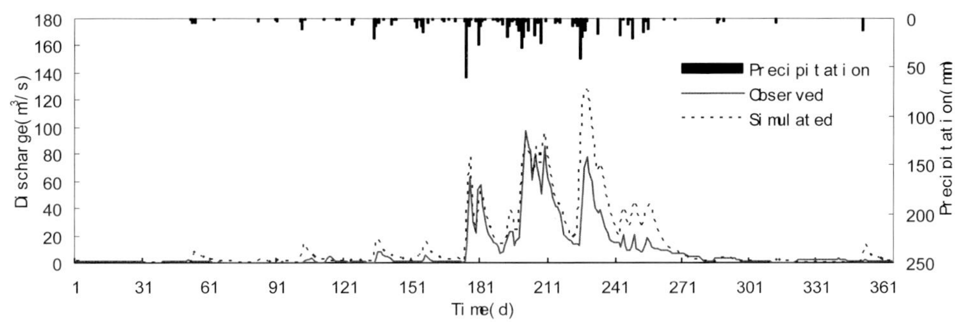

Fig. 3 The simulated flow of hybrid runoff model *versus* the observed flow in 1979.

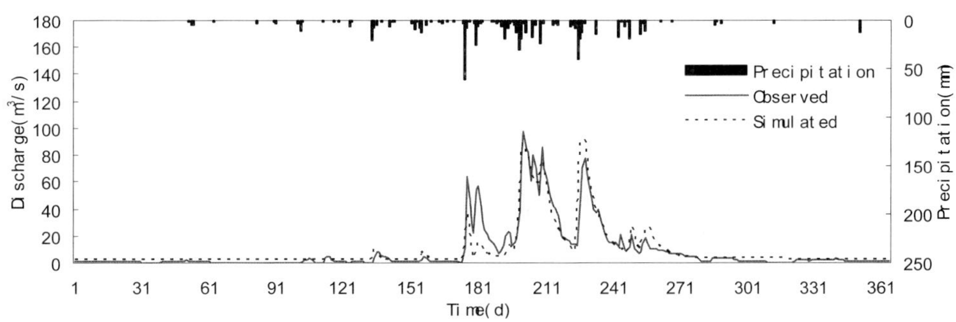

Fig. 4 The simulated flow of IHMS-VSWSC *versus* the observed flow in 1979.

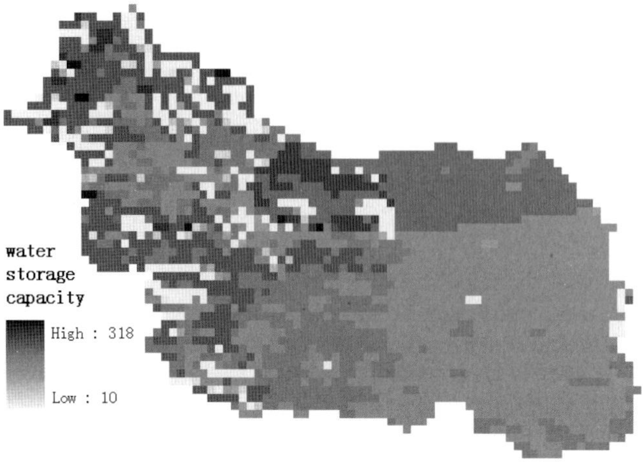

Fig. 5 The spatial distribution of water storage in Dianzi watershed.

CONCLUSIONS

The spatial distributed model of water storage capacity took into account topography, vegetation types, soil types, soil water parameter and other relative factors' influence on the spatial distribution of watershed water storage capacity. On the basis of the spatial distributed model of water storage capacity, the authors built IHMS-VSWSC, which had a simple structure and could apply multi-resource data, described the soil water change of grid vegetation root system layer with grid saturation excess runoff mechanism, and avoided the direct description of the complicated soil water flow process.

The authors applied IHMS-VSWSC and the hybrid runoff model to simulate the daily hydrological process in Laoha River Dianzi watershed. Simulation results indicated that both models are adequate for hydrological simulation in the Dianzi watershed. The average annual Nash-Sutcliffe coefficient of the daily runoff process simulation of IHMS-VSWSC reaches 0.80, which was higher than the hybrid runoff model by 13%. Except in low flow years, the error of runoff by IHMS-VSWSC was controlled in a relatively small range, which indicated that the simulation result of IHMS-VSWSC was better than that of the hybrid runoff model. Building of IHMS-VSWSC would help to solve problems such as watershed hydrological process simulation, water resource evaluation, agricultural planning and drought monitoring and forecasting.

Acknowledgements This work described in this paper was supported financially by the National Key Basic Research Program of China (Grant no. 2006CB400502) and the Special Basic Research Fund for Methodology in Hydrology (Grant no. 2007FY140900), Ministry of Science and Technology, China, the 111 Project under Grant B08048, Ministry of Education and State Administration of Foreign Experts Affairs, China, and also the Program for Changjiang Scholars and Innovative Research Team in University (Grant no. IRT0717).

REFERENCES

Ao, T., Ishidaira, H., Takeuchi, K., Kiem, A. S., Yoshitari, J., Fukami, K. & Magome, J. (2006) Relating BTOPMC model parameters to physical features of MOPEX basins. *Hydrol.* **320**, 84–102.

Bao, W. (1995) *Loess Zone Basin Water and Sand Simulation Concept Model and Application.* Hohai University Press, Nanjing, China.

Beven, K. J. (1992) The future of distributed models: model calibration and uncertainty prediction. *Hydrol. Processes* **6**, 279–298.

Beven, K. & Kirby, M. J. (1979) A physically based variable contributing area model of basin hydrology. *Hydro Sci. Bull.* **24,** 43–69.

Hu, C. Q. (1993) Computational method of runoff generation in semihumid and semiarid regions. In: *Proc., Symp. on Hydrological Information and Forecasting* (ed. by Hydrology Bureau of the Yangtze River Water Resources Commission), 57–62. China Water Power Press, Beijing, China.

Hu, H., Tang, Q., Lei, Z. & Yang, S. (2004) Runoff-evaporation hydrological model for arid plain oasis, 1, the model structure. *Adv. Water Sci.* **15**(2), 141–145.

Jia, Y., Wang, H. & Yan, D. (2006) Distributed model of hydrological cycle system in Heihe River basin I. *Model Development Verification* **37**(5), 534–542.

Liang, X. & Xie, Z. (2001) A new surface runoff parameterization with subgrid-scale soil heterogeneity for land surface models. *Adv. Water Resour.* **24**(9–10), 1173–1193.

Liu, Z., Ni, G., Lei, Z. & Wang, L. (2006) Distributed hydrological model of small and moderate size watersheds in the Loess Plateau. *J. Tsinghua Univ. (Natural Sciences)* **46**(9), 1546–1550.

Mo, X., Liu, S., Lin, Z. & Zhao, W. (2004) Simulating temporal and spatial variation of evapotranspiration over the Lushi basin. *J. Hydrol.* **285**, 125–142.

Takeuchi, K., Ao, T. & Ishidaira, H. (1999) Introduction of block-wise use of TOPMODEL and Muskingum-Cunge method for the hydro-environmental simulation of a large ungauged basin. *Hydrol. Sci. J.* **44**(4), 633–646.

Wang, G. (1998) The attempt of improvement of the Xin'anjiang model. *Hydrology, Water Conservation Information Center Monograph, extra edition*. 23–27.

Wigmosta, M. S., Vail, L. W. & Lettenmaier, D. P. (1994)A distributed hydrology-vegetation model for complex terrain. *Water Resour. Res.* **30**(6), 1665–1903.

Zhao, R. (1984) *Watershed Hydrologic Modeling*. China Water Power Press, Beijing, China.

Zheng, G., Bao, W. & Qu, S. (2002) Evaluation of effect of artificial rain induced runoff Increase on Longyangxia reservoir. *J. Hohai University (Natural Sciences)* **30**(2), 58–62.

A hybrid model based on set pair analysis and NNB for annual runoff prediction

JULIANG JIN[1,2], WENSHENG WANG[3], CHENGGUO WU[4] & YULIANG ZHOU[2]

1 *Chengdu Institute of Plateau Meteorology, CMA, Chengdu 610072, China*
jinjl66@126.com

2 *School of Civil Engineering, Hefei University of Technology, Hefei 230009, China*

3 *School of Hydraulic Engineering, Sichuan University, Chengdu 610065, China*

4 *Key Lab of Northwest Water Resources and Environment Ecology of MOE, Xi'an University of Technology, Xi'an 710048, China*

Abstract As an important part of watershed water security management, prediction of watershed annual runoff can be used to offer a scientific basis of reasonably developing and utilizing watershed water resources and preferably programming of the sustainable development of watershed society and economy. In order to deal with a weak dependence of the watershed annual runoff process and the complex relationship between annual runoff process and its influencing factors, and to avoid the uncertainty problems during the process of selecting and resolving explicit mathematics functions with parametric statistics prediction methods: (1) autocorrelation technology can be used to determine the characteristic vectors of the nearest neighbour bootstrap method (NNB); (2) set pair analysis (SPA) method can be used to quantitatively compute the comparability between the present and historical annual runoff characteristic vectors from the view of identicalness, discrepancy and contrast; (3) the weighted average of the subsequent annual runoff of the nearest neighbour characteristic vectors can be regarded as the predicted value of the present annual runoff; and (4) the hybrid model of SPA-NNB based on set pair analysis and nearest neighbour bootstrap was suggested. The application results of SPA-NNB show that the physical concepts of SPA-NNB are distinct, its computation method is visual, its prediction precision is high, the scheme of modelling SPA-NNB is simple, effective and ideal, so SPA-NNB can be applied to different prediction of hydrological time series with plenty of representative historical samples.

Key words annual runoff; nearest neighbour bootstrap method; set pair analysis; similarity prediction; SPA-NNB hybrid model

INTRODUCTION

Prediction of watershed annual runoff, which can provide a scientific basis for reasonably developing and optimally utilizing watershed water resources and preferably programming the sustainable development of watershed society and economy, is an important part of watershed water security management (Loucks, 2000). As a consequence of hydro-meteorological, physical geographical, vegetative covering, human activities, etc. the annual runoff process is a complex and dynamic system of high-dimensional, nonlinear and weak dependence. So far, the research of prediction of annual runoff is still at the positive exploratory stage (Jin *et al.*, 2002b; Wang *et al.*, 2006; Zhang *et al.*, 2006). At present, there are two common prediction methods for annual runoff: the first is the causative prediction method, which is based on the evolutional laws of atmospheric circulation, weather process and the causative dynamic model of physical conditions of the watershed underlying surface; it is an important developing research direction of annual runoff prediction (Peel, 2002; Morassutti, 2008). However, because of complex statistical characteristics in time and space of annual runoff, this method is still difficult at present (Zhang *et al.*, 2006). The second is the statistical prediction method, mainly including the time series analysis method, the multivariate statistical method and the similarity prediction method (Yan, 1998; Coulibaly, 2000; Jin *et al.*, 2002a; Mazvimavi *et al.*, 2005; Sen & Altunkaynak, 2006). Because of weak dependence of annual runoff process, there is a complex and nonlinear relationship between annual runoff and its influencing factors, so it is difficult to raise the prediction precision with the above two prediction methods (Jin *et al.*, 2002a; Wang *et al.*, 2006).

Similarity prediction, which has been proved effective by hydro-meteorological prediction practices, is a nonparametric prediction method, and its main principle, according to the principle of similar reasons causing similar results, is selecting one or more samples from the historical samples, which are the most similar with the present annual runoff, and then regard it as the predicted present consequence (Yan, 1998; Jin *et al.*, 2002a; Wang *et al.*, 2006). The nearest neighbour bootstrap (NNB) prediction model, which has been successfully applied to prediction of hydrological time series, is a visual and operable similarity prediction method (Wang *et al.*, 2001; Jin *et al.*, 2002c). The key problem of using the NNB hybrid model is how to determine the number of the nearest neighbour characteristic vectors and their weights of NNB model. A method to identify the only one nearest neighbour characteristic vector was proposed by Wang *et al.* (2006) using set pair analysis (SPA) (Zhao, 2000).

To further improve the annual runoff prediction method proposed by Wang *et al.* (2006), on the basis of the characteristics of annual runoff series, using autocorrelation technology to determine the characteristic vectors of NNB model, using methods based on average and standard deviation to carry out grade symbol processing of the characteristic vectors, and then regarding the weighted average of the subsequent annual runoff of the nearest neighbour K (K>1) characteristic vectors as the predicted value of the present annual runoff, leading to a SPA-NNB hybrid model based on SPA and NNB for annual runoff prediction is proposed in this paper.

SPA-NNB HYBRID MODEL

The establishment of SPA-NNB model includes three steps as follows:

(a) To determine the characteristic vectors of SPA-NNB with autocorrelation analysis technology. The autocorrelation coefficient R_k of the kth delayed order of annual runoff historical samples $\{x_i|i = 1, 2, \ldots, n\}$ is (Wang *et al.*, 2007):

$$R_k = \sum_{i=k+1}^{n}(x_i - \bar{x})(x_{i-k} - \bar{x}) / \sum_{i=1}^{n}(x_i - \bar{x})^2 \quad (1)$$

where average $\bar{x} = \sum_{i=1}^{n} x_i / n$; n is sample size; k = 1, 2, …, n_k<[n/4]. According to sampling distribution theory of R_k, under the condition of confidence level 1−α, when the autocorrelation coefficient satisfies:

$$R_k \notin [(-1 - u_{\alpha/2} \cdot (n-k-1)^{0.5})/(n-k), (-1 + u_{\alpha/2} \cdot (n-k-1)^{0.5})/(n-k)] \quad (2)$$

the dependence of the kth delayed order of time series x_i is obvious, otherwise it is not obvious (Wang *et al.*, 2007). The quantile value $u_{a/2}$ in equation (2) can be obtained from normal probability distribution. The characteristic vectors D_i = (x_{i-p}, x_{i-p+1}, …, x_{i-1}) of SPA-NNB are composed of delayed items of x_i, and x_i is called the successor of D_i, where i = p+1, …, n and p is the number of maximum delayed order. The dimension of characteristic vectors is p (Wang *et al.*, 2001, 2006; Jin *et al.*, 2002c).

(b) To establish set pairs ($H(D_{n+1},D_i)$, i = p+1, …, n) between the characteristic vectors D_{n+1} = (x_{n-p+1}, x_{n-p+2} ,…, x_n) of the present annual runoff x_{n+1} and the characteristic vectors of historical samples D_i, respectively. The main principle of SPA is (Zhao, 2000) to qualitatively and comparatively analyse the comparability of the two discussed sets D_{n+1} and D_i from the view of identicalness, discrepancy and contrary, and to compute the connection degree using equation (3):

$$u_{D_{n+1}-D_i} = a_i + b_i I + c_i J \quad (3)$$

where a_i, b_i and c_i, which are non-negative and satisfying a_i+b_i+c_i = 1, are the identicalness degree, discrepancy degree and contrary degree of SPA, respectively. I is coefficient of discrepancy degree, and its value ranges from −1 to 1. When the value of I is −1 or 1, representing b_i is certain. As I approaching to 0, the uncertainty of b_i is intensifying. J is coefficient of contrary degree, whose value is usually −1. i = p+1, ..., n. As the value of

$u_{D_{n+1}-D_i}$ becomes greater, it represents that the possibility of the present annual runoff determined by the successor value x_i of D_i is also becoming greater (Wang *et al.*, 2006).

In order to quantitatively compute equation (3) conveniently, the characteristic vectors can be carried out by grade symbol processing as follows:

$$z_{i,j} = \begin{cases} 1, & x_{i-j} \le \bar{x}_j - 0.44 s_j \\ 2, & \bar{x}_j - 0.44 s_j < x_{i-j} < \bar{x}_j + 0.44 s_j \\ 3, & x_{i-j} \ge \bar{x}_j + 0.44 s_j \end{cases} \tag{4}$$

where average $\bar{x}_j = \sum_{i=p+1}^{n} x_{i-j} / (n-p)$, standard deviation $s_j = [\sum_{i=p+1}^{n} (x_{i-j} - \bar{x}_j)^2 / (n-p-1)]^{0.5}$, $i = p+1, \ldots, n, j=1, 2, \ldots, p$. If x_{i-j} obeys the normal probability distribution, the possibility of $z_{i,j}$ assigned grade 1, grade 2, grade 3, respectively, defined by equation (4), is equal. In this paper, these three grades are corresponding to low-flow state, moderate-flow state, high-flow state of annual runoff, respectively. So it is obvious that the statistical physical meaning of equation (4) is more obvious than that of equations (3) and (4) used for grade symbol processing (Wang *et al.*, 2006).

Now these two sets in the set pair $H(D_{n+1},D_i)$ can be written as $D_{n+1} = (z_{n+1,p}, \cdots, z_{n+1,2}, z_{n+1,1})$ and $D_i = (z_{i,p}, \cdots, z_{i,2}, z_{i,1})$, respectively, and equation (3) can be quantitatively computed by comparing the value of their corresponding vectors. If the corresponding vectors are at the same grade, consecutive grade or distant grade, it represents the identicalness, discrepancy or contrast of $H(D_{n+1}, D_i)$, respectively, and then compute the connection degree of $u_{D_{n+1}-D_i}$, $i=p+1, \ldots, n$ according to equation (3). It is obvious that connection degree $u_{D_{n+1}-D_i}$ ranges from −1 to 1. If discrepancy of characteristic vectors D_{n+1} and D_i becomes greater, $u_{D_{n+1}-D_i}$ is closer to approaching to −1, representing that these two characteristic vectors are more dissimilar. If discrepancy of characteristic vectors D_{n+1} and D_i is becoming smaller, $u_{D_{n+1}-D_i}$ is closer to approaching to 1, representing that these two characteristic vectors are more similar. So, the connection degree $u_{D_{n+1}-D_i}$ is a relative discrepancy degree function of variable fuzzy set of similarity of characteristic vectors D_{n+1} and D_i, and the relative membership function of this fuzzy set can be described as in Chen (2005).

$$v_{n+1,i} = 0.5 + 0.5 u_{D_{n+1}-D_i} \quad (i=p+1,\ldots,n) \tag{5}$$

In equation (5), the relative membership function $v_{n+1,i}$ of the variable fuzzy set is constituted by the quantitative analysis about the value proximity of the characteristic vectors D_{n+1} and D_i from the view of identicalness, discrepancy and contrary, and the calculation process is visual, simple and universal. But when constituting $v_{n+1,i}$ of the variable fuzzy set in Chen (2005), the determination of variable attraction area I_{ab}, variable range area I_{cd} and the dot-matrix M in the variable attraction area (which membership function value is 1) are all needed, so the computation process of $v_{n+1,i}$ in Chen (2005) is very complicated.

(c) To establish SPA-NNB hybrid model for annual runoff prediction. The determination of K of characteristic vectors number, which is the number of the nearest neighbours to the characteristic vectors D_{n+1} of the predicted present annual runoff, is experiential, where commonly satisfying $K \le (n-p)^{0.5}$ (Wang *et al.*, 2001). Choose K characteristic vectors which are the most similar (that is the values of connection degree $u_{D_{n+1}-D_i}$ are the greatest) to D_{n+1} from $n-p$ characteristic vectors of historical samples, then regard the weighted average, whose weights are the relative membership $v_{n+1,i}$ corresponding to the subsequent of the characteristic vectors with $u_{D_{n+1}-D_i}$, as the predicted present value of annual runoff:

$$\hat{x}_{n+1} = \sum_{i=k_1}^{k_K} [(v_{n+1,i} / \sum_{i=k_1}^{k_K} v_{n+1,i}) x_i] = \sum_{i=k_1}^{k_K} w_i x_i \tag{6}$$

where w_i is the weight (that is the normalization value of relative membership $v_{n+1,i}$) of the subsequent x_i of characteristic vector D_i, $i = k_1, \ldots, k_K$ is the sequence number of the K nearest neighbour characteristic vectors.

CASE STUDY OF SPA-NNB

Data

Now we establish the SPA-NNB model of the annual runoff prediction with the 55 annual runoff historical samples $\{x_i | i = 1,2, \ldots, 55\}$ from 1943 to 1997 of the Cuntan Station in the upper reaches of the Yangtze River of China (Wang *et al.*, 2006). The Cuntan station is the input station of the Three Gorges hydropower station.

Selection of dimension *p*

Compute the autocorrelation coefficients R_k of the first 10 delayed orders of the historical annual runoff series, the corresponding upper limit $R_{2,k}$ and lower limit $R_{1,k}$ in equation (2), and their results are shown in Table 1. For the weak dependence of the annual runoff series, the confidence level is 60%. As shown in Table 1, the dependence of R_1, R_2, R_3 and R_5 are obvious, according to the confidence level of 60%. R_4, relative to autocorrelation coefficient above six delayed orders, is also more obvious. In order to conveniently compare and analyse SPA-NNB with the models of Wang *et al.*, 2006), the values of characteristic vectors of the SPA-NNB are the same as Wang *et al.* (2006), that is, D_i=(x_{i-5}, x_{i-4}, x_{i-3}, x_{i-2}, x_{i-1}), and the corresponding subsequent value is x_i, $i = 6, \ldots,$ 55 and the dimension p of characteristic vectors and maximum delayed order p are all 5.

Table 1 Autocorrelation coefficient values of annual runoff series and their upper and lower limit values (α = 40%).

k	1	2	3	4	5	6	7	8	9	10
$R_{1,k}$	–0.13	–0.13	–0.14	–0.14	–0.14	–0.14	–0.14	–0.14	–0.14	–0.15
R_k	0.32	0.22	0.13	–0.06	–0.14	0.04	0.04	–0.02	0.01	0.06
$R_{2,k}$	0.10	0.10	0.10	0.10	0.10	0.10	0.10	0.10	0.10	0.10

RESULTS

Choose the annual runoff of 1998 as the present annual runoff x_{n+1}, and form 50 set pairs $H(D_{n+1}, D_i)$ i = 6–55. The results after grade symbol processing of these characteristic vectors can be computed according to equation (4), the values of I and J are 0.5 and −1, respectively, which are consistent with Wang *et al.* (2006), and the values of connection degree $u_{D_{n+1}-D_i}$, $i = 6, \ldots, 55$ can be computed according to equation (3), as are shown in Table 2. In this case study, n and p are 55 and 5, respectively, so choose 7 as the nearest neighbour number K of characteristic vectors. The time series numbers of the 7 characteristic vectors chosen from 50 characteristic vectors in Table 2, which are the most similar to D_{n+1} (that is, the values of connection degree $u_{D_{n+1}-D_i}$ are the greatest), are 38, 31, 19, 54, 32, 39 and 53. Then the value x_{n+1} of annual runoff prediction of 1998 is 11 100 m^3/s. And for the same reason, choose the annual runoff values of 1999, 2000, 2001, 2002 and 2003 as the annual runoff $x(n+1)$, then the predicted values x_{n+1} of annual runoff of these years can be computed, as shown in Table 3. To validate the reliability of the prediction accuracy of SPA-NNB, the corresponding results of annual runoff prediction of SPA and fuzzy optimization model in Wang *et al.* (2006) are also listed in Table 3.

Table 2 Computed results of symbol processed characteristic vectors of annual runoff and their connection degree values.

Number of time series i	Characteristic vector denoted in grade symbol $z_{i,i-5}$	$z_{i,i-4}$	$z_{i,i-3}$	$z_{i,i-2}$	$z_{i,i-1}$	Value of the subsequent x_i (m^3/s)	Computed value of connection degree a_i	b_i	c_i	$u_{D_{n+1}-D_i}$
6	1	1	2	2	3	13200	0.20	0.60	0.20	0.30
7	1	2	2	3	3	14600	0.00	0.60	0.40	–0.10
8	2	1	3	3	3	10900	0.40	0.00	0.60	–0.20
9	2	3	3	3	2	11300	0.20	0.20	0.60	–0.30
10	3	3	3	2	2	12400	0.00	0.60	0.40	–0.10
11	3	3	2	2	3	10700	0.00	0.60	0.40	–0.10
12	3	2	2	3	2	14100	0.00	0.80	0.20	0.20
13	2	2	3	2	3	12000	0.20	0.40	0.40	0.00
14	2	3	2	3	3	11200	0.20	0.20	0.60	–0.30
15	3	2	3	3	2	11000	0.00	0.60	0.40	–0.10
16	2	3	3	2	2	10500	0.20	0.40	0.40	0.00
17	3	3	2	2	2	9400	0.00	0.80	0.20	0.20
18	3	2	2	1	1	10300	0.40	0.60	0.00	0.70
19	2	2	1	1	1	11800	0.80	0.20	0.00	0.90
20	2	1	1	1	3	12100	0.80	0.00	0.20	0.60
21	1	1	1	3	3	10900	0.40	0.20	0.40	0.10
22	1	1	3	3	2	12800	0.20	0.40	0.40	0.00
23	1	3	3	2	3	13200	0.00	0.40	0.60	–0.40
24	3	3	2	3	3	11900	0.00	0.40	0.60	–0.40
25	3	2	3	3	3	11000	0.00	0.40	0.60	–0.40
26	2	3	3	3	2	13400	0.20	0.20	0.60	–0.30
27	3	3	3	2	3	8960	0.00	0.40	0.60	–0.40
28	3	3	2	3	1	10300	0.20	0.40	0.40	0.00
29	3	2	3	1	1	9350	0.40	0.40	0.20	0.40
30	2	3	1	1	1	8790	0.80	0.00	0.20	0.60
31	3	1	1	1	1	10500	0.80	0.20	0.00	0.90
32	1	1	1	1	2	13000	0.60	0.40	0.00	0.80
33	1	1	1	1	3	10800	0.60	0.20	0.20	0.50
34	1	1	1	3	2	10300	0.40	0.40	0.20	0.40
35	1	1	3	2	1	9900	0.40	0.40	0.20	0.40
36	1	3	2	1	1	10000	0.40	0.40	0.20	0.40
37	3	2	1	1	1	9910	0.60	0.40	0.00	0.80
38	2	1	1	1	1	11200	1.00	0.00	0.00	1.00
39	1	1	1	1	2	11900	0.60	0.40	0.00	0.80
40	1	1	1	2	3	10300	0.40	0.40	0.20	0.40
41	1	1	2	3	1	11100	0.40	0.40	0.20	0.40
42	1	2	3	1	2	10700	0.20	0.60	0.20	0.30
43	2	3	1	2	2	11900	0.40	0.40	0.20	0.40
44	3	1	2	2	3	9490	0.20	0.60	0.20	0.30
45	1	2	2	3	1	10600	0.20	0.60	0.20	0.30
46	2	2	3	1	2	10800	0.40	0.40	0.20	0.40
47	2	3	1	2	2	11800	0.40	0.40	0.20	0.40
48	3	1	2	2	3	11600	0.20	0.60	0.20	0.30
49	1	1	2	3	2	10800	0.20	0.60	0.20	0.30
50	2	2	3	2	2	10100	0.20	0.60	0.20	0.30
51	2	3	2	2	1	11000	0.40	0.40	0.20	0.40
52	3	2	2	1	2	8240	0.20	0.80	0.00	0.60
53	2	2	1	2	1	10000	0.60	0.40	0.00	0.80
54	2	1	2	1	1	9450	0.80	0.20	0.00	0.90
55	1	2	1	1	1	8660	0.60	0.40	0.00	0.80

Table 3 Predicted results of annual runoff using SPA-NNB and other models.

Year i	Characteristic vectors					The nearest 7 neighbour characteristic vectors							Annual runoff (m^3/s)		Relative error of prediction (%)		
	$z_{i,i-5}$	$z_{i,i-4}$	$z_{i,i-3}$	$z_{i,i-2}$	$z_{i,i-1}$	1	2	3	4	5	6	7	Observed	SPA-NNB	M1	M2	M3
1998	2	1	1	1	1	38	31	19	54	32	39	53	12500	11100	–11.2	–10.4	–16.0
1999	1	1	1	1	3	33	32	20	39	40	6	21	11600	11700	0.9	–6.9	12.1
2000	1	1	1	3	2	34	21	49	41	40	7	45	11500	11200	–2.6	–10.4	–5.2
2001	1	1	3	2	2	22	35	6	42	49	50	8	11000	11200	1.8	3.6	16.4
2002	1	3	2	2	2	16	23	36	43	47	51	14	9380	11400	21.5	27.9	36.5
2003	3	2	2	2	1	18	17	12	28	29	37	51	10600	10600	0.0	0.9	–2.8

Note: M1 denotes SPA-NNB model; M2 and M3 denote SPA and fuzzy optimization model of Wang *et al.* (2006), respectively.

Table 3 shows: (a) If the permissive error of the annual runoff prediction is 20% of the measured value, the testing approved rate of the SPA-NNB hybrid model is 83.3%, so the model can be an important reference in practical long-term forecasting of annual runoff (Li, 1993). (b) In the SPA-NNB model, the similarity of the characteristic vectors D_{n+1} of the present annual runoff and the characteristic vectors D_i of the historical samples was quantitatively computed from the view of identicalness, discrepancy and contrast at the same time, which can express prediction results more than the fuzzy optimization model, whose similarity is expressed only by connection degree (Wang *et al.*, 2006). In the SPA-NNB model, the prediction value of the annual runoff is the weighted average of the subsequent values which correspond to the nearest $K(K > 1)$ neighbour characteristic vectors, while in the SPA model of Wang *et al.* (2006), the information of only one nearest neighbour characteristic vector is used, so the SPA-NNB model is more satisfying in analysing the similar information of historical samples, and the prediction precision is higher. (c) In the SPA-NNB model, the prediction of annual runoff is based on distance similarity of annual runoff characteristic vectors and historical samples, so the computation process is visual, and the precision is high.

CONCLUSIONS

(a) In order to deal with the weak dependence of annual runoff process and the complexity between annual runoff and its influencing factors, and to avoid the uncertain problems in the process of determining and solving the explicit mathematical function with parametric statistical prediction model, the NNB method of annual runoff prediction through nonparametric statistical approach was discussed in this paper. Aiming at the key problems of using the NNB model and characters of annual runoff prediction problem, using autocorrelation technology to determine the characteristic vectors of NNB model, using the SPA model to quantitatively compute the similarity of characteristic vectors of the annual runoff prediction and historical samples, and regarding the weighted average of the subsequent values of the $K(K > 1)$ nearest neighbour characteristic vectors as the predicted value of the annual runoff, and then the SPA-NNB hybrid model based on SPA and NNB for annual runoff prediction was established. (b) The results of the SPA-NNB model applied to the annual runoff prediction of the Cuntan station in the Yangtze River of China show: (i) the physical conception of SPA-NNB hybrid model is clear, the computation process is visual, and its prediction precision is high; (ii) the SPA-NNB model is simple and effective, and the prediction technology given in this paper is complete, so SPA-NNB model can be applied to different prediction of hydrological and meteorological time series if there are plenty of representative historical samples.

Acknowledgements The authors would like to acknowledge the funding support of the National Natural Science Foundation of China (no. 70771035, 50779042), the National Key Technologies

R&D Program of China during the 11th Five-year Plan Period (no. 2006BAB14B02), and the Opening Foundation of Chengdu Institute of Plateau Meteorology, China Meteorological Administration (LPM2008018).

REFERENCES

Chen, S. Y. (2005) *Theories and Methods of Variable Fuzzy Sets in Water Resources and Flood Control System.* Dalian University Press, Dalian, China.

Coulibaly P., Anctil F., Rasmussen P. *et al.* (2000) A recurrent neural networks approach using indices of low-frequency climatic variability to forecast regional annual runoff. *Hydrol. Processes* **14**, 2755–2777.

Jin, J. L., Fu, Q., Wei, Y. M. *et al.* (2002a) Shepard interpolation model for predicting annual runoff. *J. Yangtze River Scientific Research Institute* **19**(1), 52–55.

Jin, J. L., Wei, Y. M. & Ding, J. (2002b) Application of projection pursuit threshold regressive model for predicting annual runoff. *Scientia Geographica Sinica* **22**(2), 171–175.

Jin, J. L., Wei, Y. M. & Ding, J. (2002c) Nearest neighbour bootstrap model for predicting daily flow process. *Acta Simulata Systematica Sinica* **14**(11), 1494–1496.

Li, H. L. (1993) *Hydrology Forecasting*. China Water Power Press, Beijing, China.

Loucks, D. P. (2000) Sustainable water resources management. *Water Int.* **25**(1), 3–10.

Mazvimavi, D., Meijerink, A. M. J., Savenije, H. H. G. *et al.* (2005) Prediction of flow characteristics using multiple regression and neural networks: a case study in Zimbabwe. *Physics Chemistry of the Earth* **30**(11/16), 639–647.

Morassutti M. P. (2008) Australian runoff scenarios from a runoff–climate model. *Int. J. Climatol.* **12**(8), 797–813.

Peel, M. C., McMahon, T. A., Finlayson, B. L. *et al.* (2002) Implications of the relationship between catchment vegetation type and the variability of annual runoff. *Hydrol. Processes* **16**, 2995–3002.

Sen, Z & Altunkaynak, A. (2006) A comparative fuzzy logic approach to runoff coefficient and runoff estimation. *Hydrol. Processes* **20**, 1993–2009.

Wang, H. F., Huang, W. J., Wang, W. S. *et al.* (2006) Cuntan station of the Yangtze River annual runoff forecasting with set pair analysis method. *J. Heilongjiang Hydraulic Engineering College* **33**(4), 3–5.

Wang, W. S., Jin, J. L., Li, Y. Q. *et al.* (2007) *Stochastic Simulation Technology for Hydrology and Water Resource*. Sichuan University Press, Chengdu, China.

Wang, W. S., Xiang, H. L. & Ding, J. (2001) Predication of hydrology and water resources with nearest neighbour bootstrapping regressive model. *Int. J. Hydroelectric Energy* **19**(2), 8–10, 14.

Yan, H. S., Xie, Y. Q. & Cao, J. (1998) *Nonlinear Statistical Forecasting Methods and Applications*. Yunnan Science and Technology Press, Kunming, China.

Zhang, J. S., Liu, L. Y., Cheng, Z. S. *et al.* (2006) *Statistical Hydrology*. Yellow River Water Conservancy Press, Zhengzhou, China.

Zhao, K. Q. (2000) *Set Pair Analysis and Its Elementary Application.* Zhejiang Science and Technology Press, Hangzhou, China.

Analysing inter-annual variability of the water-energy balance in the upstream watershed of the Panjiakou Reservoir

XIANGYU XU & DAWEN YANG
State Key Laboratory of Hydro-Science and Engineering & Department of Hydraulic Engineering, Tsinghua University, Beijing 100084, China
xu-xy07@mails.tsinghua.edu.cn

Abstract The Panjiakou Reservoir is the largest reservoir on the Luan River, which is a major surface water source, but reservoir inflow has decreased significantly in recent decades. The physical aspect of the hydrological cycle in the upstream of the reservoir is studied to understand the hydrological changes and provide a sound scientific basis of water resources management. In the present research, a distributed hydrological model (i.e. GBHM) and a simple water-energy balance model based on the Budyko hypothesis were used. Using the climate and discharge data of 1980–1991, the present study analyses the inter-annual variability of water balance. The daily discharges at different locations are simulated by the GBHM model, and the actual evapotranspiration is estimated using both the GBHM and water-energy balance model. Comparing the results, both models give similar results that show the same complementary relationship between actual and potential evapotranspiration on an annual time scale.

Key words water-energy balance; distributed hydrological model; Budyko hypothesis; the Panjiakou Reservoir

INTRODUCTION

The North China Plain is an important domestic foodstuff-producing region which plays an important economic role. The Haihe River basin, the major river in the region, has a mean annual water yield of about 42 billion m^3, but the water resource per capita is only 305 m^3, which is less than 1/7 of the national average and 1/24 of the world average. Moreover, in recent decades, water shortage has been the main problem in North China due to intensive economic development. Climate change and human activity are the major impacting factors to the regional water balance and water resources. Apart from social and economic factors, the physical aspects of the hydrological cycle require better understanding in order to provide a sound scientific analysis of water resources.

The Luan River basin, in the northeast of the Haihe River Basin, is the major surface water source for Beijing and Tianjin and the surrounding regions of Hebei province. The Panjiakou Reservoir is the largest reservoir on the Luan River and provides drinking water for Tianjin city (a city of several million people), but it has been facing a water shortage in the last 10 years. In order to provide a sound scientific analysis of water resources in the upstream of the reservoir, it is significant to understand the physical aspects of the hydrological cycle in this region. Two methods can be used for hydrological analysis: the bottom-up approach and the top-down approach. The former is based on a prior perception, whereas a model is calibrated and it is emphasized that the model structure is inferred from the data.

The bottom-up approach in hydrological analysis uses the "process-based" or "physically-based" distributed hydrological model, which can predict overall catchment water balance components based on process information acquired on a smaller spatial and temporal scale. Freeze & Harlan (1969) first championed this approach in their "blueprint for a physically-based, digitally-simulated hydrologic response model". Since then, especially aided by increased process understanding, digital terrain attributes and the explosion of computing power, a large number of distributed models have been developed to simulate hydrological processes to estimate water balance in different watersheds. The System Hydrologic European (SHE) model (Abbott *et al.*, 1986) is a typical distributed model which simulates the entire hydrological cycle by discretizing the discrete grids form the computational units and the vertical axis into layers. Since then, a number of distributed hydrological models have been developed: IHDM (Morris, 1980), SLURP (Kite, 1995), SWAT (Arnold *et al.*, 1996), GBHM (Yang *et al.*, 1998, 2002), WEP (Jia *et al.*,

2001), etc. These and many other models have been applied to interpret and predict the impacts of land-use change and climate variability since they relate model parameters directly to physically observable land surface characteristics.

The top-down approach in hydrological analysis was introduced by Klemes (1983), it has been defined as the route that "starts with trying to find a distinct conceptual node directly at the level of interest (or higher) and then looks for the steps that could have led to it from a lower level". It features an attempt to predict overall catchment response and catchment functioning based on an interpretation of the observed response at the catchment scale. Budyko hypothesis is one example of this method. Based on the Budyko hypothesis, much work has been done to derive the Budyko curve (Fu, 1981; Choudhury, 1999; Zhang *et al.* 2004). Moreover, the Budyko hypothesis has been widely applied to analyse the variability of annual water balances (Budyko, 1974; Milly, 1994; Wolock & McCabe, 1999).

In the present research inter-annual variability of water-energy balances in the upstream watershed of the Panjiakou Reservoir was analysed using these two methods to better understand the mechanism of hydrological changes in this watershed.

STUDY AREA AND DATA AVAILABLE

The Luan River basin is located in the region delimited by geographical coordinates 40.4°–42.6°N and 115.5°–118.9°E with the elevation ranging from about 150 to 2000 m above sea level. The Panjiakou Reservoir is the largest reservoir on the Luan River, the upstream watershed of the Panjiakou Reservoir was selected as the study area in the present research (see Fig. 1).

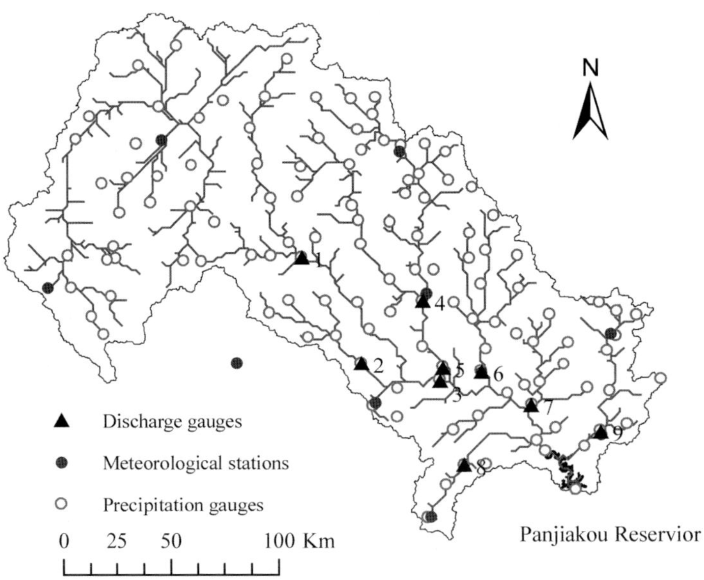

Fig. 1 The study area and locations of discharge gauges, meteorological stations and precipitation gauges.

The catchment extent and topography are represented using the digital elevation model (DEM) of a 100 m resolution. A land use map of 100 m resolution was from the Data Center for Resources and Environmental Sciences of Chinese Academy of Sciences. Based on the available data, the land use has been regrouped into 7 categories including: water body, urban-area, forest, irrigated cropland, upland, grassland and shrub. Seasonal change of vegetation is expressed by the monthly leaf area index (LAI) derived from monthly NDVI (NDVI data is obtained from the DAAC website of GSFC/NASA: http://daac.gsfc.nasa.gov/DAAC_DOCS/). The soil type data used here are based on the FAO-UNESCO's 1:5 million scale Soil Map of the World (FAO, 2003).

Daily meteorological data are obtained from the China Meteorological Administration and additional daily precipitation data are obtained from the Ministry of Water Resources. The daily meteorological data consists of precipitation, temperature, average wind speed, relative humidity, sunshine duration, etc. There are 10 meteorological stations and 121 precipitation gauges in this region (see Fig. 1). From the two data sets, a time period from 1980 to 1991 is selected according to the availability of records in these meteorological stations and precipitation gauges. The gridded meteorological data required as input to the distributed hydrological model is interpolated from the gauge data. Precipitation is interpolated using an angular-distance weighting method. In the same way, the wind speed, relative humidity and sunshine hours are also interpolated into each grid. The temperatures (maximum, minimum and mean) are interpolated using an elevation-corrected angular-direction weighting method. The daily potential evapotranspiration is calculated using the wind speed, relative humidity, sunshine hours and temperature (Shuttleworth, 1993).

The discharge data are collected for the same period of 1980–1991 from the "Hydrological Year Book" published by the Hydrological Bureau of the Ministry of Water Resources. In this study, seven gauges located on the tributaries and two gauges located on the mainstream (see Fig. 1) are selected for analysis.

METHODOLOGY

The ratio of evapotranspiration to precipitation in the study area is nearly 70–80%, therefore, estimation of evapotranspiration is important for understanding the catchment water balance. The physically-based hydrological model can estimate hourly evapotranspiration at hillslope scale, and through a bottom-up analysis the hourly evapotranspiration on each hillslope can be integrated into annual and long-term mean annual evapotranspiration at the catchment scale. The long-term mean annual and annual evapotranspiration of a catchment can be estimated using a water-energy balance model. Comparing the annual evapotranspiration, the inter-annual variability of water-energy balances in the upstream watershed of the Panjiakou Reservoir was analysed.

The distributed hydrological model–GBHM model

Since the GBHM model has been successfully applied in the non-humid regions of China, such as the Yellow River Basin (Yang *et al.*, 2004, 2006; Cong *et al.*, 2009), it is chosen as a bottom-up tool to simulate the hydrological processes and the evapotranspiration in the upstream watershed of the Panjiakou Reservoir in this paper. Here a 2-km grid is used for discretization of the study catchment. The procedures of model construction contain catchment subdivision, sub-grid parameterization, physically-based hydrological simulation on hillslopes and kinematic wave flow routing.

A total of 97 sub-catchments were divided for the study area. For each sub-catchment simulated in the GBHM model, the geomorphologic property of stream-hillslope formation is used to represent the catchment topography and it is also assumed that a large grid is comprised of a set of hillslopes located along the streams. The hillslope is the fundamental computational unit for hydrological simulation. The runoff generated from one grid is the sum of the hillslope responses including both the surface and subsurface runoff. The vertical flux, the actual evapotranspiration, is the total evapotranspiration simulated from all of the hillslopes. The soil moisture content state variable is taken as the area-averaged soil moisture of all the hillslopes.

The actual evapotranspiration is estimated from the potential evapotranspiration, which is calculated by the Penman equation (Penman, 1948). Considering seasonal variation of LAI, root distribution and soil moisture availability, it is computed individually from the canopy water storage, root zone and soil surface.

The parameters used in the model include vegetation parameters, land surface parameters, soil-water properties, river parameters, a snow-melting parameter and a groundwater parameter. Since most of the parameters have their physical meaning, they can be estimated through field tests. However, for such a large basin it is impossible to measure all parameters. Therefore, this study specifies the model parameters by referring to the existing database and handbooks. A 6-year test run from 1980 to 1985 was carried out to calibrate the model parameters. One of the calibrated parameters in this model is the snowmelt factor in the temperature-based snowmelt equation. Another calibrated parameter, the hydraulic conductivity of the groundwater, is calibrated by checking the base flow in different sub-basins.

The water-energy balance model

Water balance and energy balance are usually used to study the spatio-temporal characteristics of water resource in a region. Evapotranspiration is the key component linking water balance with energy balance. The primary factors controlling the long-term mean annual evapotranspiration are the local interaction of fluctuating water supply (precipitation) and demand (potential evapotranspiration). Budyko (1974) assumed that actual evapotranspiration is controlled by both water and energy availabilities and considered that the ratio of annual evapotranspiration and annual precipitation is the function of the ratio of annual precipitation and annual net radiation.

Based on the Budyko hypothesis, Fu (1981) derived the equation to calculate the actual evapotranspiration at the long-term time scale, which is given by:

$$\frac{E}{P} = 1 + \frac{E_0}{P} - \left[1 + \left(\frac{E_0}{P}\right)^{\varpi}\right]^{1/\varpi} \quad \text{or} \quad \frac{E}{E_0} = 1 + \frac{P}{E_0} - \left[1 + \left(\frac{P}{E_0}\right)^{\varpi}\right]^{1/\varpi} \tag{1}$$

where ϖ is a constant. This parameter has been shown to have a significant regional pattern and represents the land surface conditions of catchments, especially, vegetation cover, soil properties, and topography (Yang *et al.*, 2007). This model is simple enough to estimate actual annual evapotranspiration, only given the value of annual precipitation and potential annual evapotranspiration.

RESULTS AND DISCUSSIONS

River discharge

Ten discharge gauges in 10 different sub-catchments are selected for evaluating the GBHM model. The years from 1980 to 1985 were selected as the calibration period and the years from 1986 to

Table 1 Evaluation of the discharge simulation results of different periods.

No.	Gauge	River	Area(km^2)	R^2		RE(%)	
				period(1)	period(2)	period(1)	period(2)
1	Guojiatun	Luan River	14340	0.74	0.73	–42.0	0.7
2	Boluonuo	Xingzhou River	1378	0.87	0.55	2.9	23.8
3	Sandaohezi	Luan River	17100	0.84	0.73	21.3	5.1
4	Xiahenan	Yimatu River	2404	0.81	0.55	7.4	25
5	Hanjiaying	Yixun River	6848	0.84	0.7	4.1	19.6
6	Chengde	Wulie River	2460	0.92	0.65	0.0	0.8
7	Xiabancheng	Laoniu River	1648	0.87	0.73	36.3	16.1
8	Liying	Liu River	628	0.63	0.77	–5.0	7.2
9	Kuancheng	Pu River	1796	0.82	0.78	22.0	1.9
10	Panjiakou	Luan River	33700	0.91	0.79	2.0	1.2

Notes: period (1) represents the calibration period from 1980 to 1985 and period (2) represents the verification period from 1986 to 1991.

1991 were selected as the verification period for the daily discharge simulation. The Nash-Sutcliffe efficiency coefficient and the relative error between the annual simulated discharge and the observed one were selected as the quantitative evaluation index of the simulated hydrographs. Table 1 shows the information about the gauges and the simulation results of the daily discharge for both periods. Figure 2 shows the daily discharge hydrograph at three different gauges.

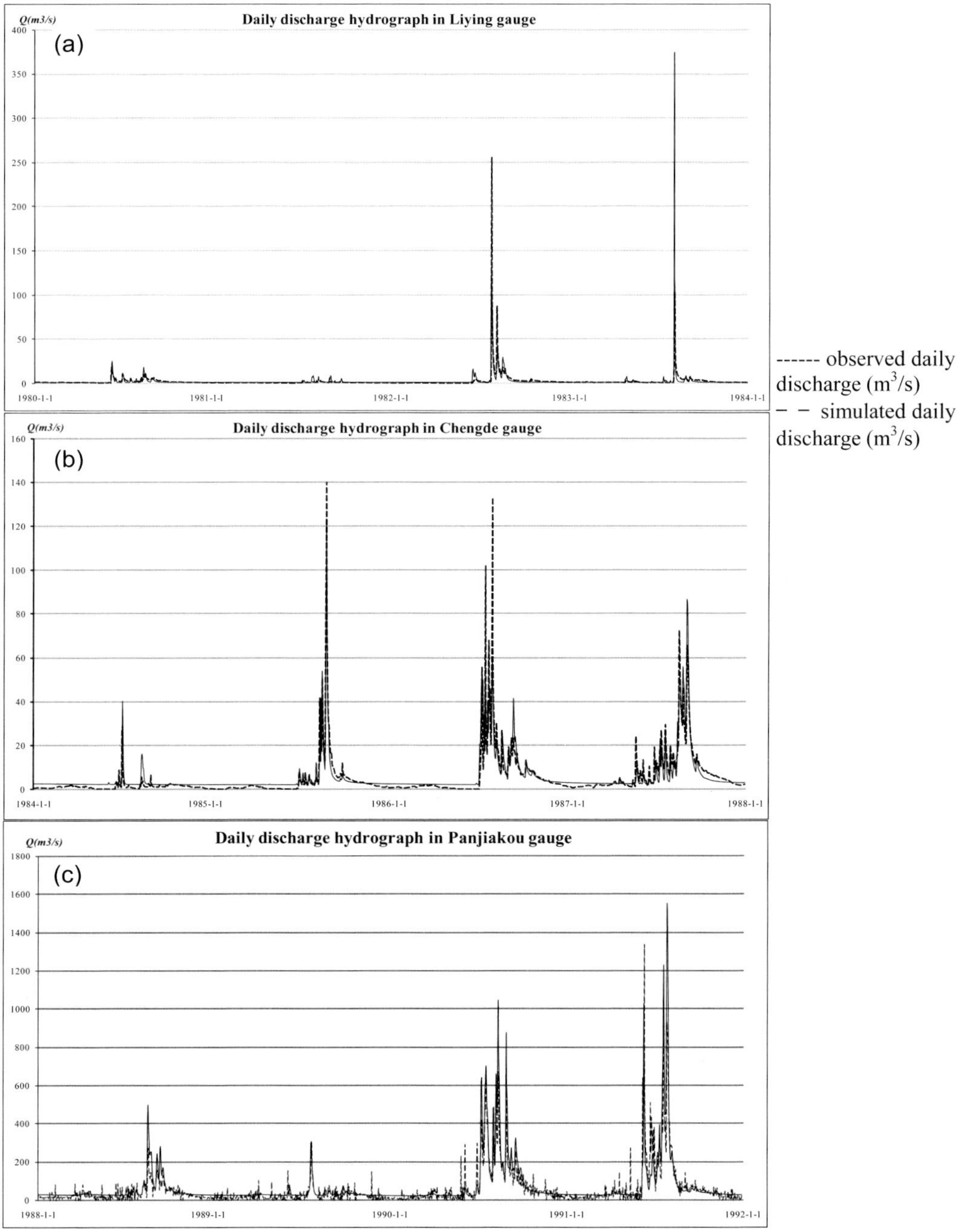

Fig. 2 Comparison of the observed and simulated daily discharge hydrograph in (a) Liying gauge (1980–1983), (b) Chengde gauge (1984–1987) and (c) Panjiakou gauge (1988–1991).

In the calibration period, the total relative runoff errors (the ratio of the difference between simulated and observed annual runoff to the observed annual runoff) have been calculated for the drainage areas above the gauges in each sub-catchment. This error ranges from 0.02% to 42.0%. In most sub-catchments, the relative error is less than 8%. The Nash coefficient for the whole basin is 0.91 and more than 0.80 in most sub-catchments. In the verification period (1986–1991), the total relative runoff errors range from 0.7% to 25.0%, and less than 8% in five sub-catchments. Nash's coefficient for the whole basin is 0.79 and more than 0.70 in most sub-catchments.

It can be seen that the simulated river discharges in the upstream have a larger error than in the downstream for both the calibration period and the verification period. The simulation error can be introduced by the uncertainty in the parameter calibration, the error of the hydrological simulation, and the error from input data. The main reason for the disagreement between simulation and observation hydrographs in the upstream basins could be due to the effect of frozen and thawing processes. It can be seen that the first peak of the hydrograph in a year was not generally simulated well.

Analysing the mean annual water-energy balance

Table 2 shows the comparison between the long-term mean values of actual evapotranspiration estimated from the annual precipitation and runoff data by neglecting the change in water storage, simulated by the GBHM model and the water-energy balance model. It can be seen that the values of the actual evapotranspiration by different models have a good agreement. Considering the value estimated from the long-term water balance by neglecting the change of water storage as the "true value" (which is denoted as the "observation"), the relative error for the simulated annual evapotranspiration are within ±5% by both the GBHM and water-energy balance models. This implies that both the detailed process-based distributed hydrological model and the simple lumped water-energy balance model can appropriately simulate the catchment long-term water balance.

Table 2 Comparison of the long-term water balance simulated by the GBHM and the water-energy balance model.

No.	Gauge	River	*P* (mm)	*R* (mm)	Simulated actual *E* (mm)			Relative error *RE* (%)	
					E_{wb}	E_{budyko}	E_{gbhm}	RE_1	RE_2
1	Guojiatun	Luan River	356.8	21.7	335.1	319.3	340.7	–4.72	1.67
2	Boluonuo	Xingzhou River	453.1	59.8	393.2	393.3	411.5	0.00	4.63
3	Sandaohezi	Luan River	375.5	23.8	351.7	336.6	351.5	–4.29	–0.07
4	Xiahenan	Yimatu River	420.7	29.8	390.9	372.5	389.7	–4.72	–0.32
5	Hanjiaying	Yixun River	426.8	30.0	396.7	409.5	393.7	3.21	–0.77
6	Chengde	Wulie River	503.4	62.1	441.2	444.2	452.3	0.69	2.51
7	Xiabancheng	Laoniu River	524.2	71.1	453.2	463.4	454.9	2.27	0.37
8	Liying	Liu River	595.3	96.6	498.6	516.0	501.0	3.47	0.47
9	Kuancheng	Pu River	533.5	85.7	447.7	464.4	452.5	3.72	1.07
10	Panjiakou	Luan River	430.0	45.2	384.8	395.1	381.8	2.68	–0.76

Notes: (1) E_{wb}, E_{budyko}, E_{gbhm} represent mean values of actual evapotranspiration calculated by long-term water balance, the water-energy balance model and the GBHM model, respectively. (2) RE_1 and RE_2 are the relative errors of the long-term mean values of actual evapotranspiration calculated using the water-energy balance model and the GBHM model, respectively, compared with the value estimated by long-term water balance.

Analysing the annual water-energy balance

In order to analyse the annual water-energy balance, Fig. 3 shows the inter-annual variability of the actual evapotranspiration in each sub-catchment from 1980 to 1991 simulated by the water-energy balance model and the GBHM model. It can be seen that both results have a very similar temporal pattern for the 10 sub-catchments. Results show that both the simple lumped water-energy balance model and the physically-based distributed hydrological model can predict catchment annual evapotranspiration and its inter-annual variability.

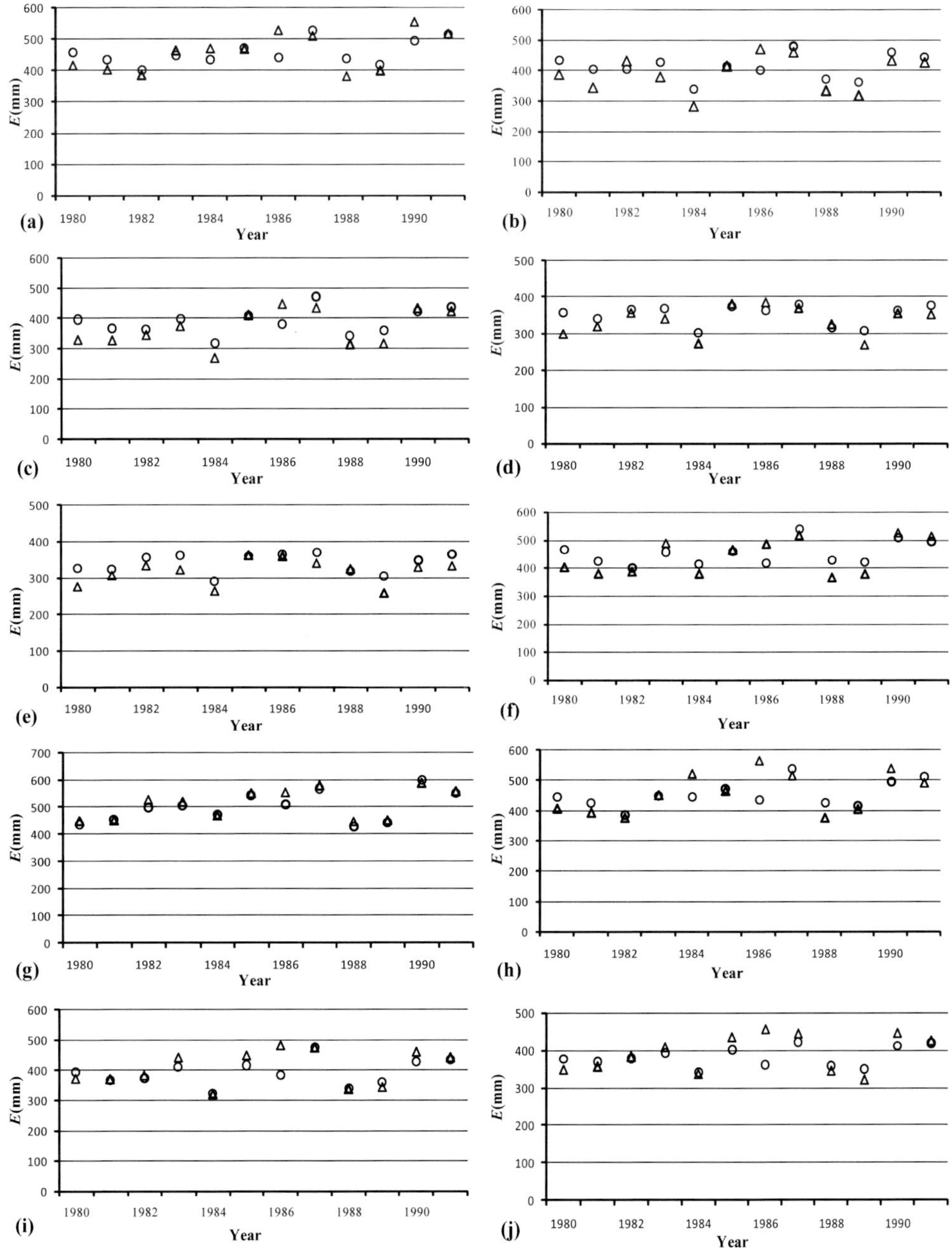

○ ---- actual annual evapotranspiration (E) predicted by GBHM model;
Δ ---- actual annual evapotranspiration (E) predicted by water-energy balance model

Fig. 3 Comparison of the annual evapotranspiration estimated by water-energy balance model and GBHM model ((a)–(i) represent the sub-catchment of the upstream of the Buoluonuo station, Chengde station, Hanjiaying station, Kuancheng station, Liying station, Xiabancheng station, Xiahenan station, Guojiatun station, and Sandaohezi station, respectively. (j) represents the whole watershed).

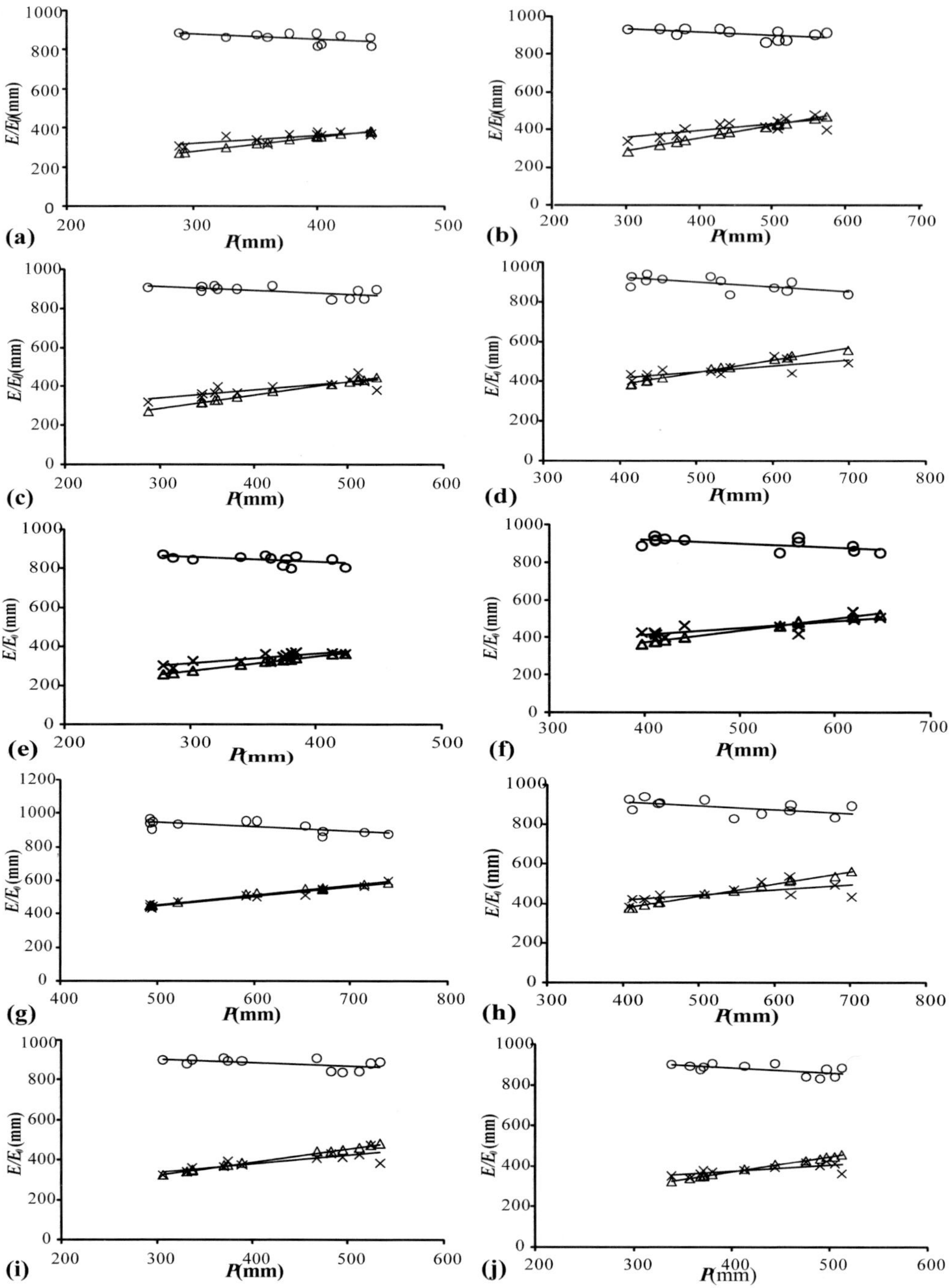

○ ---- potential annual evapotranspiration E_0 ;
x ---- actual annual evapotranspiration (*E*) predicted by GBHM
Δ---- actual annual evapotranspiration (*E*) predicted by water-energy balance model

Fig. 4 Comparison of the annual evapotranspiration estimated by water-energy balance model and GBHM model and the observed one. (a)–(i) represent the 9 sub-catchments, respectively, (j) represents the whole watershed.

According to Yang *et al.*'s study (2006), the actual evapotranspiration at annual and catchment scale is mainly controlled by precipitation rather than potential evapotranspiration in the non-humid regions of China, and the relationship between the actual and potential evapotranspiration should be complementary. The actual and potential evapotranspiration is usually plotted against the precipitation for validating the complementary relationship of evapotranspiration. Figure 4 shows the complementary relationship between the actual and potential evapotranspiration, in which the actual evapotranspiration is simulated by the GBHM and the water-energy balance model. From Fig. 4, it is known that annual catchment evapotranspiration simulated both by the GBHM and the water-energy balance model followed the complementary relationship with potential evapotranspiration.

CONCLUSIONS

In this study a distributed physically-based hydrological model (GBHM) and a simple water-energy balance model derived from Budyko hypothesis were used to understand the water-energy balances at mean annual and annual scale in 9 sub-catchments and the whole basin of the Luan River in northern China. At annual and long-term time scales, a consistent result has been achieved by the two models, which show the catchment annual water balance pattern can be predicted successfully by both a simple lumped model and a detailed distributed model in the study area and both models can be applied in the upstream watershed of Panjiakou Reservoir to evaluate the water resources on an annual scale. Furthermore, existence of the complementary relationship between the actual and potential evapotranspiration at the annual time scale in the study areas has been confirmed by both the distributed hydrological model and the lumped model.

Acknowledgement This study was partially supported by the key research program of the National Science & Technology Pillar Program in the Eleventh Five-year Plan Period (no. 2006BAB14B02) and by the Ministry of Water Resources (no. 20081012).

REFERENCES

Abbott, M. B., Bathurst, J. C. & Cunge, J. A. (1986) An Introduction to the European Hydrological System – Systeme Hydrologique Europeen, "SHE", 1, History and philosophy of a physically-based, distributed modeling system. *J. Hydrol*, **87**, 45–59.

Arnold, J. G., Williams, J. R. & Maidment, D. R. (1995) Continuous-time water and sediment-routing model for large basins. *J. Hydraul. Engng, ASCE* **121**(2), 171–183.

Budyko, M. I. (ed.) (1974) *Climate and Life*, translated from Russian by D. H. Miller. Academic, San Diego, California, USA.

Choudhury, B. J. (1999) Evaluation of an empirical equation for annual evaporation using field observations and results from a biophysical model. *J. Hydrol.* **216**, 99–110.

Cong, Z., Yang, D., Gao, B., Yang, H. & Hu, H. (2009) Hydrological trend analysis in the Yellow River basin using a distributed hydrological model, *Water Resour. Res.* **45**, doi:10.1029/2008WR006852.

FAO (2003), Digital soil map of the world and derived soil properties. Land and Water Digital Media Series Rev. 1.

Freeze, R. A. & Harlan, R. L. (1969) Blueprint for a physically-based digitally-simulated hydrological response model. *J. Hydrol.* **9**, 237–258.

Fu, B. P. (1981) On the calculation of the evaporation from land surface. *Scientia Atmospherica Sinica* **5**(1), 23–31 (in Chinese).

Gutman, G. & Ignatov, A. (1998) The derivation of the green vegetation fraction from NOAA/AVHRR data for use in numerical weather prediction models. *Int. J. Remote Sens.* **19**(8), 1533–1543.

Jia, Y. W., Ni, G. H. & Kawahara, Y. (2001) Development of WEP Model and its application to an Urban Watershed. *Hydrol. Processes* **15**, 2175–2194.

Kite, G. W. (1995) The SLURP model, In: *Computer Models of Watershed Hydrology* (ed. by V. P. Singh). Water Resources Publications, Highlands, Colorado, USA.

Klemes, V. (1983) Conceptualization and scale in hydrology. *J. Hydrol.* **65**, 1–23.

Milly, P. C. D. (1994) Climate, soil water storage, and the average annual water balance. *Water Resour. Res.* **30**, 2143–2156.

Morris, E. M. (1980) Forecasting flood flows in grassy and forested basins using a deterministic distributed mathematical model. In: *Hydrological Forecasting* (Proc. Symp. On the Application of recent Developments in Hydrological Forecasting in the Operation of Water Resource Systems, Oxford, UK), 247–265. IAHS Publ. 129. IAHS Press, Wallingford, UK.

Penman, H. L. (1948) Natural evaporation from open water, bare and grass. *Proc. R. Soc. Lond., Ser.* A **193**, 120–145.

Shuttleworth, W. J. (1993) Evaporation. In: *Handbook of Hydrology*, (ed. by D. R. Maidment), 4.1–4.53. McGraw-Hill, New York, USA.

Sivapalan, M., Zhang, L., Vertessy, R. & Bloschl, G. (2003) Downward approach to hydrological prediction. *Hydrol. Processes* **17**(11), 2101–2111.

Wolock, D. M., & McCabe G. J. (1999) Explaining spatial variability in mean annual runoff in the conterminous United States, *Clim. Res.* **11**, 149–159.

Xia, J., Liu, M. Y. & Jia, S. F. (2005) Water Security Problem in North China: Research and Perspective. *J. Pedosphere* **15**(5), 563–575.

Yang, D., Herath, S. & Musiake, K. (1998) Development of a geomorphology-based hydrological model for large catchments, *Ann. J. Hydraulic Engng, JSCE* **42**, 169–174.

Yang, D., Oki, T., Herath, S. & Musiake, K. (2002) A geomorphology-based hydrological model and its applications. In: *Mathematical Models of Small Watershed Hydrology and Applications* (ed. by V. P. Singh & D. K. Frevert), 259–300. Water Resources Publications, Littleton, Colorado, USA.

Yang, D., Li, C., Hu, H., Lei, Z., Yang, S., Kusuda, T., Koike, T. & Musiake, K. (2004) Analysis of water resources variability in the Yellow River of China during the last half century using historical data. *Water Resour. Res.* **40**, W06502, doi:10.1029/2003WR002763.

Yang, D., Sun, F., Liu, Z., Cong, Z. & Lei, Z. (2006) Interpreting the complementary relationship in non-humid environments based on the Budyko and Penman hypotheses. *Geophys. Res. Lett.* **33**, L18402, doi: 10.1029/2006GL027657.

Yang, D., Sun, F., Liu, Z., Cong, Z., Ni, G. & Lei, Z. (2007) Analyzing spatial and temporal variability of annual water-energy balance in non-humid regions of China using the Budyko hypothesis. *Water Resour. Res.* **43**, W04426, doi: 10.1029/2006WR005224.

Zhang, L., Hickel, K., Dawes, W. R., Chiew, F. H. S., Western, A. W. & Briggs, P. R. (2004) A rational function approach for estimating mean annual evapotranspiration. *Water Resour. Res.* **40**, W02502, doi: 10.1029/2003WR002710.

Hydrological modelling for the Zhangweinan River basin with intensive human activities

XITIAN CAI, BAOLIN SU & ZONGXUE XU
Key Laboratory of Water and Sediment Sciences, Ministry of Education, College of Water Sciences, Beijing Normal University, Beijing 100875, China
subl@bnu.edu.cn

Abstract One of the basins with most intensive human activities in China and even in the world, the Zhangweinan River basin was selected as the study area to apply SWAT2000 for hydrological modelling. Measured releases from six main reservoirs, actual irrigation schedules in different sub-basins, and major water uses withdrawn from rivers and groundwater were collected and investigated. Simulated ET was calibrated by using RS-based ET at a regional scale, and simulated monthly runoff was calibrated and validated. Results indicate that hydrological modelling in such a complex river basin is possible, with R^2 and E_{NS} (Nash–Sutcliffe efficiency coefficient) of around 0.75 for the calibration period, and around 0.65 for the validation period in monthly simulations, and RS-based ET is helpful for calibrating the simulated ET and for runoff simulation. Data collection is most important, and a good understanding on basin characteristics, human activities and model structure are strongly required.

Key words hydrological model; SWAT; Zhangweinan River; human activity; water use

INTRODUCTION

Human activities, which have increased enormously in the 20th century, have altered the landscape of the Earth, hydrological processes, meteorological processes (Samaniego & Bárdossy, 2007; Solomon *et al.*, 2007), and ecosystems (Eamus *et al.*, 2006). Urbanization and industrialization are the most important human activities, which increase water demand and produce large amounts of pollutants to the environment, especially to rivers. However, agriculture, which produces food for supporting the large population in China, is an even more important human activity that influences hydrological processes and water resources, because it consumes a large amount of water for irrigation and is the largest consumer of water, and has caused severe problems such as decreasing runoff and changing of the water cycle. A typical example is that in northern China, where the increase in the amount of water taken from river reaches is the direct cause of observed runoff decrease in northern China (Ren *et al.*, 1999).

The changes of landscape on the Earth, hydrological processes, and meteorological processes make it difficult in hydrological modelling. Traditional hydrological models do not consider land cover changes and cannot explain detailed heterogeneity in space. Refsgaard *et al.* (1996) emphasized the importance of physically-based and spatially-distributed hydrological models in simulating effects of watershed changes due to human activities. A description of spatial heterogeneities on hydrological variables and land surface characteristics, efficient simulation of hydrodynamics, and coupling simulation of the natural water cycle system and the artificial water cycle system are still problematic in distributed hydrological models (Jia *et al.*, 2006). Yu *et al.* (2008) gave a review on the applications of non-point source pollution modelling in China. Owing to intensive human activities in China, it is relatively more difficult to apply the Soil and Water Assessment Tools (SWAT) in China than that in North America where the model was developed. Few researchers have studied hydrological modelling in areas with intensive human activities (Döll, 2003; Jia *et al.*, 2006; He *et al.*, 2008).

The consequence of these human impacts is that previous hydrological records may no longer be reliable guides to what will happen in the future (Sivapalan *et al.*, 2003). Northern China is one typical example of this kind of ungauged areas (Liu *et al.*, 2005). In this study, Zhangweinan River basin, one of the basins with most intensive human activities in China and even in the world, was selected as the study area to employ SWAT2000 for hydrological modelling. Evapotranspiration (ET), which usually could not be discussed in detail in previous studies due to the lack of reliable data (Immerzeel & Droogers, 2007), is calibrated at the regional scale by using remote sensing derived ET. Parameter regionalization, irrigation schedule estimation for crops and water use for industry in the model are also discussed.

METHODOLOGY

Study area description

The Zhangweinan River basin (37 700 km^2) is located in the southern part of the Hai River basin in northern China and originates in the Taihang Mountains (Fig. 1). The basin is located between 112–118°E longitude and 35–39°N latitude. The terrain of the basin is high in the western area and low in the eastern area, including 68% of the mountainous area and 32% of the plain area. Upstream of the basin are the Taihang Mountains, where most areas have elevation above 1000 m. The annual mean temperature is approximately 14°C, and annual mean precipitation is approximately 608 mm. The gauges used for runoff calibration include Guantai and Yuancun in the upstream, and Linqing and Xinji in the downstream. Land use of the basin includes 53.4% of agricultural land (mainly WWHT – winter-wheat and summer-maize rotation areas, and CORN – corn) and 20.3% of pasture area (Fig. 2). The agricultural and industrial water consumptions have made the exploitation and utilization of water resources in this region largely exceed its capacity (Table 1), and the deficit was recharged by the overexploitation of groundwater and diversion of water from the Yellow River.

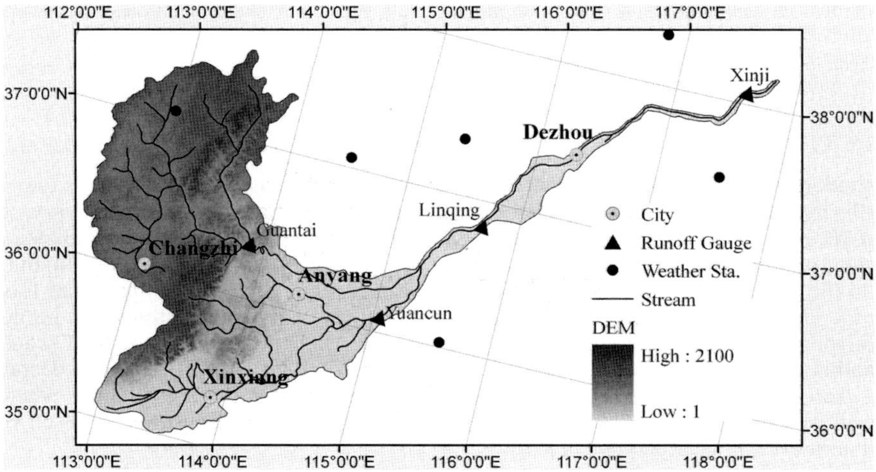

Fig. 1 Study area – the Zhangweinan River basin.

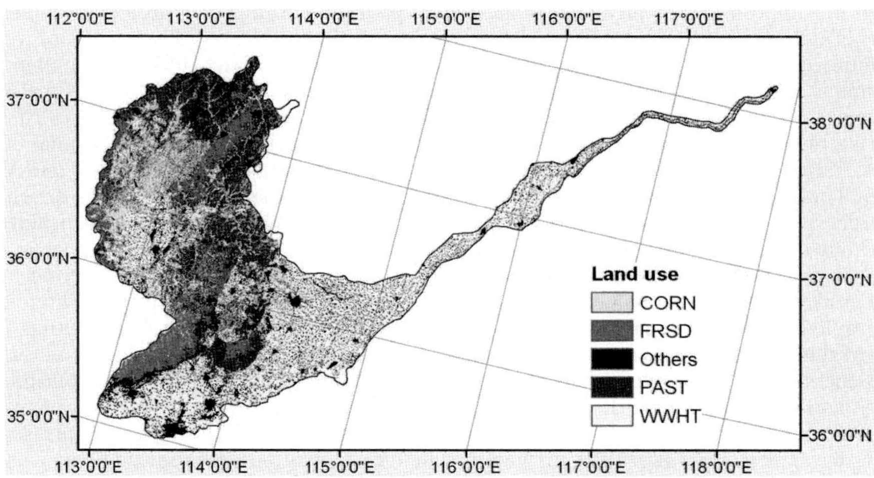

Fig. 2 The land use map of the Zhangweinan River basin (CORN, corn; WWHT, winter-wheat and summer-maize rotation; PAST, Pasture; FRSD, Deciduous Forest).

Table 1 Water resources and its utilization in the Zhangweinan River basin (unit: $\times 10^9$ m^3).

	Agriculture	Industry	Urban	Total water use	Total water resources
2000	10.68	2.27	0.88	13.83	10.21
2004	9.94	1.89	1.52	12.92	9.75

1: The data for 2000 were cited from *Comprehensive Water Resources Planning for the Haihe River Basin* (Haihe River Water Resources Commission, 2005a), and that for 2004 were cited from *Water Resources Bulletin of Haihe River Basin in 2004* (Haihe River Water Resources Commission, 2005b).

Model description

The hydrological simulation is based on SWAT2000, a public domain model on the internet. Basic theory and procedures on SWAT2000 have been described in detail in SWAT theoretical documentation and the user's manual, respectively (Neitsch *et al.*, 2001, 2002). This study only focuses on the consumptive water use by irrigation and domestic use in the study area.

Irrigation in a Hydrological Response Unit (HRU) may be scheduled by the user or automatically applied by SWAT. In addition to specifying the timing and application amount, the user must specify the source of irrigation water from one of a reach, a reservoir, a shallow aquifer, a deep aquifer, or a source outside the watershed, and the location of the water source (unless the source is outside the watershed).

Consumptive Water Use is a management tool that withdraws water from a sub-basin of the SWAT model. SWAT allows water to be withdrawn from the shallow aquifer, the deep aquifer, the reach, the reservoir or the pond within any sub-basin in the watershed. Consumptive water use is allowed to vary from month to month. For each month in the year, an average daily volume of water withdrawn from the source is specified. For reservoirs, the user may also specify a fraction of the water losses in the process of diversion.

Data description and processing

SWAT database consists of a spatial database, including Digital Elevation Model (DEM), soil and land use map; and an attribute database, including weather, hydrology, irrigation, water use, management practices, water use data, etc. (Table 2). Monthly RS-based ET was interpreted from the bi-weekly MODIS satellite data (1 km × 1 km resolution) by the Institute of Remote Sensing and Application (IRSA), Chinese Academy of Sciences, and its reliability has been examined by the expert panel of the Global Environmental Fund (GEF), and a detailed description can be found in Wu *et al.* (2008).

Table 2 Database system of the model for the Zhangweinan River basin.

Data type	Scale/station	Description	Source
DEM	1:250 000	Elevation, channel, slopes, lengths, aspect, etc.	National Fundamental Geographic Database
Land use	1:100 000	Land use classifications such as cropland, pasture, etc.	Remotely Sensed image and Survey
Soil	1:1 000 000	Soil types; physical properties, e.g. bulk density, texture, etc.	ISS, CAS[1]
Weather	10 Stations	Precipitation, temperature, relative humidity, wind speed, etc.	NMIC, CMA[2]
Precipitation	102 Stations	Precipitation	NMIC, CMA and ZRAB[3]
Reservoirs	6	Storage capacity, out flow, water supply, etc.	ZRAB
Irrigation	–	Distribution of irrigation area, water source, irrigation amount, etc.	On-site surveys and ZRAB
Management practices	–	Tillage, fertilization, pesticide, etc.	On-site surveys and ZRAB
Water Use	–	283 water intakes; groundwater pumping	On-site surveys and ZRAB

[1], Institute of Soil Science, Chinese Academy of Sciences; [2], National Meteorological Information Centre, China Meteorological Administration (CMA); [3], Zhangweinan River Administration Bureau.

The data processing mainly includes several steps. Firstly, the drainage network of the watershed was delineated based on the DEM and digitized real streams, and the whole watershed was divided into 110 sub-basins. Since the main stream in the downstream is a canal, the water only yielded near the stream can contribute to the river, which makes the downstream area very narrow (Fig. 1). Secondly, the derived sub-basin map, land use and soil maps were overlaid. Thirdly, six large/medium-scale reservoirs were specified to the sub-basins where they are located, and 409 hydrological response units (HRUs) were defined based on the land use and soil maps. Finally, weather and precipitation data were input into the model system.

RESULTS ANALYSIS

Model evaluation

In order to evaluate the performance of the model, coefficient of determination (R^2) and Nash-Sutcliffe efficiency coefficient (E_{NS}) were used. The R^2 is an indicator of strength of relationship between the observed and simulated values. $R^2 = 0$ means no relationship and $R^2 = 1$ means complete correlation. E_{NS} indicates how well the plot of the observed value *versus* the simulated value fits the 1:1 line, and values between 0.0 and 1.0 are generally viewed as acceptable levels of model performance, whereas a value <0.0 shows that the model gives a bad prediction for the observed value with unacceptable performance.

ET calibration

ET was calibrated for the crop land for the period of 2003–2004 by using the RS-based ET. Calibrating procedure is firstly checking the irrigation scheme, which requires irrigation survey for the study area; and adjusting the parameters for crop growth, especially the maximum potential leaf area index (BLAI), the fraction of growing season when leaf area index declines (DLAI), and the parameters controlling the crop growth stages (e.g. FRGRW1, LAIMX1, FRGRW2, LAIMX2, etc.); then checking crop stress level (e.g. water, temperature, etc.). The ET simulating result is shown in Fig. 3(a). It fits the RS based ET well in the growing seasons of the crop (except in May,

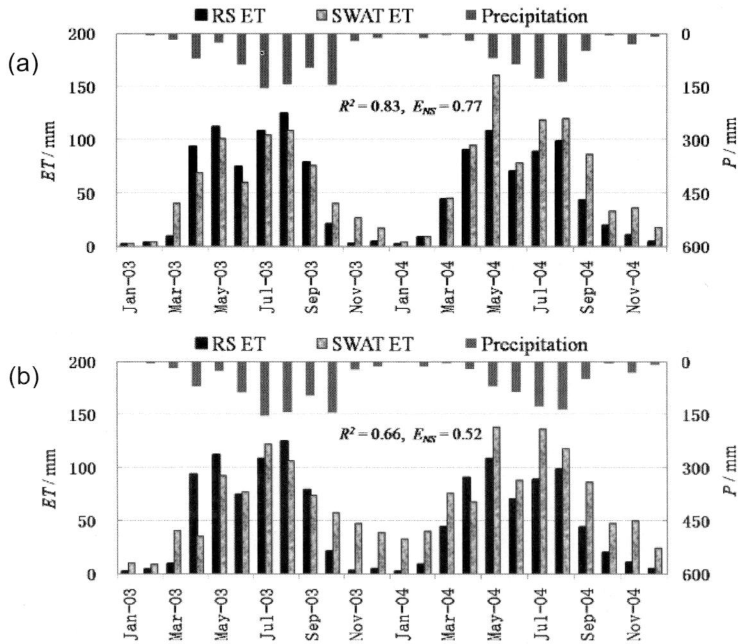

Fig. 3 Comparisons between simulated *ET* and RS derived *ET* for 11 sub-basins with predominantly land use type of WWHT, (a) crop growth parameters calibrated, (b) using default crop growth parameters of the SWAT model.

2004), and model performance is improved compared to that of no ET calibration (Fig. 3(b)); but it does not fit well from October to December, which perhaps comes from the model bias that SWAT evaporates more water from the soil than it actually does.

In this study, the irrigation scheme and crop growth parameters were set as constants during the whole simulation period. If they were set separately year by year, the results would be improved greatly. But this must be based on more detailed input data.

Runoff calibration

After ET calibration, monthly runoff was calibrated for the period 2001–2003, and validated for 2004. Some relative sensitive parameters such as CN2 – SCS runoff curve number, SOL_AWC – available water capacity of the soil layer, ALPHA_BF – baseflow alpha factor and CANMX – maximum canopy storage were adjusted. CN2 and SOL_AWC are usually adjusted so that measured and simulated surface runoff match well, and ALPHA_BF is adjusted to make baseflow acceptable. But parameter CANMX is also a very important parameter in semi-arid and semi-humid region such as the Zhangweinan River basin, because canopy storage may take a large part of a given precipitation. Therefore, different values were specified for different land use, some of which were based on previous study, and others were obtained by calibration.

Calibrated and validated results are shown in Fig. 4, and evaluation of the model performance in runoff simulation is shown in Table 3. The results showed that the trend of the hydrograph was simulated well and improved greatly compared to that of no calibration in crop growth parameters. It was acceptable in such an area with intensive human activities. It also showed that the fits at downstream gauges (Linqing and Xinji) were better than those at upstream gauges (Guantai and

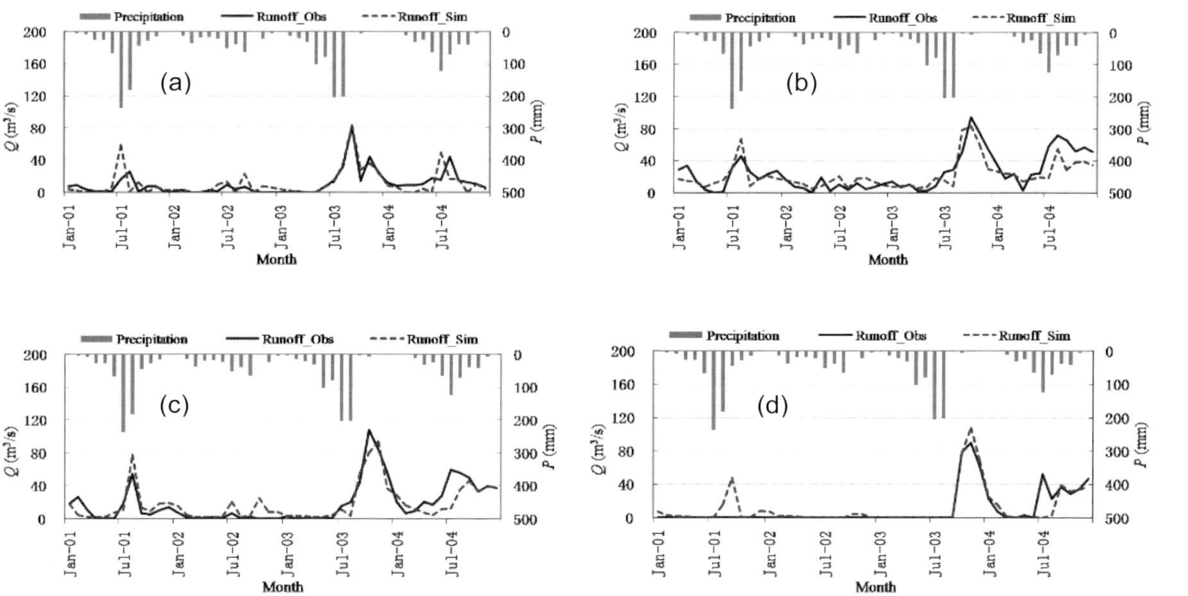

Fig. 4 Comparisons between simulated and observed runoff. (a) Guantai, (b) Yuancun, (c) Linqing, (d) Xinji.

Table 3 Evaluation of model performance in runoff simulation for the calibration and validation periods.

	Guantai Stn.		Yuancun Stn.		Linqing Stn.		Xinji Stn.	
	Calib.	Valid.	Calib.	Valid.	Calib.	Valid.	Calib.	Valid.
R^2	0.71	0.61	0.68	0.66	0.82	0.78	0.88	0.84
E_{NS}	0.61	0.47	0.68	0.59	0.81	0.75	0.81	0.77

Calib., calibration period of 2001–2003; Valid., validation period of 2004.

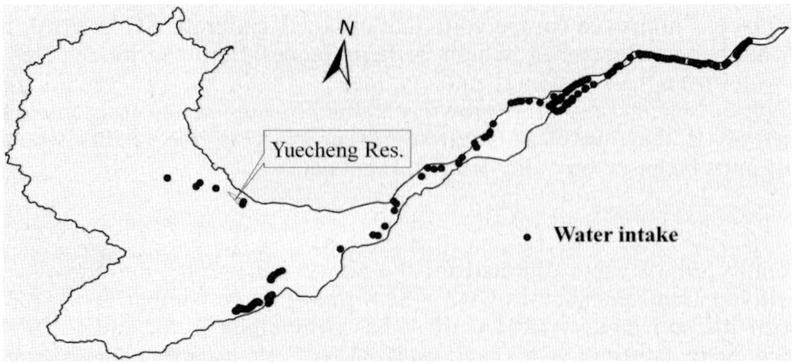

Fig. 5 The location of water intakes and the Yuecheng reservoirs.

Yuancun). This is perhaps due to the more detailed daily outflow data from the Yuecheng reservoir and water uses withdrawn from 283 water intakes (Fig. 5), most of which are located in the downstream area. The water uses in the upstream area is as complex as in the downstream area, however, few water use data were collected, and most of the water use data in the upstream area were estimated according to population, industrial and area of agricultural land use in this study, which might lead to inaccuracy in the input data.

DISCUSSION

Irrigation water uses

In order to simulate the irrigation process, it is very important to guarantee that the model can obtain an adequate amount of water from the irrigation sources.

Shallow or deep aquifer If a shallow or deep aquifer is specified as an irrigation water source, the parameter of SHALLST (Initial depth of water in the shallow aquifer) or DEEPST (Initial depth of water in the deep aquifer) become critical, because there would not be sufficient water for irrigation if SHALLST (or DEEPST) is set too low; and it would not represent the actual situation in northern China if SHALLST (or DEEPST) is set too high.

River reach If a river reach is specified as the irrigation water source, it is very important to make sure that the model can get sufficient water from that reach on the scheduled irrigation day. If runoff is too low for irrigation water, it is recommended to schedule another or other days for a given irrigation event; it will last for several days (about 8 days) for an irrigation process in a large irrigation area (Guo, 1997). Since SWAT makes the schedule for an irrigation process on a given day, it should be modified to make a schedule for a starting and an ending day to represent the irrigation duration days.

Reservoirs It is also very important to specify initial reservoir volume at the starting date for simulation and check the storage during simulating period, so that the storage in reservoirs can fit the water volume variations under real operation scheme.

Consumptive water use

Consumptive water use takes an important role in an area with intense human activities. The source of the water use can be specified by an average daily volume water removal from a pond, a reach, a shallow or a deep aquifer for the month. However, if the specified WUS (specified average daily volume water removal from a reach for the month) from a reach is determined entirely according to the survey data, water diverted from the reach may be less than the objective volume. For example (Fig. 6), according to the survey, 5 184 000 m^3 of water was removed in September, and then the WUS from a reach for September was 172 800 m^3, which was specified for each day in September. On a given day, the water available in that reach may be 259 200 m^3, which is enough to supply sufficient water; but on another day, the water available in that reach may be 86 400 m^3, which is not enough to supply insufficient water. An ideal solution is that water use data for each day are collected and input into the model. But the data availability usually does not allow us to do so, and specifying WUS for each month is still a reasonable approach. Under this situation, the WUS can be increased from 172 800 m^3 to 259 200 m^3 so that 86 400 m^3 extra water is removed on the day with high flow to compensate 86 400 m^3 deficit for another day with low flow. Perhaps it is better to have an output for total water volume removal in that month, so that it can be decided whether to increase the WUS or not.

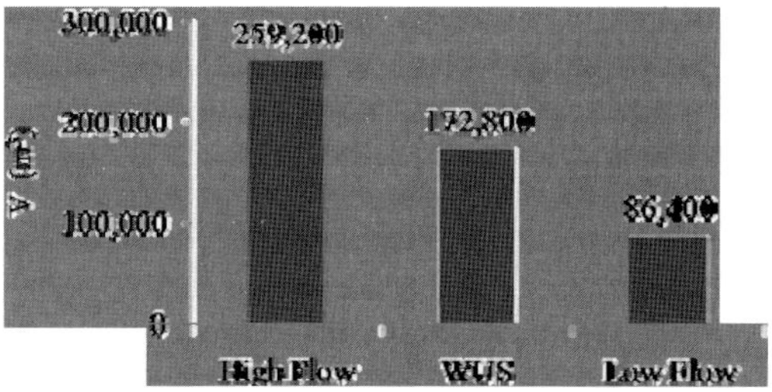

Fig. 6 Water use removal from a reach for a given day of the month. WUS, specified average daily volume water removal from a reach for the month; Low Flow, supposed low flow for a given day; High Flow, supposed high flow for a given day.

CONCLUSIONS

According to the hydrological simulation in the Zhangweinan River basin with intensive human activities, some conclusions were summarized as follows:

Firstly, on-site survey and data collection are very important for hydrological modelling in an area with intense human activities. A lot of work should be focused on the data collection and analyses, such as runoff, reservoir, irrigation, water uses, and even socioeconomic data. Secondly, if these data are available, it is possible to simulate the hydrological process in a changing environment. Thirdly, irrigation scheme and consumptive water consumption should be estimated when these data are not available, and the model structure should also be modified so that it can represent human interference more precisely. Finally, RS-based ET is helpful for calibrating ET, and therefore improving model performance in the simulation.

Acknowledgements Global Environmental Fund (GEF) should be gratefully appreciated for providing financial support for the research "Hai Basin Integrated Water Management and Environment Project". The Chinese National Meteorological Information Center, China Meteorological Administration should be appreciated for providing meteorological data. Thanks should also be given to the Zhangweinan River Administration Bureau for providing hydrological, irrigation and water use data in the Zhangweinan River basin.

REFERENCES

Döll, P., Kaspar, F. & Lehner, B. (2003) A global hydrological model for deriving water availability indicators: model tuning and validation. *J. Hydrol.* **270**, 105–134.

Eamus, D., Hatton, T., Cook, P. & Colvin, C. (2006) *Ecohydrology: Vegetation Function Water and Resource Management.* CSIRO Publishing, Collingwood, Australia.

Guo, Y. Y. (1997) *Irrigation and Water Conservancy Science* (third edition). Water Conservancy and Hydropower Press of China, Beijing, China (in Chinese).

Haihe River Water Resources Commission (2005a) *Comprehensive Water Resources Planning for the Haihe River Basin.* Haihe River Water Resources Commission, Tianjin, China (in Chinese).

Haihe River Water Resources Commission (2005b) *Water Resources Bulletin of Haihe River Basin in 2004.* Tianjin, China (in Chinese).

He, H. M., Zhou, J. & Zhang, W. C. (2008) Modelling the impacts of environmental changes on hydrological regimes in the Hei River Watershed, China. *Global Planetary Change* **61**(3–4), 175–193.

Immerzeel, W. W. & Droogers, P. (2008) Calibration of a distributed hydrological model based on satellite evapotranspiration. *J. Hydrol.* **349**, 411–424.

Jia, Y. W., Wang, H., Zhou, Z. H., Qiu, Y. Q., Luo, X. Y., Wang, J. H., Yan, D. H. & Qin, D. Y. (2006) Development of the WEP-L distributed hydrological model and dynamic assessment of water resources in the Yellow River Basin. *J. Hydrol.* **331**(3–4), 606–629.

Liu, S. X., Xia, J., Mo, X. G. (2005) Advances in Predictions in Ungauged Basins. *Water Resour. Hydropower Engng* **36**(2), 9–12 (in Chinese).

Neitsch, S. L., Arnold, J. G., Kiniry, J. R. & Williams, J. R. (2001) *Soil and Water Assessment Tool Theoretical Documentation* (Version 2000). Grassland, Soil and Water Research Laboratory, Agricultural Research Service, Temple, Texas, USA.

Neitsch, S. L., Arnold, J. G., Kiniry, J. R., Srinivasan, R. & Williams, J. R. (2002) *Soil and Water Assessment Tool User's Manual* (Version 2000). Grassland, Soil and Water Research Laboratory, Agricultural Research Service, Temple, Texas, USA.

Refsgaard, J. C., Storm, B. & Abbott, M. B. (1996) Comment on'A discussion of distributed hydrological modeling'. In: *Distributed Hydrological Modeling* (ed. by M. B. Abbott & J. C. Refsgaard), 279–287. Kluwer Academic Publishers, Dordrecht, The Netherlands.

Ren, L. L., Wang, M. R., Li, C. H. & Zhang, W. (2002) Impacts of human activity on river runoff in the northern area of China. *J. Hydrol.* **261**(1–4), 204–217.

Samaniego, L. & Bárdossy, A. (2007) Relating macroclimatic circulation patterns with characteristics of floods and droughts at the mesoscale. *J. Hydrol.* **335**(1–2), 109–123.

Sivapalan, M., Takeuchi, K., Franks, S. W., Gupta, V. K., Karambiri, H., Lakshmi, V., Liang, X. & McDonnell, J. J., Mendiondo, E. M., O'Connell P. E., Oki, T., Pomeroy, J. W., Schertzer, D., Uhlenbrook, S. & Zehe, E. (2003) IAHS Decade on Predictions in Ungauged Basins (PUB), 2003–2012: shaping an exciting future for the hydrological sciences. *Hydrol. Sci. J.* **48**(6), 857–880.

Solomon, S., Qin, D., Manning, M., Chen, Z., Marquis, M., Averyt, K. B., Tignor, M. & Miller, H. L. (eds) (2007) *Climate Change 2007: The Physical Science Basis. Contribution of Working Group I to the Fourth Assessment Report of the Intergovernmental Panel on Climate Change*. Cambridge University Press, Cambridge, UK.

Wu, B. F., Xiong, J., Yan, N. N., Yang, L. D. & Du, X. (2008) ETWatch for monitoring regional evapotranspiration with remote sensing. *Adv. Water Sci.* **19**(5), 671–678 (in Chinese).

Yu, T., Meng, W., Ongley, E., Zheng, B. H. & Deng, Y. X. (2008) Problems and recommendations for non2point source pollution identification in China. *Acta Scientiae Circumstantiae* **28**(3), 401–407 (in Chinese).

Estimating parameter uncertainty in hydrological models using the bootstrap method

ZHANLING LI[1,2], QUANXI SHAO[2], ZONGXUE XU[1] & CHONG-YU XU[3]

1 *Key Laboratory of Water and Sediment Sciences, Ministry of Education, College of Water Sciences, Beijing Normal University, Beijing 100875, China*
zongxuexu@vip.sina.com

2 *CSIRO Mathematical and Information Sciences, Private Bag 5, Wembley, Western Australia 6913, Australia*

3 *Department of Geosciences, University of Oslo, PO Box 1022 Blindern, N-0315 Oslo, Norway*

Abstract Uncertainty issues in hydrological modelling have been given much attention due to their great impact on model prediction. This study mainly investigates parameter uncertainty in two hydrological models. One is the conceptual WASMOD model with five parameters, and the other is the semi-distributed SWAT model in which nine sensitive parameters are identified for model calibration. A non-parametric technique, the bootstrap method, is employed for the assessment of parameter uncertainty in two models. The upper reaches of the Heihe River basin is chosen as the study area. Results show that, comparatively, the normality assumption is basically satisfied for parameters in the simple WASMOD model, although they are still a little skewed, while in the more complex SWAT model the normality assumption is violated in most cases (six out of nine parameters are proved not to be normally distributed). A 95% confidence interval for each parameter is obtained for the purpose of quantifying its uncertainty. These results cannot be popularized directly to other cases since the ranges of model parameters might change as scale or resolution changes and may still be dependent on model implementation and boundary conditions.

Key words parameter uncertainty; WASMOD; SWAT; Heihe River; bootstrap

INTRODUCTION

Hydrological models are common tools for estimating the runoff dynamics and the water balance for regions of interest at various scales. However, their prediction results are fraught with uncertainty due to some data errors and imperfect model structure. In general, there are four important sources of uncertainties in hydrological models: input data, output data used for model calibration, model parameters and imperfect model structure. All uncertainties have been addressed recently (Beven & Binley, 1992; Seibert, 1997; Bates & Campbell, 2001; Engeland *et al.*, 2005; Mugunthan & Shoemaker, 2006; Yang *et al.*, 2007a,b).

The main focus of this study is on the assessment of model parameter uncertainty. Several methods are available for parameter estimation and uncertainty analysis, such as first-order approximation (Vrugt & Bouten, 2002), sequential uncertainty fitting (SUFI) (Abbaspour *et al.*, 2004), Generalized Likelihood Uncertainty Estimation (GLUE) (Beven & Binley, 1992) and the Bayesian method (Bates & Campbell, 2001; Engeland *et al.*, 2005; Gallagher & Doherty, 2007). In this paper, a non-parametric technique, the bootstrap method (Efron, 1979), is employed for model parameter uncertainty analysis. Advantages of this approach are ease of description and implementation without a requirement for an external program. It does not require the normality assumption to calculate the parameter's confidence interval (Efron, 1979; Davison & Hinkley, 1997; Srinivas & Srinivasan, 2005).

MODELS

Two hydrological models were selected in this study. One is the conceptual WASMOD model developed by Xu (2002) with five calibrated parameters. The primary equations of the model are presented in Table 1. The other is the commonly used semi-distributed SWAT model. It is a continuous time model at river basin scale and can be applied to a wide range of environmental conditions, including not only humid and semi-humid, but also arid and semi-arid regions. The most sensitive nine parameters are identified for SWAT model calibration. In order to reduce the computation demands and resolve the problems of non-identifiable parameters, the concept of

Table 1 Principal equations of WASMOD model with snow module.

Component	Equation
Snowfall	$sn_t = p_t\{1-\exp[-(c_t-a_1)/(a_1-a_2)]^2\}^+$
Rainfall	$r_t = p_t - sn_t$
Snowmelt	$m_t = sp_{t-1}\{1-\exp[-(c_t-a_2)/(a_1-a_2)]^2\}^+$
Snow storage	$sp_t = sp_{t-1} + sn_t - m_t$
Actual evapotranspiration	$e_t = \min\{ep_t[1-a_3^{w_t/\max(ep_t,1)}], w_t\}$ or $e_t = \min\{w_t[1-\exp(-a_3 \times ep_t)], ep_t\}$
Slow flow	$s_t = a_4(sm_{t-1}^+)^{b_1}$
Fast flow	$f_t = a_5(sm_{t-1}^+)^{b_2}(n_t + m_t)$
Active rainfall	$n_t = p_t - ep_t\{1-\exp[-p_t/\max(ep_t,1)]\}$
Total runoff	$d_t = s_t + f_t$
Water balance	$sm_t = sm_{t-1} + r_t + m_t - e_t - d_t$

p_t, monthly precipitation; c_t, monthly air temperature; ep_t, monthly potential evapotranspiration; w_t, available water with the equation of $w_t = sm_{t-1}^+ + p_t$; b_1 = or 2; b_2 = 1 or 2; the superscript plus means for example $x^+ = \max(x,0)$.

"aggregate parameter" put forward by Yang *et al.* (2007a) is used in the SWAT model. Accordingly, not the original but the corresponding aggregate parameters are calibrated in SWAT model (Table 3). These two models are applied in the upper reaches of the Heihe River basin, Yingluoxia watershed. The model performance is estimated using the efficiency coefficient R^2 proposed by Nash & Sutcliffe (1970).

STUDY AREA AND DATA DESCRIPTION

Yingluoxia watershed with an area of 10 009 km^2 lies in the upper reaches of the Heihe River basin in Northwest China (between 98°–101°30′E and 38°–42°N) (Qi & Luo, 2006) (Fig. 1). The watershed is characterized as inland climate with annual precipitation of 400 mm and annual potential evapotranspiration of 1600 mm (Wang *et al.*, 2003). Inter-annual variability in precipitation is large, with nearly 70% concentrated in the rainy season (from June to September). Annual mean runoff depth is 160 mm/year with weak intra annual variability. The area of permanent snow or ice cover is 59 km^2, accounting for 0.59% of the total area (Zhao & Zhang, 2005).

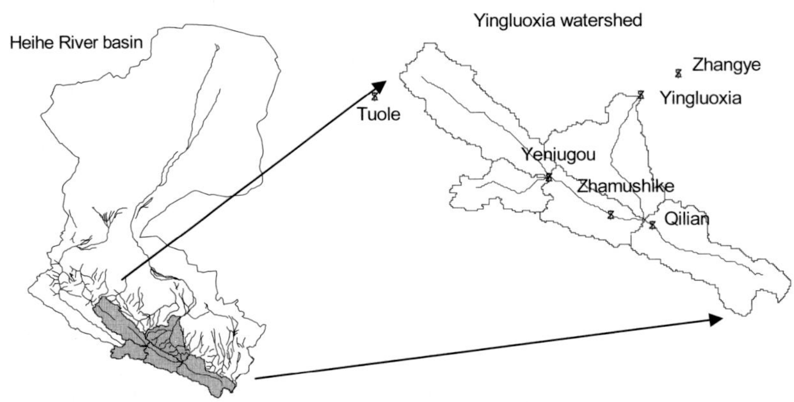

Fig. 1 Yingluoxia watershed and the location of hydro-meteorological stations.

The following datasets are used: digital elevation model with 950 m spatial resolution; soil map and land use map in 1995 at the scale of 1:1 000 000; daily meteorological data (e.g. daily maximum and minimum air temperature, wind speed and relative humidity) at Yeniugou, Qilian, Tuole and Zhangye stations, precipitation data at Yeniugou, Zhamushike, Qilian, Tuole, Zhangye and Yingluoxia stations, and runoff data at Yingluoxia station from 1985 to 2000 in which runoff data in 1988 and 1989 were not available. The first 5 years (1985–1989) is used as a warm-up period, and the following 7 years (1990–1996) is used for calibration. The last four years (1997–2000) is used for validation. All the aforementioned datasets are provided by "Environmental & Ecological Science Data Center" for West China (http://westdc.westgis.ac.cn/), National Science Foundation of China and "Digital River Basin" (http://heihe.westgis.ac.cn/).

BOOTSTRAP METHOD

The bootstrap method developed by Efron (1979) is a resampling technique that can be used to estimate properties of estimators, which then become quite popular in applications (Efron, 1979; Davison & Hinkley, 1997; Srinivas & Srinivasan, 2005). A brief description is given below:

Assumed that we have data $\{X(t),Y^{obs}(t)\}$ ($t = 1, \ldots, N$) and the hydrological model is written as:

$$Y(t) = f[X(t),\theta] \tag{1}$$

where $X(t)$ is the collection of all input data; $Y^{obs}(t)$ is the observed data; θ is the parameter vector $\theta = (\theta_1, \ldots, \theta_m)$ with m being the number of model parameters. Let:

$$\varepsilon(t) = Y^{obs}(t) - \hat{Y}(t) = Y^{obs}(t) - f[X(t),\hat{\theta}] \tag{2}$$

be the residuals, in which $\hat{\theta}$ is the estimate of θ. Let F be the joint population distribution of parameter θ and F_i the marginal distribution of parameter θ_i ($i = 1, \ldots, m$). Both joint and marginal distributions are unknown. The bootstrap is given as follows. Randomly sampling with replacement from the residuals $\varepsilon(t)$ to yield $\varepsilon^*(t)$. Compute the bootstrap sample:

$$Y^*(t) = f[X(t),\hat{\theta}] + \varepsilon^*(t) \tag{3}$$

and recompute the estimator $\hat{\theta}^*$ for each sample. This corresponds to $\hat{F}$ being the empirical distribution function of the parameters. The superscript "*" denotes something calculated from the approximate bootstrap distribution $\hat{F}_i$.

RESULTS AND DISCUSSION

Model performance

The monthly WASMOD model shows good performance with R^2 values reaching as high as 0.942 and 0.928 for calibration and validation periods, respectively. The baseflow simulates well during both periods and the peak flow also shows agreeable simulation results except for those in 1993 and 1998. The simulated hydrograph (not shown due to space limitation) seems to be acceptable compared with the observed one as a whole, and the error of simulation is relatively small.

The SWAT model can predict not only monthly, but also daily data. The monthly R^2 reaches 0.87 and 0.85 for calibration and validation periods. Daily R^2 reaches 0.70 and 0.65 for calibration and validation periods, with best performance with $R^2 = 0.76$ in 1990 and worst performance with $R^2 = 0.61$ in the dry year 1992. The lower R^2value may attribute to the bad fitness of spring runoff in which snow and ice melting contributes a lot to runoff amount.

Parameter estimation and its uncertainty

Figures 2 and 3 show the marginal distributions of calibrated parameters in WASMOD and SWAT models, respectively. Tables 2 and 3 show the results of normal test for the parameters and their uncertainty ranges. All five parameters in WASMOD model can be regarded as normally distributed based on the results of the K-S test (Table 2 and Fig. 2). In the SWAT model, nearly all parameter densities are more or less skewed (Fig. 3). Results of normal tests also show that only three aggregate parameters, a__CN2.mgt, v__ALPHA_BF.gw and v__SMFMX.bsn, accept the assumption of normal distribution (Table 3). This fact indicates that the normal distribution is not always necessarily a good approximation for parameters, especially in complex hydrological models, and therefore quantities such as confidence intervals that rely on those assumptions will also be suspect. In this study, 95% confidence intervals for parameters, obtained by ordering the parameter samples and then identifying the 2.5% and 97.5% thresholds values based on their marginal distribution, are used to indicate their uncertainty ranges.

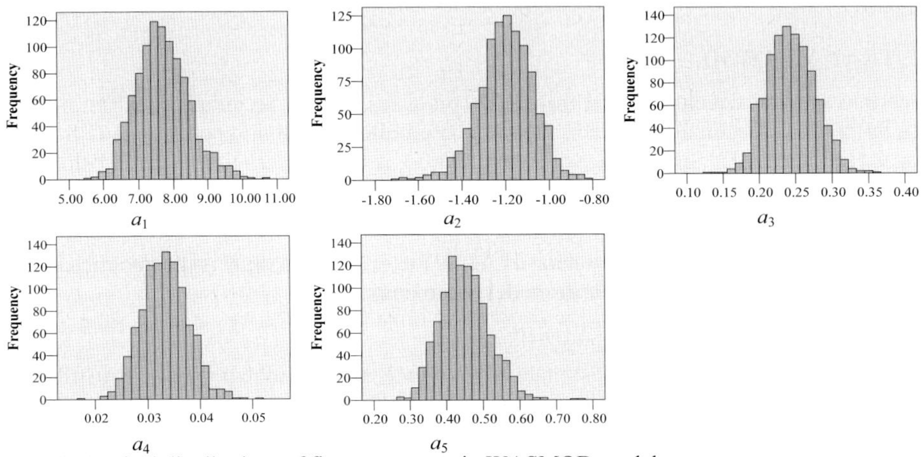

Fig. 2 Marginal distributions of five parameters in WASMOD model.

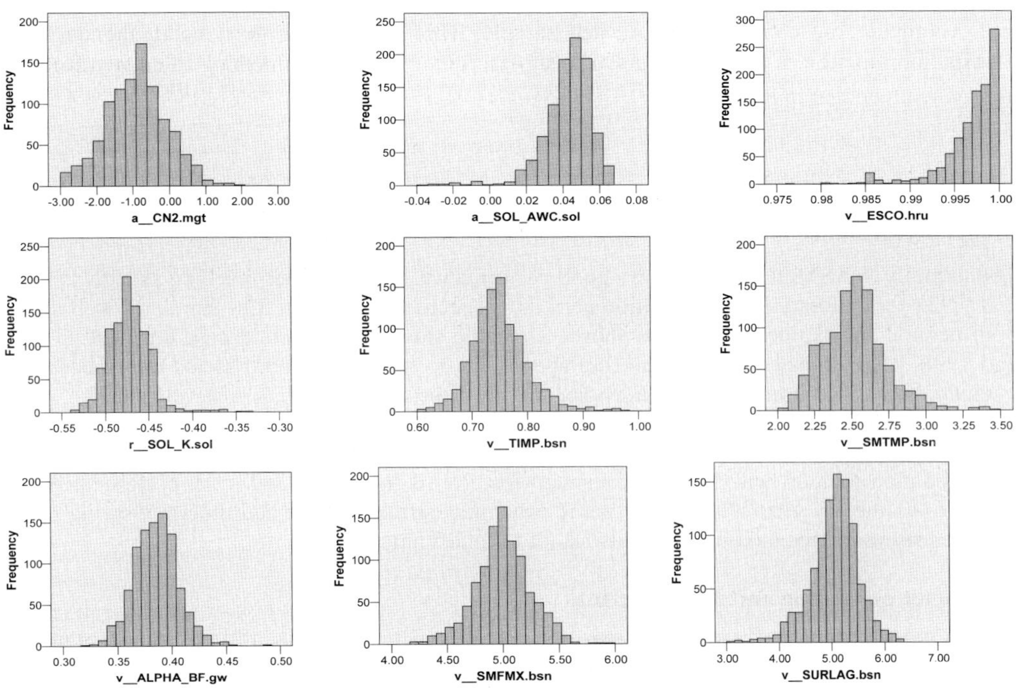

Fig. 3 Marginal distributions of nine aggregate parameters in SWAT model.

Table 2 Normal test for five parameters and their uncertainty ranges in WASMOD model.

Parameter	Kolmogorov-Smirnov Z	Asymp. Sig.(2-tailed)	95% confidence interval
a_1	1.07	0.21	(6.387, 9.400)
a_2	0.99	0.28	(–1.461, –0.969)
a_3	0.74	0.65	(0.182, 0.308)
a_4	0.63	0.83	(0.025, 0.042)
a_5	1.08	0.19	(0.331, 0.587)

Table 3 Normal test for nine aggregate parameters and their uncertainty ranges in SWAT model.

Aggregate parameter	Kolmogorov-Smirnov Z	Asymp. Sig.(2-tailed)	95% confidence interval
a__CN2.mgt	0.75	0.63	(–2.568, 0.722)
a__SOL_AWC.sol	3.31	0.00	(0.015, 0.063)
v__ESCO.hru	5.42	0.00	(0.986, 1.000)
r__SOL_K.sol	1.49	0.02	(–0.515, –0.424)
v__TIMP.bsn	1.83	0.00	(0.654, 0.859)
v__SMTMP.bsn	1.92	0.00	(2.145, 3.022)
v__ALPHA_BF.gw	0.96	0.32	(0.345, 0.427)
v__SMFMX.bsn	1.23	0.10	(4.449, 5.490)
v__SURLAG.bsn	1.86	0.00	(4.04, 5.88)

Aggregate parameters are constructed according to Yang's work (Yang *et al.*, 2007a, b). "a__", 'v__", "r__" means an absolute increase, a replacement, and a relative change to the initial parameter value, respectively.

In the WASMOD model, the 95% confidence intervals (CI) for a_1 and a_2 range from 6.39 to 9.40 and –1.46 to –0.97 (Table 2). Parameters a_1 and a_2 are two threshold temperature parameters controlling snowfall and snowmelt. Parameter a_3 controls the relationship between potential evapotranspiration and actual evapotranspiration with 95% CI of 0.18–0.31 and the smaller the value for a_3, the greater the evaporation losses at all moisture storage states. Parameter a_4 controls the proportion of runoff that appears as "base flow", with 95% CI ranging from 0.03 to 0.04 and a_5 controls the proportion of fast runoff with 95% CI ranging from 0.33 to 0.59. The simulation of both base flow and peak flow presents agreeable results in most years, leading to the whole hydrograph simulated well in total.

In the SWAT model, the 95% CI for a__CN2.mgt is (–2.57, 0.72) (Table 3). The increase in CN2 reflects higher surface runoff than that in the previous simulation. SURLAG, the surface runoff lag coefficient, has the same 95% CI as the parameter v__SURLAG.bsn, varying from 4 to 6, for v__ denotes "replacement". The changes in temperature related parameters (TIMP, SMTMP and SMFMX) demonstrate that temperature factors have a significant influence on snow melt and further on runoff. Their 95% CI are from 0.65 to 0.86, 2.15 to 3.02 and 4.45 to 5.49 for parameters v__TIMP.bsn, v__SMTMP.bsn, and v__SMFMX.bsn, respectively. ALPHA_BF has a great effect on groundwater recharge, and the 95% CI of its aggregate parameter varies from 0.34 to 0.43. Compared with others, the range of v__ESCO.hru is very narrow, from 0.986 to 1. ESCO is the soil evaporation compensation factor. Its higher estimated optimum value indicates that the soil evaporation probably plays a weaker role in the whole evaporation process, which is possibly related to the low air temperature and dense plant cover. Another reason leading to its marginal distribution near to the boundary of its prior interval possibly attributes to the complicated and strong correlations among parameters and the compensation effect among parameters.

CONCLUSIONS

Both WASMOD and SWAT models show good performance in simulating runoff in the upper reaches of the Heihe River basin, especially in normal years.

The marginal distributions of model parameters obtained by using the bootstrap method show that parameters in the WASMOD model are approximately normally distributed; while in the SWAT model, six of nine parameters reject the null hypothesis that the distributions are the same as normal, and the rest are also more or less skewed. Therefore, the assessment of parameter uncertainty by using normal approximation seems not so suitable, especially in complex hydrological models. In parameter uncertainty ranges, the 95% confidence intervals for parameters are obtained by ordering the parameter samples and then identifying the 2.5% and 97.5% thresholds values in terms of their marginal distribution. Different parameters have different uncertainty ranges. Such ranges are expected to be reduced with more data and more information provided. However, such results cannot be popularized to other studies directly due to the fact that the ranges of model parameters might change with changing scale or resolution and may still be dependent on model implementation and boundary conditions.

Acknowledgements This study has been financially supported by the "Jingshi Scholar" Leading Professor Programme, Beijing Normal University, and the China Scholarship Council, as well as the support provided by CSIRO Mathematical and Information Sciences.

REFERENCES

Abbaspour, K. C., Johnson, C. A. & van Genuchten, M. T. (2004) Estimating uncertain flow and transport parameters using a sequential uncertainty fitting procedure. *Vadose Zone J.* **3**, 1340–1352.

Bates, B. C. & Campbell, E. P. (2001) A Markov chain Monte Carlo scheme for parameter estimation and inference in conceptual rainfall-runoff modelling. *Water Resour. Res.* **37**, 937–947.

Beven, K. & Binley, A. (1992) The future of distributed models – model calibration and uncertainty prediction. *Hydrol. Processes* **6**, 279–298.

Davison, A. C. & Hinkley, D. V. (1997) *Bootstrap Methods and their Application.* Cambridge University Press, Cambridge, UK.

Efron, B. (1979) Bootstrap methods: another look at the jackknife. *Ann. Stat.* **7**, 1–26.

Engeland, K., Xu, C. Y. & Gottschalk, L. (2005) Assessing uncertainties in a conceptual water balance model using Bayesian methodology. *Hydrol. Sci. J.* **50**(1), 45–63.

Gallagher, M. & Doherty, J. (2007) Parameter estimation and uncertainty analysis for a watershed model. *Environ. Modelling Software* **22**, 1000–1020.

Mugunthan, P. & Shoemaker, C. A. (2006) Assessing the impacts of parameter uncertainty for computationally expensive groundwater models. *Water Resour. Res.* **42**, W10428, doi:10.1029/2005WR004640.

Nash, J. E. & Sutcliffe, J. V. (1970) River flow forecasting through conceptual models: Part 1. A discussion of principles. *J. Hydrol.* **10**(3), 282–290.

Qi, S. Z. & Luo, F. (2006) Land-use change and its environmental impact in the Heihe River Basin, arid northwestern China. *Environ. Geol.* **50**(4), 535–540.

Seibert, J. (1997) Estimation of parameter uncertainty in the HBV Model. *Nord. Hydrol.* **28**, 247–262.

Srinivas, V. V. & Srinivasan, K. (2005) Matched block bootstrap for resampling multiseason hydrologic time series. *Hydrol. Processes* **19**, 3659–3682.

Vrugt, J. A. & Bouten, W. (2002) Validity of first-order approximations to describe parameter uncertainty in soil hydrologic models. *Soil Sci. Soc. Am. J.* **66**, 1740–1751.

Wang, G. X. Liu, C. M. & Huang, Y. B. (2003) The theory of SWAT model and its application in Heihe Basin. *Progress Geography* **22**, 79–86 (in Chinese).

Xu, C. Y. (2002) WASMOD–The Water And Snow balance MODelling system. In: *Mathematical Models of Small Watershed Hydrology and Applications* (ed. by V. P. Singh & D. K. Frevert), Chap 17. Water Resources Publications, LLC, Chelsea, Michigan, USA.

Yang, J., Reichert, P. & Abbaspour, K. C. (2007a) Bayesian uncertainty analysis in distributed hydrological modelling: A case study in the Thur River basin (Switzerland). *Water Resour. Res.* **43**, W10401, doi:10.1029/2006WR005497.

Yang, J., Reichert, P., Abbaspour, K. C. & Yang, H. (2007b) Hydrological modelling of the Chaohe Basin in China: Statistical model formulation and Bayesian inference. *J. Hydrol.* **340**(3–4), 167–182.

Zhao, D. & Zhang, W. (2005) Rainfall–runoff simulation using the VIC-3L model over the Heihe River mountainous basin, China. In: *International Geoscience and Remote Sensing Symposium* (IGARSS), V6, 25th Anniversary IGARSS 2005: IEEE International Geoscience and Remote Sensing Symposium, 4391–4394.

Analysis of the dominant climatic factors of evaporation change over the main basins in mainland China based on Budyko and Bouchet hypotheses

HANBO YANG, DAWEN YANG, ZHENTAO CONG & ZHIDONG LEI
State key Laboratory of Hydro-Science and Engineering & Department of Hydraulic Engineering, Tsinghua University, Beijing 100084, China
yhb99@mails.tsinghua.edu.cn

Abstract Actual evaporation (E) has different change trends over various regions. It is important to understand why actual evaporation changes and what the dominant climatic factors are. Based on the Budyko and Bouchet hypotheses, we established a water-energy coupled balance model that can quantify the effects of radiation, precipitation and temperature on regional actual evaporation. Furthermore, the analysis, according to this model and meteorological data from about 700 stations across mainland China during 1961–2005, revealed the regional change of actual evaporation and its dominant climatic factors. The results show that temperature is the key factor in the Song-Liao River basin; while precipitation is the major factor in the Northwest River basin, the Hai River basin, the Yellow River basin, and the Southwest Rivers basin. In addition, net radiation controls E in the Huai River basin, the Yangtze River basin, the Southeast Rivers basin, the Pearl River basin, and Hainan province.

Key words climate change; evaporation; dominant factors; Budyko hypothesis; Bouchet hypothesis

INTRODUCTION

Actual evaporation is an important component of the surface fluxes coupling the land surface with the atmosphere and water balance with heat balance at the surface, which is remarkably influenced by climate change. During the past century (1906–2005), the global average surface temperature has increased by about 0.74°C (IPCC, 2007), which is called global warming. It is expected to lead to the atmosphere transporting higher amounts of water vapour. Accompanying global warming, a large amount of observations show that global average solar radiation reaching the ground surface has a declining ratio of 0.5 W/(m^2 year^{-1}) (Stanhill & Cohen, 2001), which is named global dimming. Global warming indicates an accelerating water cycle; however, declining solar radiation implies a weakening energy supply for evaporation, which is expected to lead to decreasing actual evaporation. Additionally, the regional change of actual evaporation has been reported widely (e.g. Jiang *et al.*, 2005; Gao *et al.*, 2007). Why the regional actual evaporation changes under climate change has therefore become a topical research subject in recent years.

Budyko (1974) proposed that actual evaporation was mainly controlled by available water and energy (Budyko hypothesis), which can be expressed as the water–energy balance equation. This equation is a useful tool for estimating regional actual evaporation (Zhang *et al.*, 2004; Yang *et al.*, 2008a), which is especially a function of precipitation and potential evaporation for a catchment over a long-term time scale (Budyko, 1974). Potential evaporation is an indicator of available energy for evaporation, which is considered as the forcing of evaporation occurring. Sometimes, pan evaporation is used to measure potential evaporation, a regional change in which has been observed during the past dozens of years over China (Thomas, 2000; Wang *et al.*, 2007). However, when potential evaporation drives actual evaporation, actual evaporation will also modify near surface atmospheric conditions (e.g. air temperature, wind speed, and humidity), which results in a change in potential evaporation. This feedback of potential evaporation to actual evaporation was documented as the complementary relationship between potential and actual evaporations (Bouchet, 1963). In other words, the change in potential evaporation is not only the result of the change in actual evaporation, but also the cause. In addition, another forcing to actual evaporation change, precipitation, also has a regional pattern of change trends (Zhai *et al.*, 2005).

In the hydrological cycle, water balance and energy balance are the most important principles, which can be expressed as:

$$P = E + R + \Delta S \tag{1}$$

and

$$R_n = \lambda E + H + G \tag{2}$$

where P, E, and R are precipitation, actual evaporation, and runoff, respectively; ΔS is a change of water storage; R_n, λE, H, and G are net radiation, latent heat, sensible heat, and soil heat fluxes, respectively. On a long-term time scale, ΔS and G can be ignored, leading to:

$$P = E + R \tag{3}$$

and

$$R_n = \lambda E + H \tag{4}$$

According to equations (3) and (4), we can consider the changes in precipitation P and net radiation R_n as inducing changes in the hydrological cycle.

This study aims to establish a model for evaluating evaporation change, based on the Budyko and Bouchet hypotheses; and then to analyse regional changes of actual evaporation, in order to explore why it changes, and capture the dominant factors resulting in these changes.

DATA AND METHOD

In this study, daily meteorological data collected from 740 stations during 1961–2005, including air temperatures, sunshine duration, wind speed, relative humidity, and precipitation, were collected from the National Meteorological Administration of China. Also a digital elevation model (DEM) of 1-km resolution from USGS (the US Geological Survey) was used to capture the topography. To analyse the regional patterns of actual evaporation, China was divided into 11 basins (see Fig. 1). Because of no observation data from Taiwan province, the study region is limited to the mainland.

The procedures for calculating basin average climatic variables are: (1) a 10-km grid data set covering the study areas was interpolated from the gauge data; (2) the basin average values were then calculated. Air temperature was interpolated using an inverse distance weighted technique, taking into consideration the impact of elevation. Other variables were interpolated using an inverse distance weighted technique.

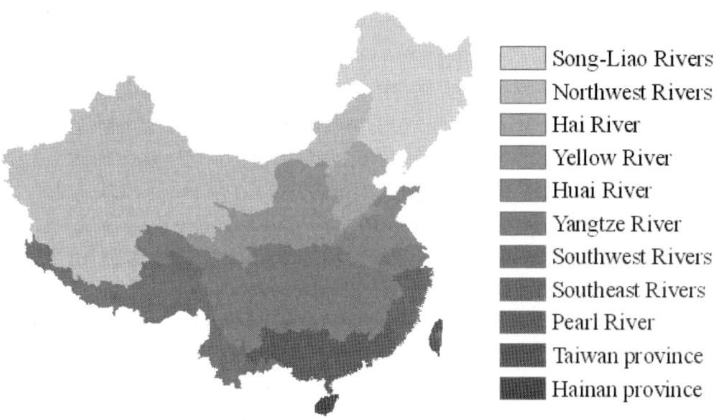

Fig. 1 Study basins in this study.

To qualify the hydrological response to climate change, we first introduced the mean annual water-energy balance equation:

$$E = \frac{E_0 P}{\left(P^n + E_0^n\right)^{1/n}} \tag{5}$$

with n being a parameter representing basin characteristics (Yang *et al.*, 2008a). And to describe the feedback of E_0 to E, we brought in the complementary relationship between potential and actual evaporation (Bouchet, 1963), which can be expressed as:

$$E + E_0 = 2\,E_w \tag{6}$$

where E_w was defined as the wet environment evaporation (Brutsaert & Stricker, 1979), which can be calculated by the Priestley-Taylor equation (Priestley & Taylor, 1972):

$$E_w = \alpha\frac{\Delta}{\Delta+\gamma}\left(R_n - G\right)/\lambda \tag{7}$$

where Δ is the slope of the saturated vapour pressure *versus* temperature curve, γ is a psychometric constant, λ is the latent heat of vaporization of water, and α is a parameter (with a standard value of 1.26) that indicates the capacity of available energy ($R_n - G$) to transform latent heat (Eagleson, 2002). Further, the differential forms of equations (5) and (6) can be derived as:

$$\frac{dE}{dt} = \frac{\partial E}{\partial P}\cdot\frac{dP}{dt} + \frac{\partial E}{\partial E_0}\cdot\frac{dE_0}{dt} \tag{8}$$

and

$$\frac{dE}{dt} + \frac{dE_0}{dt} = 2\frac{dE_w}{dt} \tag{9}$$

Equation (8) includes an assumption that the change of basin characteristics (i.e. land cover change) can be neglected, i.e. $dn/dt = 0$. According to equation (9), by neglecting the soil heat flux, dE_w/dt can be derived as:

$$\frac{dE_w}{dt} = \alpha\frac{\Delta}{\Delta+\gamma}/\lambda\cdot\frac{dR_n}{dt} + \alpha R_n/\lambda\cdot\frac{d\left(\frac{\Delta}{\Delta+\gamma}\right)}{dT}\cdot\frac{dT}{dt} \tag{10}$$

Combining equations (8), (9) and (10), we can derive a model which expresses the responses of actual evaporation to the changes in net radiation, precipitation, and air temperature as:

$$\frac{dE}{dt} = a\cdot\frac{dR_n}{dt} + b\cdot\frac{dP}{dt} + c\cdot\frac{dT}{dt} \tag{11}$$

And the coefficients a, b, and c were calculated by the equation:

$$\begin{cases} a = 2\dfrac{\partial E}{\partial E_0}\cdot\alpha\dfrac{\Delta}{(\Delta+\gamma)\cdot\lambda}/\left(1+\dfrac{\partial E}{\partial E_0}\right) \\ b = \dfrac{\partial E}{\partial P}/\left(1+\dfrac{\partial E}{\partial E_0}\right) \\ c = 2\dfrac{\partial E}{\partial E_0}\cdot\alpha R_n/\lambda\dfrac{\partial\left(\dfrac{\Delta}{\Delta+\gamma}\right)}{\partial T}/\left(1+\dfrac{\partial E}{\partial E_0}\right) \end{cases} \tag{12}$$

These coefficients describe the sensitivity of actual evaporation to climate factors.

This model includes two parameters, i.e. n from equation (5) and α from equation (7). It is worthy of notice that the parameter α has a regional variability, which increases with the latitude increasing and the distance from ocean decreasing (Yang *et al.*, 2008b). Following this, the value for each basin was given (see Table 1). Another parameter, the parameter n, can be estimated according to basin and vegetation characteristics (Yang *et al.*, 2008a). And in this study, according to equation (5), n was estimated for each basin by using mean annual precipitation, potential evaporation, and runoff. Further, according to equation (12), the coefficients a, b, and c were estimated, using the mean annual values of meteorological variables (see Table 1).

Table 1 change of regional actual evaporation over main basins in Mainland China during 1961–2005.

Num.	Basin	n	α	(a) · dR_n/dt (mm/year^2)	(b) · dP/dt (mm/year^2)	(c) · dT/dt (mm/year^2)	dE/dt (mm/year^2)
1	Song-Liao River	1.4	1.2	–0.18	–0.52	**0.61**	–0.08
2	Northwest Rivers	2.0	1.0	–0.02	**0.40**	0.10	0.47
3	Hai River	1.6	1.2	–1.11	**–1.34**	0.52	–1.93
4	Yellow River	1.6	1.1	–0.29	**–0.68**	0.40	–0.57
5	Huai River	1.4	1.2	**–2.05**	–0.36	0.38	–2.04
6	Yangtze River	1.1	1.0	**–1.17**	0.16	0.21	–0.80
7	Southwest Rivers	1.1	1.0	–0.21	**0.56**	0.29	0.63
8	Southeast Rivers	1.1	1.0	**–3.35**	0.10	0.35	–2.90
9	Pearl River	1.1	1.0	**–2.74**	0.21	0.18	–2.35
10	Hainan Province	1.1	1.0	**–2.93**	–0.03	0.40	–2.57

RESULTS AND DISCUSSION

We calculated the change rate of net radiation, precipitation, and air temperature by regress analysis (the results are shown in Table 1), and then the effects of net radiation, precipitation, and air temperature on actual evaporation, as quantified as (a) · dR_n/dt, (b) · dP/dt, and (c) · dT/dt. Consequently we evaluated the change of regional actual evaporation dE/dt according to equation (11) (see Table 1).

Table 1 shows the effect of changing climatic variables on actual evaporation from 1961 to 2005. Declining net radiation was found to have a widespread negative effect on actual evaporation, and the effect is roughly larger in the southern basins than in the northern basins. Regarding the effect of changing precipitation, it is negative in some basins, and positive in other basins. In addition, this effect has a significant regional pattern, i.e. it is larger in the northern basins than in the southern basins. Regarding the effect of warming climate, it leads to a positive effect on actual evaporation over all basins, which is larger in the northern basins (except the Northwest Rivers basin) than in the southern basins (except Hainan province).

Actual evaporation has a change, as a result of climatic variable changes. The Northwest Rivers and the Southwest Rivers basins have an upward actual evaporation trend, while the other basins have a declining actual evaporation. This is consistent with the conclusions in some previous studies. Wang *et al.* (2005) documented the Yangtze River basin as having an increasing actual evaporation during 1961–2000. Besides, using a distributed hydrologic model, Gao *et al.* (2008) found an increasing actual evaporation in the Yarlung Zangbo River basin, as a part of the Southwest Rivers basin.

Furthermore, it can be found that different climatic factors dominate the change of actual evaporation in different basins, namely net radiation in the Huai River basin, the Yangtze River basin, the Southeast Rivers basin, and the Pearl River basin and Hainan province; precipitation in the Northwest Rivers basin, the Hai River basin, the Yellow River basin, and the Southwest Rivers basin; and air temperature in the Song-Liao River basin. Declining net radiation is the leading cause of changes of actual evaporation in most southern basins, and changing precipitation is the leading cause in most northern basins.

CONCLUDING REMARKS

Solar radiation and precipitation are important factors in the hydrological cycle, where changes have taken place. To evaluate their impact on actual evaporation, we established a model for evaluating the regional hydrological response to changes in net radiation, precipitation, and air temperature, based on the Budyko and Bouchet hypotheses. According to this model, we analysed the change of actual evaporation over the main basins in mainland China. The results show that most basins have a decreasing actual evaporation. Furthermore, leading causes for the declining actual evaporation were not identical in different basins. The actual evaporation change was controlled by changing precipitation in most southern basins (the Northwest Rivers basin, the Hai River basin, and the Yellow River basin); and it was dominated by declining net radiation in most northern basins (the Huai River basin, the Yangtze River basin, the Southeast River basin, and the Pearl River basin and Hainan province).

Additionally, it is worthy of notice that this study neglected the effect of land cover change on actual evaporation. In future work, an improvement, establishing a model describing the relationship between the parameter n of the mean annual water-energy balance equation and land cover, is expected.

Acknowledgements This research was partially supported by the public-benefit project of the Ministry of Water Resources PR China (no. 200801012) and the National Natural Sciences Foundation of China (no. 50909051).

REFERENCES

Bouchet, R. J. (1963) Evapotranspiration reelle evapotranspiration potentielle, signification climatique. In: *Evaporation*. IAHS Publ. 62, 134–142. IAHS Press, Wallingford, UK.

Brutsaert, W. & Stricker, H. (1979) An advection-aridity approach to estimate actual regional evapotranspiration. *Water Resour. Res.* **15**(2), 443–450.

Budyko, M. I. (1974) *Climate and Life*. Academic Press, New York, UK.

Eagleson, P. S. (2002) *Ecohydrology: Darwinian Expression of Vegetation Form and Function*. Cambridge University Press, Cambridge, UK.

Gao, B, Yang, D. W., Liu, Z. Y. & Zhu, C. B. (2008) Application of a distributed hydrological model for the Yarlung Zangbo River and analysis of the river runoff. *J. China Hydrol.* **28**(3), 40–44 (in Chinese).

Gao, G., Chen, D. L., Xu, C. Y. & Simelton, E. (2007) Trend of estimated actual evapotranspiration over China during 1960–2002. *J. Geophys. Res.* 112, D11120, doi:10.1029/2006JD008010.

IPCC (2007) *Observation: Surface and Atmospheric Climate Change* (ed. by K. E. Trenberth & P. D. Jones). Cambridge University Press, Cambridge, UK.

Jiang, T., Su, B. D., Wang, Y. J. & Qin, X. N. (2005) Trends of temperature, precipitation and runoff in the Yangtze River Basin from 1961 to 2000. *Adv. Clim. Change Res.* **1**(2), 65–68.

Priestley, C. H. B. & Taylor, R. J. (1972) On the assessment of surface heat flux and evaporation using large-scale parameters. *Mon. Weather Rev.* **100**, 81–92.

Stanhill, G. & Cohen, S. (2001) Global dimming: a review of the evidence for a widespread and significant reduction in global radiation with discussion of its probable causes and possible agricultural consequences. *Agr. Forest Meteorol.* **107**, 255–278.

Thomas, A. (2000) Spatial and temporal characteristics of potential evapotranspiration trends over China. *Int. J. Climatol.* **20**, 381–396.

Wang, Y., Jiang, T., Bothe, O. & Fraedrich, K. (2007) Changes of pan evaporation and reference evapotranspiration in the Yangtze River basin. *Theor. Appl. Climatol.* **90**, 13–23.

Wang, Y. J., Jiang, T., Xu, C. Y. & Shi, Y. F. (2005) Trends of evapotranspiration in the Yangtze River Basin in 1961–2000. *Adv. Clim. Change Res.* **1**, 99–105.

Yang, H. B., Yang, D. W., Lei, Z. D. & Sun, F. B. (2008a) New analytical derivation of the mean annual water-energy balance equation. *Water Resour. Res.* **44**, W03410, doi:10.1029/2007WR006135.

Yang, H. B., Yang, D. W., Lei, Z. D., Sun, F. B. & Cong, Z. T. (2008b) Regional variability of the complementary relationship between actual and potential evapotranspiration. *J. Tsinghua Univ.* (Sci. & Tech.) **48**(9), 1413–1416 (in Chinese).

Zhai, P. M., Zhang, X. B., Wan, H. & Pan, X. H. (2005) Trends in total precipitation and frequency of daily precipitation extremes over China. *J. Climate* **18**, 1096–1108.

Zhang, L., Hickel, K., Dawes, W. R., Chiew, F. H. S. & Briggs, P. R. (2004) Arational function approach for estimating mean annual evapotranspiration. *Water Resour. Res.* **40**, W02502, doi: 10.1029/2003ER002710.

A unit hydrograph model with spatial variations of rainfall based on an artificial neural network

XIANG ZHANG[1,3], JUN XIA[2], XI CHEN[3], WENHUA ZHANG[1] & LIPING ZHANG[1]

1 *State key Laboratory of Water Resources and Hydropower Engineering Science, Wuhan University, Wuhan 430072, China*
zhangxiang@whu.edu.cn

2 *Key Laboratory of Water Cycle and Related land Surface Processes, IGSNRR, CAS, Beijing 100101, China*

3 *State key Laboratory of Hydrology-Water Resources and Hydraulic Engineering, Hohai University, Nanjing 210098, China*

Abstract Impacts of watershed morphologic and topographic structures, and spatial-temporal variations of rainfall on the unit hydrograph (UH) have been a key issue for investigation of watershed flow routing. In this paper we developed a new method to derive UH by considering the spatial-temporal variations of rainfall. The new method is based on the S-curve approach from a cumulative curve of UH. A nonlinear function with two parameters was used to fit the S-curve. These two parameters were found to be related to the barycentre of storm and the intensity of rainfall in the basin which describe the spatial-temporal variations of rainfall. The nonlinear relationship between two parameters of the S-curve function and the barycentre of storm and the intensity of rainfall is established by using an artificial neural network (ANN). Therefore, the unit hydrograph derived from the new method integrates with changing positions of storm barycentre and intensity of rainfall, and thus has characteristics of the Sherman unit hydrograph and a function of the description of the temporal-spatial variations of rainfall. This method is successfully applied in the Juhe catchment of the Yangtze basin, China. The simulation results show that this method is effective in watershed flow routing.

Key words unit hydrograph; *S*-curve; Barycentre of storm; artificial neural network

INTRODUCTION

During the research of surface flow routing, L. K. Sherman brought up the concept of the unit hydrograph for the first time in 1932, and analysed the empirical unit hydrograph according to the observed rainfall and runoff information. More than half a century has passed, and the Sherman Unit Hydrograph is widely used as an extremely effective and key method in surface runoff routing for hydrological forecasting and computation, although its principles of multiplication and superposition do not fully meet the fact, and are affected by the uneven distribution of the rainfall caused by the spatial and geographic conditions. Rui Xiaofang (1999) pointed out that the unit hydrograph is just the hydrological response of the basin when the system-analysis idea is used to explain the runoff routing. Early in the 1930s to 1940s, hydrologists and physical geographers used empirical statistics analysis, e.g. unit hydrograph peak, peak time-lag, etc. and basin topography parameters, such as the empirical relationship between the basin slope and area, to establish the main features of the unit hydrograph in order to determine the unit hydrograph in ungauged basins. In recent years, Rui Xiaofang & Shi Peng (2002) wrote about the geomorphologic instantaneous unit hydrograph based on geomorphologic and hydrodynamic dispersion. Rui Xiaofang (2003) proposed the method of determining the geomorphologic unit hydrograph by means of probability density functions of path length and slope. Xie Ping & Sun Zhi (2003) established the geomorphologic instantaneous unit hydrograph based on linear dynamic wave. Nash (1958), starting from the linear reservoir assumption, provided the instantaneous unit hydrograph, and considered the impact on unit hydrograph by net-rain intensity via nonlinear amendment of instantaneous unit hydrograph parameter, with the classification treatment method adopted in the impact of storm centre. Based on mathematic expression of the Nash model (similar to Γ distribution) and applied by the maximum entropy principle, Zhang Ming (2002) suggested a new method of Nash model parameter estimation, the maximum information entropy method. Shi Peng & Rui Xiaofang (2003) put forward the method that the geomorphic parameters are firstly extracted by DEM and then the Nash model parameters are calculated. Using the stochastic differential equation, Sun Yingna & Rui Xiaofang (2005) discussed the calculation method of confluence on circumstance of stochastic input, with the feature of having considered the impact on confluence process by uncertain factors.

The studies above indicate that the spatial-temporal variation of rainfall and landscape feature, as well as their impacts on unit hydrograph by uncertain factors, are always the difficult problems in the calculation of flow routing in basin. This paper, based on the empirical Sherman unit hydrograph, firstly deduces the *S*-curve method affected by the position of barycentre of storm and rainfall intensity, then a new *S*-curve equation is given. The relationships between the parameters of the *S*-curve and the position of the barycentre of storm as well as the rainfall intensity are established in this paper by using an artificial neural network model. Using the method above the unit hydrograph can be determined, provided that the position of the storm centre and the rainfall intensity are obtained.

S-CURVE AND ARTICIAL NEURAL NETWORK

The variation of the unit hydrograph is continuous, and is caused by several comprehensive factors. Therefore there is no choice but to amend the concept and assumption of the unit hydrograph by means of a unit hydrograph classification method. This is because the analysed unit hydrograph does not adapt to an appropriate equation, and for this reason we suggest the following method.

Our idea is, firstly we transform the analysed unit hydrograph into an *S*-curve, draw the curve $S(t)$ point in the rectangular coordinate system with ordinate S and abscissa t, choose an *S*-curve equation, calculate its two parameters after linearization, set up a relevant relationship among the storm centre, the net-rainfall and the parameters, and lastly, transform the *S*-curve into a period unit hydrograph, thus we can work out any period unit hydrograph given any storm centre and rainfall intensity.

Deduction of the formula of basin and period unit hydrograph

The general *S*-curve equation is given by

$$S = \frac{1}{a + be^{-t}} \tag{1}$$

where a and b are the parameters, e is the base of natural logarithm and t is time.

From equation (1), we know that, when $t = 0$, $S = 1/a + b$; when $t \to \infty$, $S \to 1/a$. Obviously, this kind of *S*-curve is not fit for the basin period unit hydrograph because the water volume of the unit hydrograph already exists before the net-rainfall starts. If we want the initial water volume of the unit hydrograph to be 0, the only solution is to make time go to infinity, which is obviously unreasonable. So we propose a new *S*-curve equation which goes through the origin of coordinates, and it is:

$$S = \frac{1}{1 + mt^{-n}} \tag{2}$$

where m and n is the two parameters of new *S*-curve.

From equation (2), we know, when $t \to 0$, $S \to 0$; when $t \to \infty$, $S \to 1$. It corresponds to the *S*-curve defined by the empirical unit hydrograph that can be obtained by the difference between $S(t)$ and $S(t - \Delta t)$, i.e.

$$U(t) = S(t) - S(t - \Delta t) \tag{3}$$

where $U(t)$ is the value of unit hydrograph at t; Δt is the time interval.

Substituting equation (2) in equation (3), we get:

$$U(t) = \frac{1}{1 + mt^{-n}} - \frac{1}{1 + m(t - \Delta t)^{-n}} = \frac{m\left[(t - \Delta t)^{-n} - t^{-n}\right]}{(1 + mt^{-n})\left[1 + m(t - \Delta t)^{-n}\right]} \tag{4}$$

Denoting $\Delta t = t_2 - t_1$, then substituting Δt in equation (4), we get:

$$U(t_2) = \frac{m\left\{\left[t_2 - (t_2 - t_1)\right]^{-n} - t_2^{-n}\right\}}{(1 + mt_2^{-n})\left\{1 + m\left[t_2 - (t_2 - t_1)\right]^{-n}\right\}} = \frac{m(t_2^{-n} - t_1^{-n})}{(1 + mt_2^{-n})(1 + mt_1^{-n})} \tag{5}$$

As mentioned above, the unit hydrograph is the result affected by several comprehensive factors, and the main factors are spatial distribution of rainfall and net-rainfall intensity, which have an impact on unit hydrograph by the parameters of m and n in equation (5). Therefore, we can set up the relationships by analysing the different position of the barycentre of the storm, the net-rainfall intensity and the corresponding unit hydrograph parameters of m and n, i.e.:

$$\begin{aligned} n &= \varphi(L_c, I) \\ m &= f(n) = f(\varphi(L_c, I)) \end{aligned} \tag{6}$$

or

$$\begin{aligned} m &= \varphi(L_c, I) \\ n &= f(m) = f(\varphi(L_c, I)) \end{aligned} \tag{7}$$

where L_c is the barycentre of the storm, I is the net-rainfall intensity.

From the causing and forming theory of the basin runoff, the unit hydrograph consists of the water volume in the basin. The geographic centre of a storm cannot reflect the modality of the forming unit hydrograph fully, but the barycentre of the storm can reflect the main water volume forming the unit hydrograph, which corresponds to the causing and forming theory of basin runoff. Therefore, we choose the barycentre of the storm instead of the geographic centre of the storm.

Firstly, we choose the flow direction as ordinate axis y, the vertical direction of the flow as abscissa axis x, then the coordinate of the barycentre of storm is defined as:

$$\begin{aligned} x_c &= \frac{\sum p_i x_i}{\sum p_i} \\ y_c &= \frac{\sum p_i y_i}{\sum p_i} \qquad i = 1,2,\ldots\ldots,k \end{aligned} \tag{8}$$

where, x_c and y_c are the position of the barycentre of storm L_c, k is the number of rainfall gauge stations, x_i and y_i is the position of the ith rainfall gauge station, p_i is the rainfall rate at the ith station.

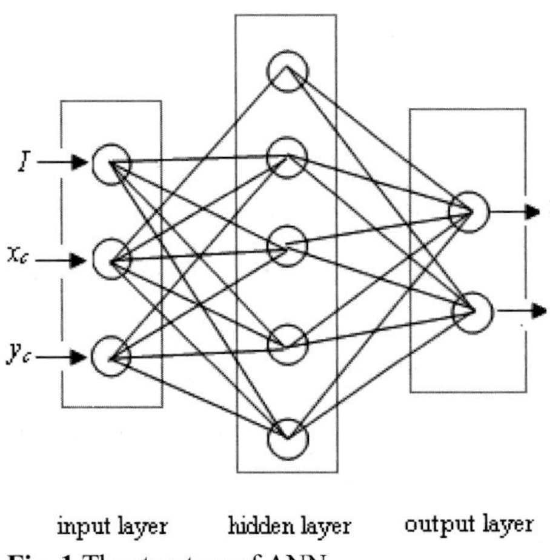

Fig. 1 The structure of ANN.

Artificial Neural Networks

Artificial Neural Networks (ANN) are widely used in the fields of nonlinear pattern recognition, intelligent control and nonlinear optimization, etc. ANN is capable of simulating the complex nonlinear relationship between the input variables and outputting variables. The results of theoretical research show one three-layer forward ANN having a finite number of nodes in the hidden layer can realize the approximation of any complex functions.

An artificial neural network is developed to simulate the nonlinear relationship expressed by formula (6) or (7). The forwarding artificial neural network trained by BP (Back Propagation) algorithm is composed of three layers, including input layer, one hidden layer and output layer (see Fig. 1). The input variables are the position of the barycentre of storm, x_c and y_c, and the net rainfall intensity, I. The output variables are the parameters of S-curve, m and n.

CASE STUDY

Brief introduction of the basin

Zuhe River, lying between E110°56′–111°70′ and N30°95′–31°78′, is the original source of Zuzhang River located in Hubei province, China. The length of the river above the outlet, Houziyan, is 226 km, and the area 2611 km^2.

The basin belongs to sub-tropical monsoon zone, and is hot and rainy in summer, cold and less-rainy in winter, higher temperature in autumn than spring, heavier rain in spring than autumn. The characteristics of climate are the great variations in temperatures in the whole year, long period of frost-free, average temperature of 15.9°C for many years, highest temperature reaching 40°C, lowest temperature reaching –13.5°C, average humidity of 78% for many years, northeast wind direction. There is abundant rainfall in the basin, with multi-annual average rainfall of 1037 mm.

In the basin there are 14 rainfall stations, Oujiadian, Guandouping, Miaoping, Xuanhu, Jiuli, Maliangping, Dianya, Fenshui, Xiakou, Yangping, Houziyan, Banqiao, Xunjian and Nilong (see Fig. 2). Data collected during 1978–1987 included daily rainfall and sub-daily rainfall from 14 rainfall stations and the evaporation data from two evaporation gauging stations near the basin. The outlet of the basin, Houziyan, was observed for discharge. We chose 48 flood events from the 1978–1981 and 1983–1987 calibration of the model, and four flood events in 1982 for the validation.

RESULTS

The key issue is to establish the relationships between the barycentre of the storm, x_c and y_c, the net rainfall intensity, I, and the parameters m and n. We use the data of 48 storm-flood events to analyse the parameters of the storm centre as x_c and y_c and surface net-rainfall intensity as I, to determine the corresponding empirical unit hydrograph, to transform the corresponding empirical unit hydrograph into S-curve, and lastly to get the parameters m and n by curve fitting. Table 1 shows a part of the analysed results. The parameters m and n of the S-curve and the parameters of the barycentre of the storm, x_c and y_c, and the rainfall intensity, I, present an obvious nonlinear relationship. We set up the relationships between the barycentre of storm, x_c and y_c, the net rainfall intensity, I, and the parameters, m and n by using an artificial neural network model, BP model, with five neural nodes in the hidden layer and three inputting nodes for the variables x_c, y_c and I and two outputting nodes for the variables m and n.

The four flood events in 1982 are used to make a model validation. Firstly, by analysing the rainfall data we obtained the corresponding rainfall intensity I and storm barycentre parameters, x_c and y_c. Then, they are input into the trained BP model for the determination of the parameters of S-curve, m and n, that are the outputs of the BP model. As long as the S-curve is obtained, the empirical unit hydrograph of the corresponding flood can be derived for the flow routing. Figure 3 shows the model validation results with the Nash-Sutcliffe efficiency 0.71 and correlation coefficient 0.93.

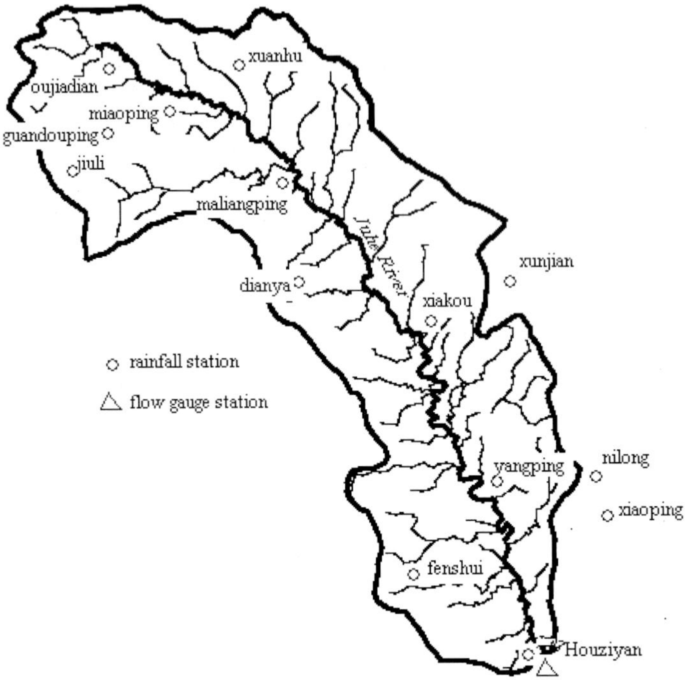

Fig. 2 Juhe River basin.

Table 1 Some values of m and n of S-curve, rainfall intensity I, and storm centre parameter x_c and y_c.

Flood events number	m	n	I	x_c	y_c
197805	6872.7	2.62	1.58	5.57	14.41
197907	796.87	2.04	3.37	5.68	14.32
198009	2983.6	2.13	0.57	5.72	15.36
198107	8685.1	2.81	1.32	5.30	15.85
198306	12847.0	2.24	1.75	5.39	15.98
198505	215.53	1.68	3.49	6.0	9.06
198706	1945.0	2.21	1.21	4.6	15.96

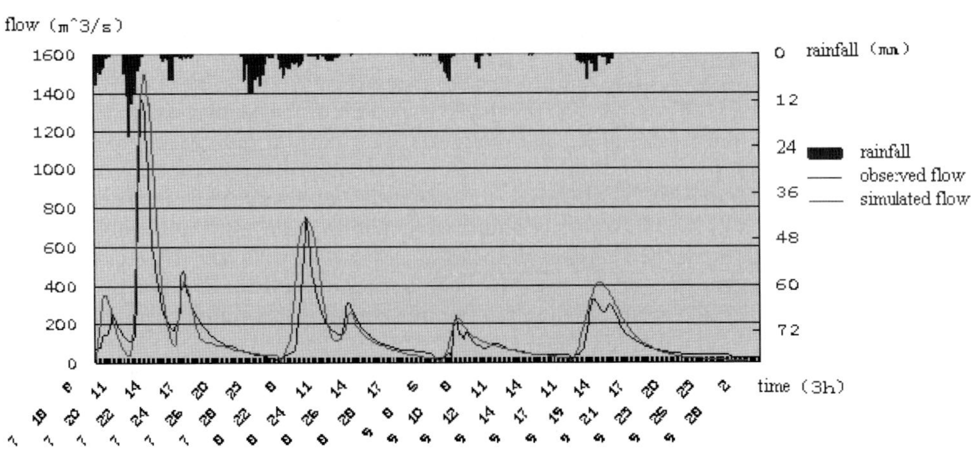

Fig. 3 Validation of hydrograph in 1982 at Houziyan in Zuhe River basin.

CONCLUSIONS

In this paper, a new *S*-curve method, considering the effect of the storm barycentre and rainfall intensity, is proposed. It firstly transforms the analysed unit hydrograph into the *S*-curve, and then uses the new *S*-curve equation to fit the obtained *S*-curve. The two parameters, m and n, can be determined by the curve fitting. In order to simulate the impact of the spatial-temporal distribution of the storm on the unit hydrograph, the relationships between the barycentre of storm, x_c and y_c, the net rainfall intensity, I, and the parameters, m and n, are discussed and simulated by an artificial neural network model, the BP model. Therefore, the method can work out the empirical unit hydrograph using the barycentre of storm and rainfall intensity, using the relationships between the parameters of *S*-curve, the barycentre of the storm and rainfall intensity. It has simple and practical characteristics like the empirical Sherman unit hydrograph, with the advantage of considering the spatial-temporal variation of the storm. In order to further improve the accuracy, more analyses about the flood data are needed, to make the relationships between the *S*-curve, the barycentre of the storm, and the rainfall intensity more precise.

Acknowledgements The authors wish to thank the following for their support: the projects of China Natural Sciences Foundation (40730632, 50579053) and the Opening Research Foundation of State Key Laboratory of Hydrology-Water Resources and Hydraulic Engineering (2006412011), Hohai University.

REFERENCES

Chang Jiang (1993) *Water Resources Commission, The Hydrographic Forecasting Methods*. China Water Power Press, Beijing, China (in Chinese).

Hubail Ajward, M. & Muzik, I. (2000) A spatially varied unit hydrograph model. *J. Environ. Hydrol.* **8**, 1–8.

Nash, J. E. (1958) The form of the instantaneous unit hydrograph. In: *Surface Water, Prevision, Evaporation* (General Assembly of Toronto, 3–14 September 1957), 114–121. IAHS Publ. 45. IAHS Press, Wallingford, UK.

Rinaldo, A., Marani, A. & Rigon, R (1991) Geomorphological dispersion. *Water Resour. Res.* **27**(4), 513–525.

Rui Xiaofang (1999) Some advances in geomorphologic instantaneous unit hydrograph theory. *Adv. Water Sci.* **10**(3), 45–350 (in Chinese).

Rui Xiaofang (2003) Study of determining geomorphologic unit hydrograph by means of probability density functions of path length and slop. *Adv. Water Sci.* **14**(5), 602–606 (in Chinese).

Rui Xiaofang & Shi Peng (2002) Study on WGIUH by means of geomorphologic and hydrodynamic dispersion. *Adv. Water Sci.* **13**(4), 439–444 (in Chinese).

Shi Peng, Rui Xiaofang & Qu Simin (2003) Determination of parameters for Nash runoff model by DEM. *J. Hohai Univ. (Natural Sciences)* **31**(4), 378–381 (in Chinese).

Sun Yingna & Rui Xiaofang (2005) Investigation on calculation method for flow concentration with stochastic input. *Adv. Water Sci.* **16**(5), 661–665 (in Chinese).

Xie Ping & Sun Zhi (2003) Geomorphic instantaneous unit hydrograph model based on linear dynamic wave. *J. Wuhan University of Hydraulic and Electric Engineering* **36**(1), 5–8 (in Chinese).

Zhang Ming (2002) Entropy method for estimating the parameters of Nash model. *J. Hydraul. Engng* **1**, 62–66 (in Chinese).

Zhang Wenhua & Zhang Xiang (2005) The suggestion for a new S-curve. *Mathematics in Practice and Theory.* **35**(6), 202–205 (in Chinese).

Real-time flood forecasting with the use of corresponding rising difference model

GUANGHUA QIN[1], SHUNJIU WANG[2] & REN MIU[1]

1 *State Key Laboratory of Hydraulics and Mountain River Engineering, School of hydraulic Engineering of Sichuan University, Chengdu 610065, China*
ghqin2000@163.com

2 *Institute of Plateau Meteorology, China Meteorological Administration, Chengdu 610071, China*

Abstract According to the Saint-Venant continuity equation, this paper presents a corresponding rising difference (CRD) model for river flow and water level prediction. The proposed model has only one parameter, which is estimated by the recursive least square method with a variable factor, compared to several parameters in other models. Two case studies from the middle reach of the Yangtse River are prepared and compared with the auto-regressive (AR) model. Based on the results, this paper suggests a better alternative for predicting river flow and water level.

Key words corresponding rising difference model; flood forecasting; recursive least square

INTRODUCTION

Research channel flood forecasting is an extremely complex problem that is influenced by a wide variety of factors. Various mathematical models have been developed to deal with these problems. Those models can be classified into two categories: physically-based hydrological (PBH) models which require the extremely complex relationships between river flow and its influencing factors to be established (Shamseldin *et al.*, 1997) and system identification (SI) models, which are based on the statistics theory (Yue Wang, 2002).

The applications of many nonlinear SI models to channel flood forecasting have become very popular in recent years, such as the artificial networks theory (French *et al.*, 1992; Maier & Dandy, 2000; Hu & Lam, 2001), chaos theory (Wilcox *et al.*, 1981), etc. Most of those new methods pay attention to the course of calculation far more than the physical causes of the hydrological variables, although they have an inherent ability to capture the nonlinear relationships involved.

In this paper, one new model of channel flood real-time forecasting has been built, named as the corresponding rising difference (CRD) model. In order to examine the potential of the CRD model in estimating river flows and water levels, the results are compared with the auto-regressive (AR) model.

THE CORRESPONDING RISING DIFFERENCE MODEL

The CRD model for water level forecasting

To the equation of continuity

$$\frac{\partial A}{\partial t} + \frac{\partial Q}{\partial x} = 0 \qquad (1)$$

where A is the cross-section area, Q is the flow; x is the direction of flow, and t is the time.

For

$$\frac{\partial Q}{\partial x} = \frac{\partial Q}{\partial A}\frac{\partial A}{\partial x} \qquad (2)$$

based on the formula of Seddon

$$C = \frac{\partial Q}{\partial A} \qquad (3a)$$

where C, the velocity of flood wave, is slope in relation curve between Q and A. So equation (3a) can be expressed as:

$$\frac{\partial Q}{\partial x} = C\frac{\partial A}{\partial x} \tag{3b}$$

takes it into equation (1), resulting in:

$$\frac{\partial A}{\partial t} + C\frac{\partial A}{\partial x} = 0 \tag{4}$$

for:

$$A = Bh \tag{5}$$

where B is the average width of water surface, h is the water depth. Taking equation (5) into equation (4), resulting in:

$$\frac{\partial h}{\partial t} + C\frac{\partial h}{\partial x} = 0 \tag{6}$$

For the partial differential equation of unsteady flow in open channels, we know if we research steady riverbeds, then:

$$\frac{\partial h}{\partial t} = \frac{\partial z}{\partial t}\cos\gamma \approx \frac{\partial z}{\partial t} \tag{7}$$

where γ is riverbed slope, Z is water level. Assumes $\cos\gamma \approx 1$, taking equation (7) into equation (6), this gives:

$$\frac{\partial z}{\partial t} + C\frac{\partial h}{\partial x} = 0 \tag{8}$$

deal with equation (8), as follows:

In the cross section of $j+1$, taking $\frac{\partial z}{\partial t}$ as central difference, taking $\frac{\partial h}{\partial x}$ as forward difference, resulting in:

$$\begin{aligned}&\frac{Z(j+1,t+\Delta t) - Z(j+1,t-\Delta t)}{2\Delta t}\\ &= -C_1\frac{[Z(j+1,t) - D(j+1)] - [Z(j,t) - D(j)]}{\Delta x}\end{aligned} \tag{9}$$

where $Z(j+1, t+\Delta t)$, $Z(j+1, t-\Delta t)$, $Z(j+1, t-\Delta t)$ are the water levels of the cross section $j+1$ at time $t+\Delta t$, $t-\Delta t$ and t, respectively, $Z(j, t)$ is the water level of the cross section j at time t. $D(j+1)$ and $D(j)$ are river bottom altitude in the cross section $j+1$ and j.

In the cross section of $j+1$, taking $\frac{\partial h}{\partial t}$ as backward difference, taking $\frac{\partial h}{\partial x}$ the same difference to equation (9), resulting in:

$$\begin{aligned}&\frac{Z(j+1,t) - Z(j+1,t-\Delta t)}{\Delta t}\\ &= -C_2\frac{[Z(j+1,t) - D(j+1)] - [Z(j,t) - D(j)]}{\Delta x}\end{aligned} \tag{10}$$

where $Z(j+1, t+\Delta t)$, $Z(j+1, t-\Delta t)$, $Z(j+1, t-\Delta t)$, $D(j+1)$ and $D(j)$ have the same meaning as in equation (9).

Then, dividing equation (10) by equation (9), results in:

$$Z(j+1,t+\Delta t) = 2\alpha_t Z(j+1,t) + (1-2\alpha_t)[Z(j+1,t-\Delta t)] \tag{11}$$

where $\alpha_t = C_1/C_2$, which is called rising difference proportion coefficient. The CRD model assumes that $2\alpha_t = \beta_t$, the CRD model for water level forecasting can be expressed as:

$$Z(j+1,t+\Delta t)=Z(j+1,t-\Delta t)+\beta_t[Z(j+1,t)-Z(j+1,t-\Delta t)] \tag{12}$$

The CRD model for river flow forecasting

Equation (4) can be expressed as equation (13) by multiplying by V, which is average velocity of the cross-section:

$$\frac{\partial Q}{\partial t}+C\frac{\partial Q}{\partial x}=0 \tag{13}$$

In equation (13), take the difference as follows:

In the cross section of *j+1*, taking $\frac{\partial Q}{\partial t}$ as the central difference, taking $\frac{\partial Q}{\partial x}$ as the forward difference, resulting in:

$$\frac{Q(j+1,t+\Delta t)-Q(j+1,t-\Delta t)}{2\Delta t}=-C_1\frac{Q(j+1,t)-Q(j,t)}{\Delta x} \tag{14}$$

where Q (j+1, $t+\Delta t$), $Q(j+1, t-\Delta t)$, $Q(j+1, t-\Delta t-$ are flows of the cross section $j+1$ at time $t+\Delta t$, $t-\Delta t$ and t, respectively, Q (j, t) is flow of the cross section j at time t.

In the cross section of *j+1*, taking $\frac{\partial Q}{\partial t}$ as backward difference, taking $\frac{\partial Q}{\partial x}$ the same difference to equation (6), resulting in:

$$\frac{Q(j+1,t)-Q(j+1,t-\Delta t)}{\Delta t}=-C_2\frac{Q(j+1,t)-Q(j,t)}{\Delta x} \tag{15}$$

where Q (j+1, $t+\Delta t$), $Q(j+1, t-\Delta t)$, $Q(j+1, t-\Delta t)$ have the same means with equation (6).

Then, divide equation (15) by equation (14), resulting in:

$$Q(j+1,t+\Delta t)=2\alpha_t Q(j+1,t)+(1-2\alpha_t)[Q(j+1,t-\Delta t)] \tag{16}$$

where $\alpha_t = C_1/C_2$, which called rising difference proportion coefficient. Assume $2\alpha_t = \beta_t$, the CRD model for flow forecasting can be expressed as:

$$Q(j+1,t+\Delta t)=Q(j+1,t-\Delta t)+\beta_t[Q(j+1,t)-Q(j+1,t-\Delta t)] \tag{17}$$

The character of the parameter β_t

12-h flow measurements were used from two hydrological stations located in Hubei Province on the Yangtse River to analyse the character of the parameter β_t. One is time series of river flows at Jianli station, the other is time series of water levels at Shashi station. The data of each station began at 08:00 h on 1 May and ended at 08:00 h on 31 October. The interval of the time series is 12 hours, i.e. Δt =12 h.

It can be seen from Fig. 1 that the curve showing the parameter β_t varying with time is irregular and disorderly. This implies that it would be incorrect if the parameter β_t was a fixed value, just like the parameters in the auto-regressive (AR) model.

Research at two gauging stations has found that the parameter β_t is relatively steady when the flow exceeds a certain extreme value. We can estimate that the parameter β_t in Fig. 1 varies from approx. 0.4–0.6, while the river flow exceeds about 10 000 m^3s^{-1}. The same finding is shown in Fig. 2. This is a very significant research, which can help to estimate parameter acceleration and forecasting accuracy improvement.

The estimation of the parameter β_t

As a time-varying parameter β_t must be revised according with increasing of information. In the conventional forecasting models the least square method is often used for parameter estimation by

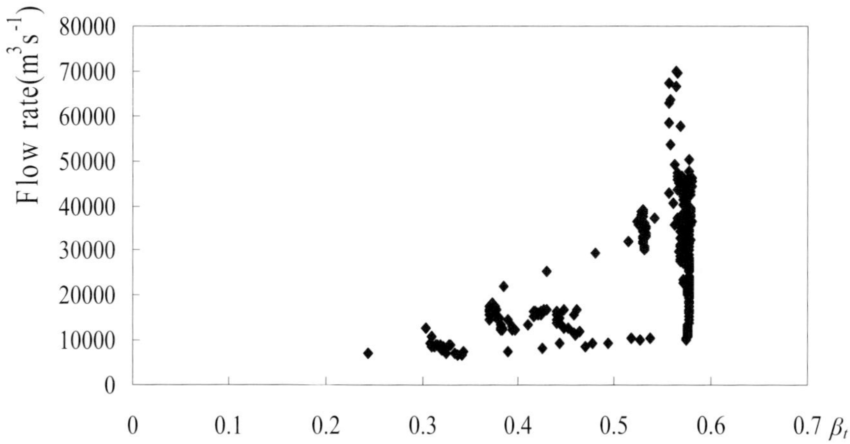

Fig. 1 Scatter plots of the parameter β_t against river flow at Jianli station in 1979.

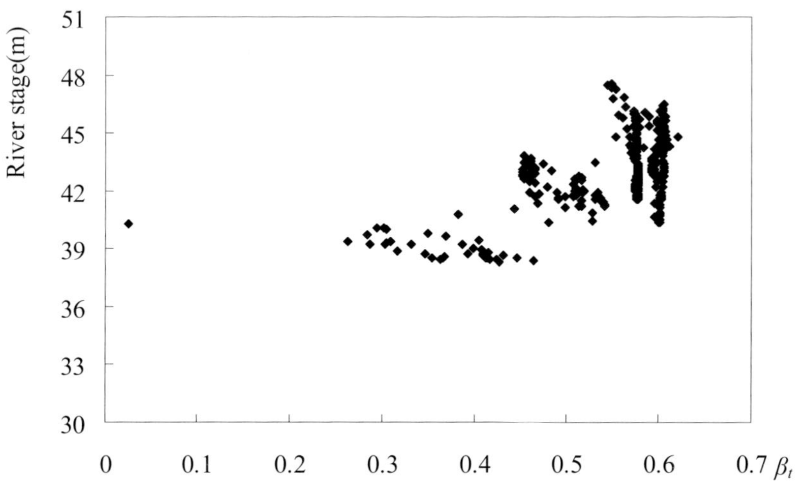

Fig. 2 Scatter plots of the parameter β_t against water level at Shashi station in 1980.

two approaches. One is the fix-method which determines the parameters empirically, the other is subsection-method which gives the different parameter in each section. Each of the above two methods has its own merits and shortcomings. The paper applied a parameter estimation which named as recursive least-square estimation with variable forgetting factors. It can revise the parameter in real time with new input data.

The steps of calculation are described as follows:

(1) set initial values of θ, P and R, where θ is parameter, P is covariance matrix and R is sum of error square;
(2) set $t = 0$;
(3) calculate forecasting value: $\hat{Y}(t+1) = X^T(t+1)\theta(t)$;
(4) input measure value: $Y(t+1)$;
(5) calculate forecasting error: $e(t+1) = Y(t+1) - \hat{Y}(t+1)$;
(6) calculate gain matrix: $K(t+1) = P(t)X(t+1)[1 + X^T(t+1)P(t)X(t+1)]^{-1}$;
(7) calculate variable forecasting factor: $\lambda(t+1) = 1 - [1 - X^T(t+1)K(t+1)e(t+1)^2]/R$;

(8) calculate covariance matrix: $P(t+1)=[P(t)-K(t+1)X^T(t+1)P(t)]/\lambda(t+1)$;

(9) revise parameter estimation: $\theta(t+1)=\theta(t)+K(t+1)[Y(t+1)-X^T(t+1)\theta(t)]$

(10) set $t=t+1$, turn to (3).

The whole iterative calculation cannot stop until the forecasting error is less than the tolerance error.

APPLICATION

Evaluation criterion of the CRD model performance

The CRD model was applied to the Yangtse River in China to evaluate its ability to predict the river flow at Jianli station and water level at Shashi station. The prediction accuracy was quantified by the following measures:

(1) Efficiency coefficient R^2

$$R^2=\frac{F_0-F_1}{F_0}\times 100\%$$

in which

$$F_0=\sum_{i=1}^{N}(Y_i-\overline{Y})^2$$

$$F_1=\sum_{i=1}^{N}(\hat{Y}_i-Y_i)^2$$

where N is the length of the time series, Y_i is observed value, $\hat{Y}_i$ is forecasting value and $\overline{Y}$ is mean value.

(2) Mean absolute error (MAE)

absolute error: $AE_i=\hat{Y}_i-Y_i$

mean absolute error: $MAE=(\sum_{i=1}^{N}|AE_i|)/N$

(3) Mean relative error (MRE)

mean error: $RE_i=\frac{Y_i-\hat{Y}_i}{Y_i}\times 100\%$

mean relative error: $MRE=(\sum_{i=1}^{N}|RE_i|)/N$

(4) Absolute error of peak value: $AE_m=|Y_m-\hat{Y}_m|$

relative error of peak value: $RE_m=\frac{Y_m-\hat{Y}_m}{Y_m}\times 100\%$

where Y_m, $\hat{Y}_m$ are observed peak value and forecasting peak value, respectively.

(5) Appearance time error of flood peak

$AT_m=TY-TY_m$

where TY, TY_m are appearance time of observed flood peak and forecasting flood peak, respectively.

(6) Up-to-standard rate (UTSR).

Up-to-standard rate represents the percentage of the output values, whose absolute value of the relative errors are less than 10%, in the total output values.

RESULTS

River flow data covering the years 1979–1988 were obtained from Yangtse River at Jianli station and water level data covering the years 1980–1988 were at Shashi station. The data of each year begin at 08:00 h on 1 May and end at 08:00 h on 31 October. The interval of the time series is 12, i.e. $\Delta t = 12h$. The flow sequence at Jianli station for all 10 years and the water level sequence at Shashi station for nine years are shown in Fig. 3 and Fig. 4, respectively.

The flow pattern sets at Jianli station were divided so that the data from 1979 to 1987 were used for calibrating, and 1988 was set aside for testing. In a similar way, the water level time series at Shashi station were divided so that the data from 1980 to 1987 were for calibrating, and that from 1988 was for testing.

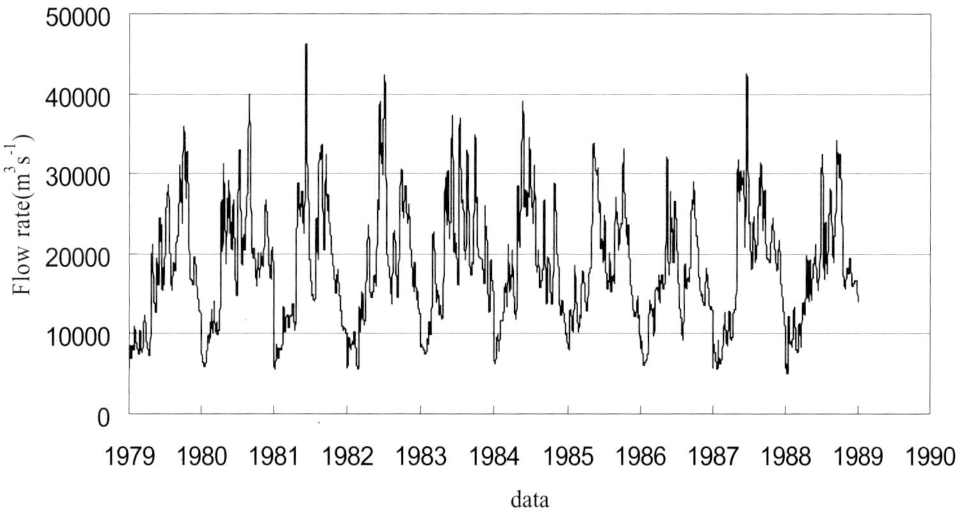

Fig. 3 Ten-year flow series of Yangtse River at Jianli station.

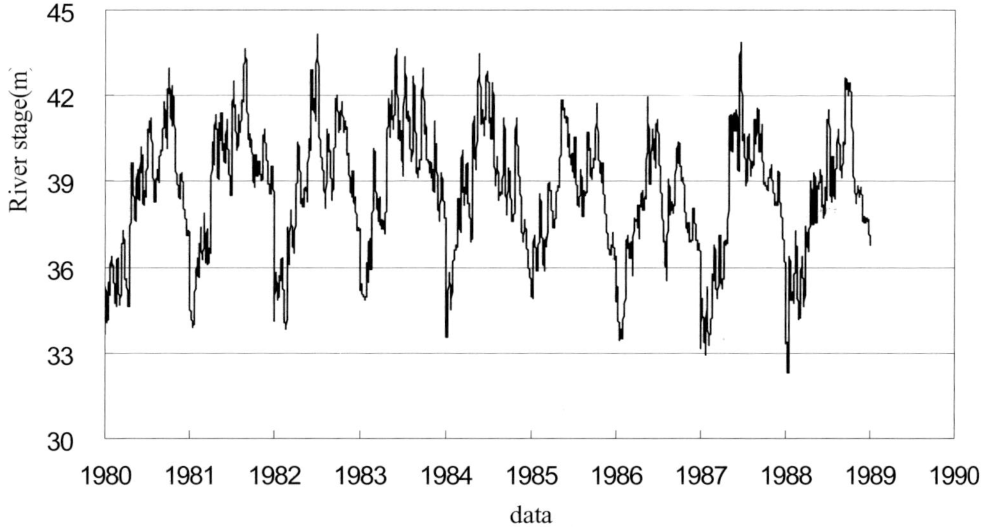

Fig. 4 Nine-year water level series of Yangtse River at Shashi station.

Table 1 Comparisons between the CRD model and AR (2) model for river flow forecasting at Jianli station in 1988.

Models	R^2 (%)	MRE (%)	REm (%)	ATm (Δt)	UTSR (%)
CRD	0.997	2.07	2.29	0	93.42
AR (2)	0.994	2.11	3.10	0	92.92

Table 2 Comparisons between the CRD model and AR (2) model for water level forecasting at Shashi station in 1988.

Models	R^2 (%)	MAE (m)	AEm (m)	ATm (Δt)	UTSR (%)
CRD	0.997	0.178	0.130	1	100
AR (2)	0.993	0.212	0.501	1	98.8

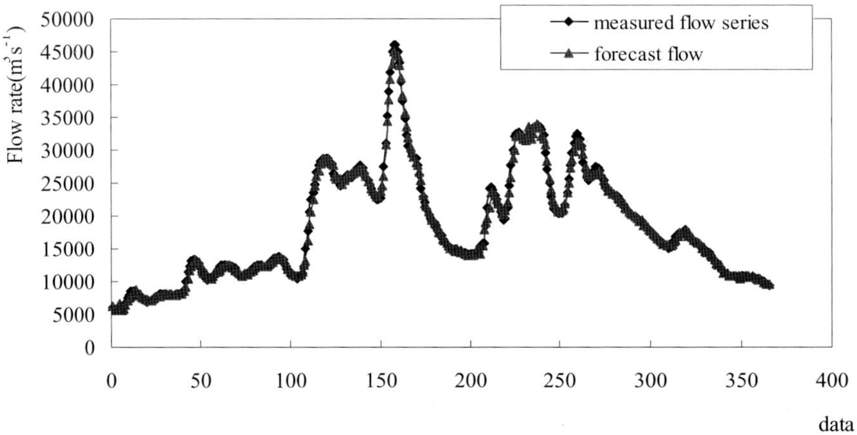

Fig. 5 Comparisons of the observed and forecasted discharge hydrographs with use of the CRD model for Jianli station in 1988.

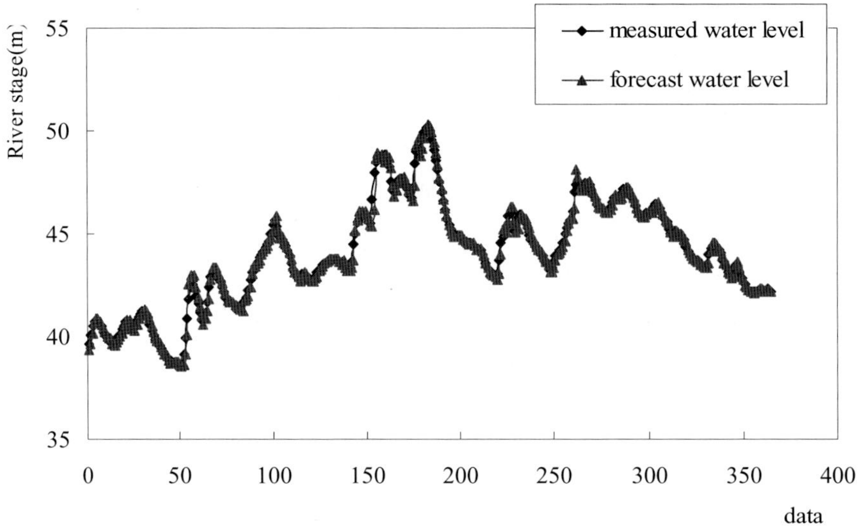

Fig. 6 Comparisons of the observed and forecasted water level hydrographs with use of the CRD model for Shashi station in 1988.

In the case of river flow and water level estimation, Tables 1 and 2 show the forecasting results of the CRD and AR (2) models. It can be observed that the errors of CRD model are reduced to a certain degree compared with the AR (2) model.

In order to indicate the differences in model performance, Figs 5 and 6 present a comparison of the observed and estimated river flow hydrographs of Jianli station in 1988 and water level hydrographs of Shashi station in 1988.

DISCUSSION AND CONCLUSIONS

A new real-time flood forecasting model was proposed for better hydrological time series modelling, which is based on Saint-Venant equations and the recursive least square method with variable forgetting factors. The applicability and potential of the CRD model in river flow and water level time series prediction were examined using data from two stations on the Yangtse River in China.

The CRD model has only one parameter β_t and need not fix on model orders. Whether the parameter β_t is estimated by offline mode or online mode, it is very simple. Moreover, the parameter β_t in equation (12) and equation (17) only relies on one section data, which fits for use in a basin that is short of data.

The key problem of applying the CRD model is parameter estimation. Farther research about the parameter β_t is still urgently needed. The implication of this investigation is that more attention should be paid to enhancing the generalization performance of the forecasting model rather than simply focusing on improving learning performance of the CRD model.

Acknowledgements The authors would like to thank the National Science Foundation Committee of China for funding the research (Project no. 50809041). The comments and suggestions made by the anonymous reviewer(s) were highly pertinent and gratefully appreciated. The project is supported by special Fund of Institute of Plateau Meteorology, Chengdu, "Impact of Climate Change of Qinghai-Tibet Plateau on Runoff Water Resource and Simulation Research" (no. BROP 200701).

REFERENCES

Bin Zhang, Rao, S. Govindaraju (2000) Prediction of watershed runoff using Bayesian concepts and modular neural networks. *Water Resour. Res.***36**, 753–762.

Cottrell, M., Girard, Y., Mangeas, M. & Muller, C. (1995) Neural modeling for time series: a statistical stepwise method for weight elimination. *IEEE Trans. on Neural Networks* **6**(6), 1355–1364.

French, M. N., Krajewski, W. F. & Cuykendall, R. R. (1992) Rainfall forecasting in space and time using a neural network. *J. Hydrol.* **137**, 1–31.

Fernando, A. K. & Jayawardena, A. W. (1998) Runoff forecasting using RBF networks with OLS algorithm. *J. Hydrol. Engng, ASCE* **3**(3), 203–209.

Hu, T. S., Lam, K. C. & Ng, S. T. (2001) River flow time series prediction with a range-dependent neural network. *Hydrol. Sci. J.* **46**(5), 729–745.

Maier, H. R., & Dandy, G. C. (2000) Neural networks for the prediction and forecasting of water resources variables: a review of modeling issues and applications. *Environ. Modeling & Software* **15**, 101–124.

Nalbantis, I. (2000) Real-time flood forecasting with the use of inadequate data. *Hydrol. Sci. J.* **45**(2), 269–283.

Pisarenko, V. F. & Lyubushin, A. A. (2005) Statistical methods for river runoff prediction. *Water Res.* **32**(2), 115–126.

Shamseldin, A. Y., O'Connor, K. M. & Liang, G. C.(1997) Methods for combining the outputs of different rainfall–runoff models. *J. Hydrol.* **197**, 203–229.

Sugihara, G. & May, R. M. (1990) Nonlinear forecasting as a way of distinguishing chaos form measurement error in time series. *Nature* **344,** 734–741.

Wilcox, B. P., Seyfried M. S. & Martison T. H., (1981) Searching for chaotic dynamics in snowmelt runoff. *Water Resour. Res.* **27**(6), 1005–1010.

Yue Sheng & Wang Chunyuan (2002) Assessment of the significance of sample serial correlation by the bootstrap test. *Water Resour. Manage.* **16**, 23–35.

Study on the effect of DEM spatial resolution and sampling algorithms on runoff simulation by BTOPMC

YU-AN WAN[1,3], FANG YIN[1], ZHAN-ZHOU LIU[1], WEI-CAI CUI[1] & TIAN-QI AO[1,2]

1 *College of Water Resources and Hydropower, Sichuan University, Chengdu 610065, China*

2 *State Key Lab of Hydraulic and Mountain River Engineering, Sichuan University, Chengdu 610065, China*
aotianqi@scu.edu.cn

3 *Jiangmen Survey and Design Institute of Water Resources and Hydropower Co., Ltd, Jiangmen, Guangdong 529000, China*

Abstract The DEM spatial resolution and sampling algorithms have a great effect on the parameters of watershed characteristics, which perhaps can result in uncertainty in hydrological modelling. In this paper, ZB algorithms, grid-minimum algorithms and grid-average algorithms are first used to obtain four different spatial resolutions DEM of Fuji-kawa basin in Japan from SRTM, and then a physically-based distributed hydrological model BTOPMC (block-wise use of TOPMODEL with Muskingum-Cunge method) is used to simulate a flood event using these data. The results show that DEM spatial resolution has a great effect on the grid-minimum algorithm and ZB algorithm. Compared to the other two algorithms, the grid-average algorithm can achieve better simulated results.

Key words DEM spatial resolution; sampling algorithms; Fuji-kawa basin; BTOPMC

INTRODUCTION

In hydrological modelling one of the most common and significant challenges faced is the reduction of uncertainty in estimation and calibration of model parameters. As one of the most important inputs of the distributed hydrological model, the DEM spatial resolution and sampling algorithms have a great impact on the parameters of watershed characteristics, which may result in uncertainty in hydrological modelling. In recent years, the most common sampling algorithms are grid-average algorithm, grid-centre algorithm and grid-minimum algorithm (Hao & Chi, 2004). Take the flow accumulation of grids into account of a weight factor, Yu *et al.* introduced the ZB algorithm (Yu *et al.*, 2001). Compared to the grid-average algorithm and grid-minimum algorithm (Yang *et al.*, 2007; Wan *et al.*, 2009), it was discovered that sampling algorithms have an important influence on the parameters of watershed characteristics. The ZB algorithm can hold stream network continuity and real watershed boundaries very well, whilst decreasing the amount of other geographical information falling off compared to other algorithms.

In this study, a physically-based distributed hydrological model BTOPMC (block-wise use of TOPMODEL with the Muskingum-Cunge method) and three sampling algorithms were first introduced, and then the effects of DEM spatial resolution and sampling algorithms on runoff simulation were investigated, attempting to find the better sampling algorithm for applying BTOPMC.

DESCRIPTION OF BTOPMC MODEL

BTOPMC is a physically-based distributed hydrological model developed for large river basins. It is a mixed model based on subdividing the entire basin into sub-basins (lumped) that consist of a number of grid cells (distributed) (Ao *et al.*, 1999; Takeuchi *et al.*, 1999; Tianqi *et al.*, 2006). Its main sub-models and corresponding parameters are briefly introduced as follows.

Topographic sub-model and its parameters

All topographic variables required in BTOPMC are specified in this component using automatic topographic analysis tools that have been integrated into the model (Ao *et al.*, 2001, 2003a). The topographic index, λ_i, of any a grid cell, *i* (Beven & Kirkby, 1979; Quinn *et al.* 1995) is calculated by equation (1):

$$\lambda_i = \ln(a_i / \tan\beta_i) \tag{1}$$

where a_i is the drainage area per contour length, $\tan\beta_i$ denotes the slope of grid cell, *i*.

The basin is then subdivided into sub-basins, and their average topographic indices $\lambda(k)$ are calculated as:

$$\lambda(k) = \frac{1}{N_p(k)} \sum_{i=1}^{N_p(k)} \lambda_i \tag{2}$$

where *k* is sub-basin code and $N_p(k)$ is the total number of grid cells belonging to sub-basin, *k*. Sub-basins are carried out using the Pfafstetter coding system (Verdin & Verdin 1999; Ao *et al.*, 2002). Both λ_i and $\lambda(k)$ are used in runoff calculation. Equations (1) and (2) demonstrate the topographic parameters in BTOPMC. These are specified by the DEM and subdivision level (Ao *et al.*, 2002, 2003b).

Runoff generation sub-model and its parameters

Runoff calculation is carried out for each grid cell by applying the assumptions and concepts of TOPMODEL to each sub-basin. The effective precipitation $R_e(i,t)$ stored in the root zone at time step *t* is represented as:

$$R_e(i,t) = R_o(i,t) - S_{r\max}(i) - E_p(i,t) \tag{3}$$

where $E_p(i,t)$ denotes evapotranspiration; $R_o(i,t)$ is total precipitation approximated by the Thiessen method; $S_{r\max}$ is maximum storage capacity of the root zone and is a parameter reflecting the effects of vegetation interception, soil moisture and pit storage. The storage in the unsaturated zone is calculated as:

$$S_{uz}(i,t) = S_{uz}(i,t-1) + R_e(i,t) - q(i,t) \tag{4}$$

where $q_v(i,t)$ is the recharge to groundwater calculated by:

$$q_v(i,t) = T_0(k)\exp[-S(i,t)/m(k)] \tag{5}$$

where T_0 is the saturated soil transmissivity in m^2/h and *m* is the decay factor (in metres) of T_0. $S(i,t)$ is the local saturation deficit in metres that can be thought of as the distance from groundwater level to ground surface. It is determined by:

$$S(i,t) = S_{bar}(k,t) + m(k)[\lambda(k) - \lambda_i] \tag{6}$$

where $S_{bar}(k,t)$ is the average saturation deficit in metres of sub-basin *k* and calculated as:

$$S_{bar}(k,t) = S_{bar}(k,t-1) + \frac{1}{N_p(k)} \sum_{i=1}^{N_p(k)} (q_b(i,t) - \frac{1}{N_p(k)} \sum_{i=1}^{N_p(k)} q_v(i,t) \tag{7}$$

where $q_b(i,t)$ is the base flow flowing into streams as calculated by:

$$q_b(i,t) = T_0(k)\exp[-S(i,t)/m(k)]\tan\beta_i \tag{8}$$

For $t = 1$, the initial value of the average saturation deficit, $S_{bar0}(k)$, which is a parameter to be calibrated, indicates $S_{bar}(k,0)$. As shown in equations (3), (5) and (7), in the runoff generation sub-model there are four parameters in total to be calibrated: $S_{r\max}(i)$, $T_0(k)$, $m(k)$ and $S_{bar0}(k)$.

Flow routing sub-model and its parameters

The Muskingum-Cunge (M-C) method (Cunge, 1969) is adopted. In this flow routing component a real time adjustment technique has been developed (Ao *et al.*, 2000) to solve the inherent negative outflow problem in the M-C method to attempt to ensure basic routing accuracy when large space-time intervals corresponding to runoff calculations are used. In flow routing calculation, the river cross-section is assumed to be rectangular and river width *B* (metres) is approximated by:

$$B(i) = C\sqrt{A(i)} \tag{9}$$

where constant $C = 10$ and A is the drainage area in square kilometres (Lu *et al.*, 1989). The equivalent Manning roughness coefficient of a grid cell is estimated as:

$$n_i = n_0(\mathrm{k})[\tan\beta_i/\tan\beta_0(\mathrm{k})]^{1/3} \tag{10}$$

where n_0 and $\tan\beta_0$ are the equivalent roughness coefficient and slope at the outlet of sub-basin k, respectively, and n_0 is a model parameter to be calibrated.

DEM SAMPLING ALGORITHMS

In this study, three DEM sampling algorithms are used to get new grid-scale DEM data. When using the grid-minimum algorithm, a new grid-scale DEM data is the minimum one among horizontal and vertical I × I original grid-scale DEM data. It is determined by:

$$\min(i,j) = \sum_{m=1}^{I}\sum_{n=1}^{I}\min\left(elev((i-1)\times I+m,(j-1)\times I+n)\right) \tag{11}$$

where min(*i*,*j*) is the elevation of new grid-scale DEM data, *I* is the multiple of expanding, *elev*(*i*,*j*) is the elevation of original grid-scale DEM data.

When using the grid-average algorithm, a new grid-scale DEM data is the average of horizontal and vertical I × I original grid-scale DEM data. It is determined by:

$$aver(i,j) = \frac{\sum_{m=1}^{I}\sum_{n=1}^{I} elev((i-1)\times I+m,(j-1)\times I+n)}{I^2} \tag{12}$$

where $aver(i,j)$ is the elevation of new grid-scale DEM.

The core of ZB algorithm (Yu *et al.*, 2001) increases the weight of elevation where the grids have large confluence (river networks), while maintaining the elevation where the grids have small confluence (basin boundaries) when the high-resolution DEM data is gathered into low-resolution DEM data, i.e. the ZB algorithm introduces the flow accumulation which includes the information of drainage basin boundaries and river networks as a weighting factor.

The ZB algorithm is calculated as follows: first of all, move the pits of the original DEM elevation, calculate its drainage networks and the flow accumulation, and then compute the elevation of large-scale DEM data according to formulas (13) and (14), keeping the flow accumulation as a weighting coefficient, while maintaining the elevation of basin boundaries, controlled by FAC.

$$FAC(i,j) = \sum_{m=1}^{I}\sum_{n=1}^{I} fac((i-1)\times I+m,(j-1)\times I+n) \tag{13}$$

$$elev(i,j) = \sum_{m=1}^{I}\sum_{n=1}^{I}[elev((i-1)\times I+m,(j-1)\times I+n)\times fac((i-1)\times I+m,(j-1)\times I+n)]\Big/FAC(i,j) \tag{14}$$

where *fac*(i,j) is the cumulative flow of original grid(i, j), *elev*(i,j) is the elevation of original grid (i,j) (unit: m).

CASE STUDY

Introduction of the basin

A case study at Fuji-kawa basin was selected to observe the effect of DEM spatial resolution and sampling algorithms on runoff simulation in BTOPMC. The Fuji-kawa basin is located in the centre region of Honshu Island in Japan; it has an area of 3570 km^2 and length of 128 km. The highest point in this basin is the famous mountain Fuji Mountain (3776 m), and about 88% of this basin is covered by forest. The Fuji-kawa basin belongs to the sub-humid region with mean annual precipitation of 2100 mm. There are hourly rainfalls from 7 gauging stations and runoff data from 4 discharge stations in the basin.

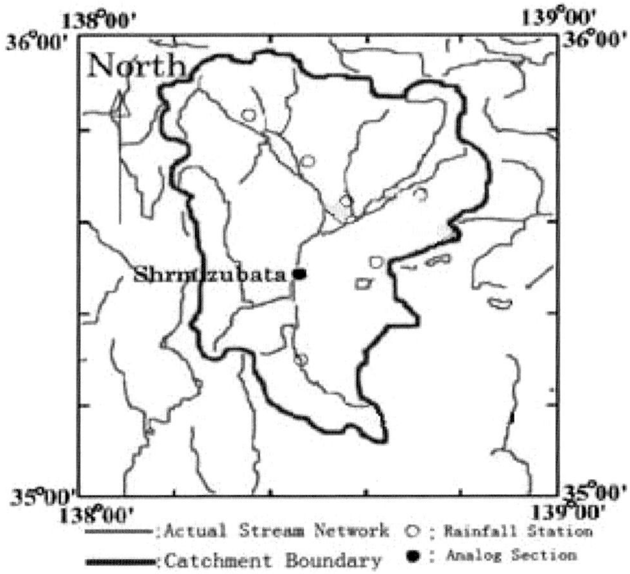

Fig. 1 Location of study basin.

Data used and processing

The original 3s resolution DEM data SRTM (Shuttle Radar Topography Mission, about 90 × 90 m) were downloaded from The CGIAR Consortium for Spatial Information (CGIAR-CSI). The DEM data used in this study are enlarged into four different resolutions (360 × 360 m, 540 × 540 m, 720 × 720 m, 900 × 900 m) by three different sampling algorithms.

The soil data are obtained from the Food and Agriculture Organization (FAO). According to the FAO ratios of soil particles provided by Yamanashi University of Japan, the soil composition of Fuji-kawa basin can be determined. The saturated hydraulic conductivity of different soil types are estimated. In this study, we suppose the land-cover and combinations of three soil types are the same over the whole research basin.

A total of 110 hours rainfall data from 7 gauging stations and runoff data from Shimizubata discharge station on 18 September 1993 to 22 September 1993 are selected in this study, which is a complete flood.

RESULTS AND DISCUSSION

The optimization is conducted by the SCE-UA algorithm (Duan *et al.*, 1994). Model parameters $S_{r\max}$, T_0, m and n_0 are calibrated for each grid cell, and S_{bar0} for each sub-basin. In BTOPMC, model performance is evaluated by the Nash-Sutcliffe Efficiency (E) measure and the Volume Ratio (V_r) of simulated to observed discharges. The optimized parameters are shown in Table 1.

Table 1 Parameter sets optimized for different cases.

BTOPMCs Parameters		Grid-average algorithm				Grid-minimum algorithm	
		900 m	720 m	540 m	360 m	900 m	720 m
T_0 (m^2/h)	Clay	1.052	7.28	49.599	12.344	1.558	3.541
	Sand	100.122	107.678	199.796	105.529	101.373	100.357
	Silt	51.021	67.804	99.786	58.633	51.437	52.001
m (m)	Clay	0.055	0.039	0.057	0.049	0.054	0.019
	Sand	0.130	0.026	0.012	0.014	0.087	0.009
	Silt	0.157	0.062	0.020	0.050	0.101	0.048
$S_{r\max}$ (m)	landcover	0.003	0.008	0.009	0.005	0.003	0.005
S_{bar0} (m)	$\lambda \le 10$	0.986	0.660	0.432	0.707	0.960	0.364
	$10 < \lambda \le 20$	0.473	0.170	0.136	0.121	0.342	0.158
n_0	landuse	0.018	0.001	0.001	0.001	0.049	0.001
	Clay	0.296	0.256	0.256	0.186	0.206	0.108
	Sand	0.162	0.078	0.169	0.081	0.246	0.249
	Silt	0.113	0.060	0.109	0.055	0.296	0.300
BTOPMC's Parameters		Grid-minimum algorithm		ZB algorithm			
		540m	360m	900m	720m	540m	360m
T_0 (m^2/h)	Clay	47.303	42.205	3.222	48.769	29.312	46.950
	Sand	199.135	196.436	101.992	198.957	195.458	199.470
	Silt	99.384	84.424	52.134	99.363	93.125	99.222
m (m)	Clay	0.063	0.051	0.003	0.028	0.049	0.041
	Sand	0.009	0.022	0.063	0.026	0.024	0.034
	Silt	0.026	0.013	0.006	0.018	0.013	0.008
$S_{r\max}$ (m)	landcover	0.007	0.003	0.003	0.018	0.012	0.003
S_{bar0} (m)	$\lambda \le 10$	0.401	0.529	0.439	0.359	0.404	0.886
	$10 < \lambda \le 20$	0.137	0.110	0.159	0.135	0.141	0.111
n_0	landuse	0.001	0.001	0.016	0.001	0.001	0.001
	Clay	0.224	0.093	0.168	0.360	0.277	0.237
	Sand	0.247	0.248	0.249	0.249	0.241	0.249
	Silt	0.299	0.300	0.299	0.300	0.296	0.300

From Table 1 it can be seen that model parameter values are dependent on DEM spatial resolution and sampling algorithms, i.e. different DEM spatial resolution and sampling algorithms lead to different parameter sets. This phenomenon appears in any of the 12 model parameters, although the degree of difference is different. This is principally considered as that different DEM spatial resolution and sampling algorithms reflect physical basin features and man-made interruption to different degrees. The physical reasonability for most of the optimized parameter values is acceptable. For example, in Table 1, in all 12 cases, clay has the lowest T_0 value, sand has the highest, and silt is between the former two soil types.

The final certainty factor (Nash%) and V_r (the ratio of total simulated discharge to the total observed discharge) are shown in Table 2, the peak time (the observed peak time t = 45) and peak flow are also given in Table 2.

Table 2 Parameter sets optimized for different cases.

Sampling algorithms	Resolution	Nash	V_r	t	Q_s/Q_o
Grid-average algorithm	900 m	96.02	1.018	45	1.047
	720 m	96.02	1.023	45	1.154
	540 m	96.02	1.005	45	1.092
	360 m	98.40	1.026	45	1.156
Grid-minimum algorithm	900 m	89.43	1.015	46	1.150
	720 m	91.29	0.977	45	1.229
	540 m	95.93	1.022	45	1.195
	360 m	96.17	1.020	45	1.158
ZB algorithm	900 m	90.79	1.003	44	1.239
	720 m	95.51	1.011	45	1.117
	540 m	96.55	1.028	45	1.123
	360 m	97.89	1.012	45	1.142

From Table 2, it can be seen that:

- As a whole, with the DEM generated by three sampling algorithms, all of the simulated results are very good. As can be seen from Table 2, the accuracy of simulation is above 90% except in one case, which means the accuracy results can meet the requirements of the A-level in hydrological forecasts. This indicates that the physically-based hydrological model BTOPMC used for simulation can achieve a satisfying result.
- Compared to the simulated results in four DEM spatial resolutions under the three sampling algorithms, it can be seen that the DEM spatial resolution has no obvious effect on the simulation result when using the grid-average algorithm, but it has a great effect when using the grid-minimum algorithm and ZB algorithm. As can be seen from Table 2, the accuracy of the simulated results become worse with the enlarging of the DEM grid scale when using the grid-minimum algorithm and ZB algorithm, but the results are more stable when using grid-average algorithm.
- Compared to the simulated results in three sampling algorithms under the same DEM spatial resolution, it can be seen that the grid-average algorithm can achieve the best simulation result, while the grid-minimum algorithm gets the worst simulated result.

The results of former research (Wan *et al.*, 2009) shows that the mean square deviation and information entropy of elevation, slope and topographic index corresponding to ZB algorithm is larger than the other two algorithms, which means the ZB algorithm can hold stream network continuity and real watershed boundaries very well, and meanwhile it can decrease the amount of other geographical information falling off compared to other algorithms. However, the information entropy becomes less as the DEM resolution becomes lower, which means with the DEM resolution reduced the geographical information of the research basin will reduce gradually. According to this conclusion, using the DEM sampled by the ZB algorithm will get better simulation results than the other two algorithms, and using high-resolution DEM data will get better simulation results than using low-resolution DEM data in theory. However, this study indicates that to maintain the value of the characteristics of the research basin better does not necessarily achieve the best simulated result. It is worthy of further study to investigate the relationship between the geographical information.

CONCLUDING REMARKS

In this study, three kinds of sampling algorithms are used, with each of the sampling algorithm there are four different spatial resolution DEM data, and then a hydrological model BTOPMC is used to simulate a flood process in Fuji-kawa basin in Japan. The results show that DEM spatial resolution has a great effect on the minimum sampling method and ZB algorithm. Compared to the other two methods, the average sampling method can achieve better simulated results.

Acknowledgements This study is supported by the Key Project of National Natural Science Foundation of China (no. 50739002).

REFERENCES

Ao, T. Q., Ishidaira, H. & Takeuchi, K. (1999) Study of distributed runoff simulation model based on block type TOPMODEL and Muskingum-Cunge method. *Ann. J. Hydraulic Engng, JSCE* **43**, 7–12 (in Japanese).

Ao, T. Q., Takeuchi, K. & Ishidaira, H. (2000) On problems and solutions of the Muskingum-Cunge routing method applied to a distributed rainfall runoff model. *Ann. J. Hydraulic Engng, JSCE* **44**, 139–144 (in Japanese).

Ao, T. Q., Takeuchi. K. & Ishidaira, H. (2001) On the method of generating artificial stream networks of large river basins and its effect on runoff simulation. *Ann. J. Hydraul. Engng, JSCE* **45**, 139–144 (in Japanese).

Ao, T. Q., Yoshitani, J., Fukami, K., Mutsuura, T., Takeuchi, K. & Ishidaira, H. (2002) Topographic analysis for distributed runoff model BTOPMC when applied to large river basins (in Japanese). *Ann. J. Hydraul. Engng, JSCE* **47**, 193–198.

Ao, T. Q., Takeuchi. K., Ishidaira, H., Yoshitani, J. & Fukami, K. (2003a) Development and application of a new algorithm for automated pit removal for grid DEMs. *Hydrol. Sci. J.* **48**(6), 985–997.

Ao, T. Q., Yoshitani, J., Takeuchi, K., Fukami, K., Mutsuura, T. & Ishidaira, H. (2003b) Effects of sub-basin scale on runoff simulation in distributed hydrological model: BTOPMC. In: *Weather Radar Information & Distributed Hydrological Modelling* (ed. by Y. Tachikawa, B. E. Vieux, K. P. Georgakakos & E. Nakakita), 227–233. IAHS Publ. 282. IAHS Press, Wallingford, UK.

Ao, T. Q., Ishidaira, H. & Takeuchi, K. *et al.* (2006) Relating BTOPMC model parameters to physical features of MOPEX basins. *J. Hydrol.* **320**(1), 84–102.

Beven, K. J. & Kirkby, M. J. (1979) A physically based, variable contributing area model of hydrology. *Hydrol. Sci. Bull.* **24**(1), 43–69.

Duan, Q., Sorooshian, S. & Gupta, V. K. (1994) Optimal use of the SCE-UA global optimization method for calibrating watershed models. *J. Hydrol.* **158**, 265–284.

Hao Zhen-chun & Chi Chen-xing (2004) The effect of spatial resolution and sampling method on the watershed features derived from DEM. *J. Glaciol. Geocryology* **26**(5), 610–616.

Quinn, P. F., Beven, K. J. & Lamb R. (1995) The ln(a/tanβ) Index: how to calculate it and how to use it within the TOPMODEL framework. *Hydrol. Processes* **9**, 161–182.

Takeuchi, K., Ao, T .Q. & Ishidaira, H. (1999) For hydro-environmental simulation of a large ungauged basin – introduction of block-wise use of TOPMODEL and Muskingum-Cunge method. *Hydrol. Sci. J.* **44**(4), 633–646.

Verdin, K. L. & Verdin, J. P. (1999) A topographical system for delineation and codification of the Earth's river basins. *J. Hydrol.* **218**, 1–12.

Wan, Yu-an, Ao, Tianqi, Liu, Zhanzhou *et al.* (2009) Research on the Impact of Different Constructing Digital watershed methods on the features of the watershed. *China Rural Water and Hydropower* **3**, 13–16.

Yang, Chuanguo, Yu, Zhongbo, Lin, Zhaohui & Hao, Zhenchun (2007) Method study of constructing digital watershed for large-scale distributed hydrological model. *Progress Geography Sci.* **26**(1), 68–76.

Yu, Z., Guo, Y., Voortman, J., White, R. & Miller, D. A. (2001) Streamflow simulation using Geographical Information System with a distributed approach. *J. Am. Water Resour. Assoc.* **37**, 957–971.

Evaluating the hydrological responses to climate change over a large basin with a hydrological–vegetation coupling model

XINGGUO MO, SUXIA LIU & ZHONGHUI LIN

Key Laboratory of Water Cycle & Related Land Surface Processes, Institute of Geographical Sciences and Natural Resources Research, Chinese Academy of Sciences, Beijing 100101, China

moxg@igsnrr.ac.cn

Abstract A physically-based distributed hydrological–vegetation coupling model, VIP, is developed to predict the responses, i.e. the relative changes, of hydrological processes to climate change in the Wuding River basin, one of the largest tributaries to the middle Yellow River, located in the Loess Plateau. It is found that the discharge is sensitive to climate change scenarios, and the response amplitudes are quite different at sub-basins. The response of discharge ranges from 11% to 25% under the scenarios of ±10% precipitation variations and 1°C temperature increment, showing that precipitation amplifies stream runoff change, but the warming mitigates the responses of runoff. Regressive analysis shows that better correlation exists between the responses of runoff and the annual runoff (basin area), with the Pearson coefficient (r) being from 0.35 to 0.61, than the correlation between the response of runoff and annual precipitation with r being from 0.29 to 0.43. It is illustrated that considering precipitation processes other than just the annual precipitation, and running the ecohydrological model at as small as a daily scale and coupling with vegetation dynamics, are crucial in exploring the responses of hydrological processes to climate change.

Key words distributed ecohydrological model; streamflow; evapotranspiration; climate change; Loess Plateau

INTRODUCTION

Climate change has gained considerable attention over the last decade, due to the concerns of unfavourable impacts on water resources, agriculture and ecosystems, as well as human welfare. According to the IPCC Fourth Assessment Report (IPCC, 2007), global mean temperature has risen approximately 0.74°C in the last 100 years. Global climate models (GCMs) predict temperature increases of 1.1–6.4°C over the period 1990–2100. Climate change is expected to considerably impact the surface hydrological processes with variabilities in different regions (Caballero *et al.*, 2007). Many studies show that there will be an alteration of seasonal variations with winter flow increasing and summer flow decreasing (Miller *et al.*, 2003; Drogue *et al.*, 2004). An alteration of precipitation regime will have a greater impact on discharge magnitude, whereas air warming will affect the seasonal flow evolution (McCabe & Hay, 1995; Zierl & Bugmann, 2005).

The distributed models, which describe the spatial patterns of various inputs and boundary conditions, such as topography, vegetation, land use, soil characteristics, rainfall and other meteorological driving forces, are taken as valuable tools for prediction of the hydrological responses under climate change. So far, most of the hydrological models do not embed vegetation dynamics as a component (Cayrol *et al.*, 2000), with few exceptions such as RHESSys (Tague & Band, 2004), TOPOG (Silberstein *et al.*, 1999), SWIM (Krysanova *et al.*, 1998, 1999). Further, only some studies use such kinds of coupling models to explore the impacts of climate change on hydrological processes. Hence, combining vegetation dynamics with a process-based hydrological model to investigate the interactions between vegetation, hydrology and climate still poses challenges.

In this study, a distributed hydrological–vegetation dynamic coupling model is designed to simulate the hydrological responses to climate changes in the Wuding River basin, a large basin on the Loess Plateau. Through evaluating the responses of hydrological processes (streamflow and ET, etc.) to climate change at the basin scale, the study will provide a scientific basis to decision makers for sound management of soil and water.

THE VIP DISTRIBUTED ECO-HYDROLOGICAL MODEL

The model used here is called VIP (Vegetation Interface Processes) model (Mo & Liu, 2001; Mo *et al.*, 2005b, 2009). It was developed under the support of geographical information data. Its ecological and hydrological processes are coupled by a one-dimensional soil–vegetation–atmosphere transfer (SVAT) scheme, a distributed runoff routing scheme and a vegetation dynamic scheme. The model framework is introduced briefly as following:

Water budget

Water budget in a grid can be described as,

$$P = E_v + E_g + R + \Delta S \tag{1}$$

where P is the precipitation; E_v and E_g are the canopy transpiration and soil evaporation, respectively; R is the runoff of overland and groundwater; ΔS is the storage change. The unit of all these variables is mm.

Water cycle and energy transfer are coupled through ET (latent heat flux) process. Neglecting solar energy fixed by photosynthesis, energy balances in canopy and soil surfaces are expressed as,

$$C_v \frac{\partial T_v}{\partial t} = R_{nv} - H_v - LE_v \tag{2}$$

$$C_{m1} \frac{\partial T_g}{\partial t} = R_{ng} - H_g - LE_g - G \tag{3}$$

where C_v and C_{m1} are the bulk heat capacity of canopy and surface soil layer (J m^{-2} K^{-1}); T_v and T_g are the canopy and soil surface temperatures (K); R_n, H, LE and G are the net radiation, sensible, latent and soil heat fluxes (W m^{-2}), with subscript v and g referring to canopy and ground, respectively.

Overland runoff is estimated with a variable infiltration curve (Mo *et al.*, 2005a), modified from Zhao (1992). The groundwater is divided into two layers and routed to the channel with the linear reservoir assumption. Routines of both overland and channel runoff are calculated with the kinematic wave equation.

Vegetation dynamics

The evolutions of foliage, stem and root biomass are expressed as:

$$\frac{dM_i}{dt} = a_i P_g - R_{g,i} - R_{m,i} - D_i \tag{4}$$

where M is the biomass of vegetation compartments (g m^{-2}); subscript $i = l$, s, r, representing the foliage, stem and root biomass of woody vegetation types, respectively; for herbaceous vegetation, $i = l$, r, representing foliage and root, respectively; P_g is the daily gross photosynthetic amount (g (CH_2O) m^{-2} day^{-1}; a is the allocation coefficient; R_m and R_g are the maintenance and growth respiration rates (g m^{-2} day^{-1}), respectively; D is the senescent rate (g m^{-2} day^{-1}); t is time (days).

The estimation of canopy photosynthesis is based on Farquhar's biochemical model. The canopy leaves are simplified into sunlit and shaded groups to account for their different photosynthetic responses to radiation and a multilayer scheme is designed to describe the radiation penetration in canopy. Leaf area index (*LAI*) is the key state variable linking the hydrological processes and vegetation dynamics. *LAI* is derived from the foliage biomass, M_l, and the specific leaf area, s (m^2 g^{-1}), namely:

$$LAI = sM_l \tag{5}$$

STUDY BASIN AND DATA

Study basin

The Wuding River, one of the largest tributaries in the middle reach of the Yellow River, is located on the Loess Plateau, covering an area of 30 261 km^2 (Fig. 1). The altitude in the basin ranges from 600 m to 1800 m with the main flatland being about 1200 m a.s.l. in the northwestern part. In its southern part, the geomorphologic characteristics consist of terraces, hill slopes and incised gullies. The main channel originates from Mount Baiyu, with a length of 491.1 km and a mean slope of 0.197%. The soil texture is distinguished as sandy loam, sandy silt, silt, sand and coarse sand, of which 55% is sandy silt. It belongs to the semi-arid climate zone. The annual precipitation amount is 376 ± 92 mm over the period from 1970 to 2000, most of which occurs in summer monsoon season, especially from July to August. The long-term mean stream flow over the period from 1970 to 2000 is about 30 $m^3 s^{-1}$ at the basin outlet (Baijiachuan site).

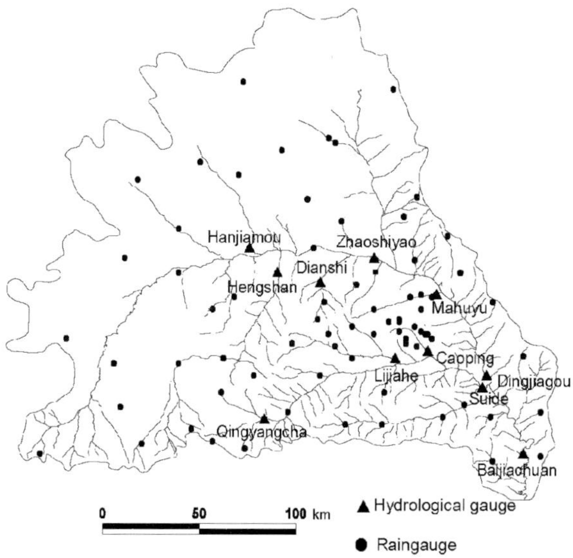

Fig. 1 Hydrological and rain gauges over the Wuding River basin.

Data

Based on the land use map published in 1990 (Wu *et al.*, 1990), the land use/cover patterns in the basin are classified into six types, namely, farmland, deciduous broadleaf forest, mixture forest, dwarf shrub, grassland and desert with ratios being 22.4%, 1%, 0.3%, 10%, 45% and 21.4%, respectively. The planted crops include maize, millet, sorghum, rice, soybean and some others. A DEM generated from contour with 1:250 000 scale is used to derive the channel pattern and flow directions of the whole basin with 1 km resolution.

Daily precipitation at 82 raingauges and meteorological data recorded at 16 climatic stations in and around the basin are used to interpolate the surface synoptic variables, which include daily atmospheric pressure, maximum and minimum air temperatures, humidity, wind velocity, precipitation and sunshine duration. Diurnal variation of air temperature and solar radiation are estimated with sine shape curves. Geographical information of vegetation types is incorporated to assign the land surface attributes spatially. Then the model is run with a 30-minute time step and 1-km spatial resolution.

Climate change scenarios

Climate change considerably affects the ecosystem water cycle and carbon assimilation processes. According to HadCM3 GCMs projections, the air temperature and precipitation over the Loess Plateau will increase about 1°C and 10% in 2035, respectively. Under climate change, the temperature of the Wuding River basin will increase along with global warming, whereas precipitation is quite uncertain. Considering the uncertainty in the climate change scenarios, as well as the spatial variability of precipitation, we set 0.5°C and 1.0°C increments of air temperature, along with 10% increment/decrement of precipitation, respectively, as the climate change scenarios. Only sensitivities of the hydrological cycle to climate change in the Wuding River basin under air temperature increase and precipitation increase/decrease scenarios are analysed here.

RESULTS

Discharge validation

The predicted daily discharges from 1988 to 1997 are validated with the measurements at the basin and six sub-basin outlets (Fig. 2). The Pearson correlation coefficient is 0.70 between the two series. The absolute relative biases of annual discharge are less than 13% between the predicted and measured discharge, except for two years. For the six sub-basins, the mean absolute relative bias of annual discharge was 14% in 1988 (Fig. 3). In this kind of semi-arid basin, rainfall–runoff process is related to rainfall intensity, soil hydraulic conductivity, as well as vegetation condition. Bias in the runoff prediction may come from parameter uncertainty in the overland runoff generation scheme.

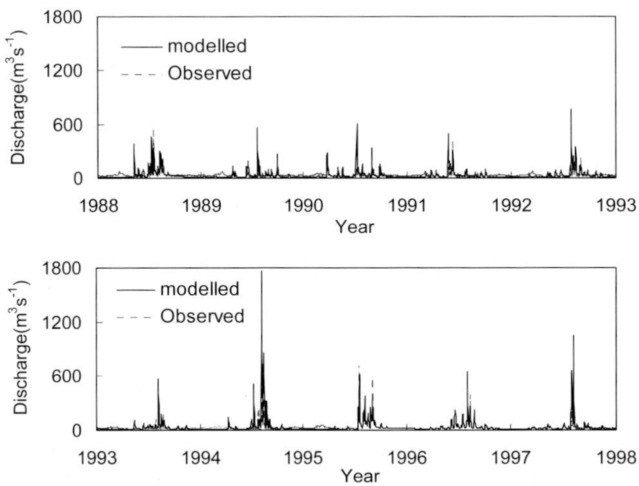

Fig. 2 Comparison of the observed and simulated daily stream discharge at the basin outlet from 1988 to 1997.

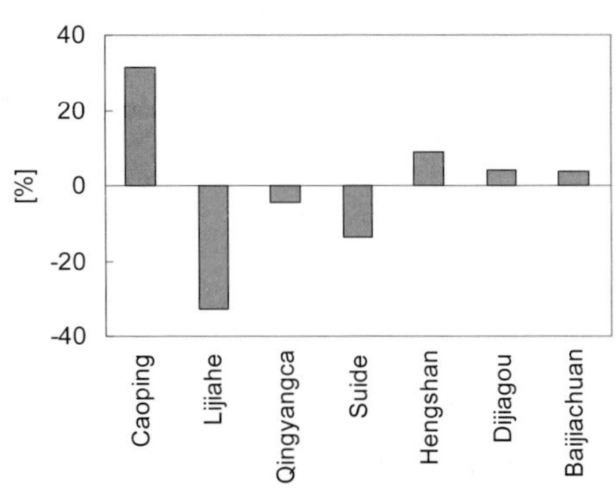

Fig. 3 Relative errors of the simulated annual discharges at the basin and its six sub-basin outlets in 1988.

Climate change effects

From the model simulation, it shows that for +10% precipitation conditions, runoff will increase by 11–20%; while for –10% precipitation conditions, runoff will reduce by 13–22% in the basin and six sub-basins (Table 1). Without precipitation change, air warming will reduce the runoff. The predicted ranges of the responses, i.e. the relative change, of runoff to climate change are comparable to literature reports, such as Jones *et al.* (2006), which reported a 1% change of annual precipitation inducing 2.1–2.5% change of runoff in Australia basins. The stream runoff usually amplifies the effect of precipitation change, whereas the warming mitigates the runoff responses. This is also found from the analysis based on observed data (communication with the Bureau of Hydrology, Yellow River Conservancy Committee).

The simulation results also show that the responses of streamflow at the entire basin and sub-basin scale are quite different, which is largely conditioned by spatial variability of climatic and physiographical characteristics between the sub-basins (Drogue *et al.*, 2004). In our case, the response variability is mainly related to the difference in temporal precipitation series.

The mechanisms of the heterogeneous hydrological responses to climate change over the large basin can be explained as follows: (1) during dry seasons, higher temperature will enhance the land surface ET rate, then more soil moisture is depleted and the recharge to groundwater is reduced, and finally the streamflow becomes lower; (2) as the response of overland runoff to daily precipitation is nonlinear and also regulated by vegetation conditions, the same percentage increment of rainfall will result in quite varied runoff responses, e.g. the response magnitudes are diverse for two sub-basins with almost the same amount of annual precipitation.

Better correlation exists between the response and runoff depth on an annual scale (basin area), with the Pearson correlation coefficient being from 0.35 to 0.61, than the correlation between the response and precipitation, with the Pearson correlation coefficient being from 0.29 to 0.43, as shown in Table 2. In such a semi-arid basin, the precipitation amount in the specific storm events and soil moisture at the beginning of the rainy season, play a more critical role in shaping the flow than the rainfall total, as demonstrated by the simulation at a daily scale.

Table 1 Area, annual precipitation and runoff amounts in 1988 and the relative changes of the stream flow to climate changes (+0.5°C, +1.0°C temperature and ±10% precipitation change) over the entire basin and its six sub-basins (Baijiachuan is the basin outlet).

	Caoping	Lijiahe	Qingyangca	Hengshan	Suide	Dingjiagou	Baijiachuan
Area (km^2)	187	807	662	2415	3893	23422	29662
P (mm $year^{-1}$)	503	435	434	423	444	419	424
Q ($10^4 m^3 year^{-1}$)	2.96	6.02	7.73	18.73	44.03	236.02	341.24
+0.5°C, +10%	0.19	0.15	0.13	0.16	0.15	0.18	0.19
+0.5°C, –10%	–0.20	–0.15	–0.13	–0.16	–0.17	–0.17	–0.20
+1.0°C,+10%	0.18	0.12	0.11	0.16	0.13	0.16	0.16
+1.0°C, –10%	–0.21	–0.17	–0.14	–0.16	–0.18	–0.18	–0.22

Table 2 The Pearson correlation coefficient between the responses and annual precipitation and between the responses and runoff amounts over the entire basin and its six sub-basins.

		Q ($10^4 m^3 year^{-1}$)	P (mm $year^{-1}$)
Responses	dT=+0.5°C, dP=+10%	0.61	0.29
	dT=+0.5°C, dP=–10%	–0.51	–0.43
	dT=+1.0°C, dP=+10%	0.35	0.37
	dT=+1.0°C, P=–10%	–0.57	–0.39

In contrast, ET from the land surface is slightly intensified (about 2%) with 1°C air temperature increment, even though the precipitation is reduced by 10%, due to the extending growing period and water vapour deficit. Under the above climate change scenarios, the relative changes of ET are much less than that of runoff caused by the precipitation increment/decrement (not shown here).

DISCUSSION AND CONCLUSIONS

Climate changes the vegetation composition in the relative long term, which significantly influences the vegetation and hydrological processes. The climate change alters the vegetation density, resulting in hydrological processes modification. To exploit the hydrological mechanisms under climate changes, eco-hydrological models, which integrate vegetation dynamics, hydrological processes and the feedback of climate are effective tools.

The paper conducted a sensitivity analysis of hydrological processes to climate change by using the eco-hydrological VIP model. It shows that the streamflow is quite sensitive to precipitation change, whereas the sensitivities are quite different at sub-basin scale. The responses range from 11% to 25% under scenarios of ±10% precipitation change and 1°C temperature increments.

Acknowledgements This study was jointly supported by the Natural Science Foundation of China grant (40671033, 40671032), the Chinese Ministry of Science and Technology for “863” agriculture high technology project (2006AA10Z228), the Chinese Ministry of Science and Technology for “973”projects (2009CB421307, 2010CB428400-4).

REFERENCES

Caballero, Y., Voirin-Morel, S., Habets, F., Noilhan, J., LeMoigne, P., Lehenaff, A. & Boone A. (2007) Hydrological sensitivity of the Adour-Garonne river basin to climate change. *Water Resour. Res.* **43**, W07448, doi:10.1029/-2005WR004192.

Cayrol, P., Kergoat, L., Moulin, S., Dedieu, G. & Chehbouni, A. (2000) Calibrating a coupled SVAT–vegetation growth model with remotely sensed reflectance and surface temperature – a case study for the HAPEX-Sahel grassland sites. *J. Appl. Met.* **39**, 2452–2472.

Drogue, G., Pfister, L., Leviandier, T. E. I., Idrissi, A., Iffly, J. F., Matgen, P., Humbert, J. & Hoffmann, L. (2004) Simulating the spatio-temporal variability of streamflow response to climate change scenarios in a mesoscale basin. *J. Hydrol.* **293**, 255–269.

IPCC (2007) The Physical Science Basis. Contribution of I to the Fourth Assessment Report of the Intergovernmental Panel on Climate Change. In: *Climate Change 2007.* (ed. by S. Solomon, D. Qin, M. Manning, Z. Chen, M. Marquis, K. B. Averyt, M. Tignor & H. L. Miller). Cambridge University Press, Cambridge, UK.

Krysanova, V., Muller-Wohlfeil, D. & Becker, A. (1998) Development and test of a spatially distributed hydrological/water quality model for mesoscale watersheds. *Ecol. Modelling* **106**, 261–289.

Krysanova, V., Wechsung, F., Becker, A., Poschenrieder, W. & Grafe, J. (1999) Mesoscale ecohydrological modelling to analyse regional effects of climate change. *Environ. Modeling & Assessment* **4**, 259–271.

Jones, R. N., Chiew, F. H. S., Boughton, W. C. & Zhang, L. (2006) Estimating the sensitivity of mean annual runoff to climate change using selected hydrological models. *Adv. Water Resour.* **29**(10), 1419–1429.

McCabe, G. J. & Hay, L. E. (1995) Hydrologic effects of hypothetical climate changes on water resources in the East River basin, Colorado. *Hydrol. Sci. J.* **40**, 303–318.

Miller, N. L., Bashford, K. E. & Strem, E. (2003) Potential impacts of climate change on California hydrology. *J. Am. Water Resour. Assoc.* **39**, 771–784.

Mo, X. & Liu, S. (2001) Simulating evapotranspiration and photosynthesis of winter wheat over the growing season. *Agric. For. Met.* **109**, 203–222.

Mo, X., Beven, K., Liu, S., Leslie, L.M. & de Roo, Ad (2005a) Long-term water budget estimation with the modified distributed model – LISFLOOD-WB over the Lushi basin, China. *Meteorol Atmos Phys.* **90**, 1–16.

Mo, X., Liu, S., Lin, Z., Xu, Y., Xiang, Y. & McVicar, T. R. (2005b) Prediction of crop yield, water consumption and water use efficiency with a SVAT-crop growth model using remotely sensed data on the North China Plain. *Ecol. Modelling* **183**, 310–322.

Mo, X., Liu, S., Chen, D., Lin, Z. & Zhao, W. (2009) Grid-scale effects on evapotranspiration and gross primary production estimation over a large loess plateau basin. *Hydrol. Sci. J.* **54**(1), 160–173.

Silberstein, R. P., Vertessy, R. A., Morris, J. & Feikema, P. M. (1999) Modelling the effects of soil moisture and solute conditions on long-term tree growth and water use: a case study from the Shepparton irrigation area, Australia. *Agric. Water Manage.* **39**, 283–315.

Tague, C. & Band, L. E. (2004) RHESSys: Regional hydro-ecologic simulation system: an object-oriented approach to spatially distributed modeling of carbon, water and nutrient cycling. *Earth Interactions* **8**, 1–42.

Wu, C. (1990) *1:1 000 000 Land Use Map of China*. Science Press, Beijing, China.

Zhao, R. J. (1992) The Xinanjiang model applied in China. *J. Hydrol.* **135**, 371–381.

Zierl, B. & Bugmann, H. (2005) Global change impacts on hydrological process in Alpine catchments. *Water Resour. Res.* **41**, W02028, doi: 10.1029 /2004WR003447.

Hydrological trends during the past 50 years in the Yangtze River basin

ZHENTAO CONG, JINGJING ZHAO, ZHENNING SHANG & DAWEN YANG
State Key Laboratory of Hydroscience and Engineering, Department of Hydraulic Engineering, Tsinghua University, Beijing 100084, China
congzht@tsinghua.edu.cn

Abstract The hydrological cycle has been highly influenced by climate changes and human activities, and it is significant to analyse the hydrological trends that occurred in past decades in order to understand past changes and predict future trends. Water resources in the Yangtze River basin have been paid more attention, especially for the Three Gorges Dam and the South-to-North water diversion project. Based on the climate data from 180 stations in the Yangtze River basin, it was found that pan evaporation showed a decreasing trend as a whole during the past 50 years with the increasing in air temperature, indicating that the evaporation paradox existed in this basin. At the same time, actual evaporation estimated by water balance decreased with increasing precipitation. Based on the Budyko curve, it can be concluded that the decrease in actual evaporation was caused by the decrease in energy.

Key words hydrological trend; Yangtze River; evaporation paradox; evaporation

INTRODUCTION

The hydrological cycle traces the largest movement of any substance on Earth (Chahine, 1992), and at the same time, it has been greatly influenced by climate changes and human activities in the past five to ten decades. Renewable fresh water is the foundation for life in terrestrial and freshwater ecosystems (Jackson *et al.*, 2001), thus it is important to analyse the hydrological trends in history to predict future changes. Components of the regional hydrological cycle include precipitation, evapotranspiration, runoff and water storage change, including variations in soil water, groundwater, ice water, snow water, canopy water and reservoir water. Runoff is given more attention because of its close relationship to water resources (McCabe *et al.*, 1997; Bronstert *et al.*, 2002).

It is reported that the surface temperature of the Earth increased at a rate of about 0.13°C per decade over the past 50 years (IPCC, 2007). One expects that global warming trends would make the air near the Earth surface drier, and result in an increase in the rate of evaporation from terrestrial open water bodies. However, many observations show that the rate of evaporation from open pans of water has been steadily decreasing all over the world over the past 50 years (Peterson *et al.*, 1995; Chattopadhyay *et al.*, 1997; Thomas, 2000; Liu *et al.*, 2004; Roderick *et al.*, 2004). The contrast between the expected and the observed is called the evaporation paradox (Roderick *et al.*, 2002).

Overall, global land precipitation has increased since the beginning of the 20th century, but the increase is neither spatially nor temporally uniform (IPCC, 2007). Over the past 50 years, the annual precipitation in China as a whole has had no obvious trend (Zhai *et al.*, 1999; Liu *et al.*, 2005). However, on a regional scale, while there has been a decreasing trend in the Yellow River Basin (Yang *et al.*, 2004) there has been an increasing trend in the middle and lower Yangtze River Basin (Zhang *et al.* 2005). During the same period, it was found that runoff had decreased in the Yellow River Basin (Fu *et al.*, 2004; Yang *et al.*, 2004; Liu *et al.*, 2008) but increased in the Yangzte River Basin (Yang *et al.*, 2005). It is disputed that the decreasing of the pan or reference evaporation predicates the decrease of the actual evaporation (Peterson *et al.*, 1995; Brutsaert *et al.*, 1998). Based on the Budyko hypothesis, Yang *et al.* (2006) suggested that a complementary relationship exists in non-humid areas where the precipitation is the controlling factor of the actual evapotranspiration and that a proportional relationship exists in humid areas where energy is the controlling factor. Quantitative analysis of evapotranspiration is still needed for a better understanding of the regional variability of the water balance.

The Yangtze River, also called Changjiang in Chinese, is the longest river in China with a length of 6321 km, and the largest one with a drainage area of 1 800 000 km^2. This river was paid more attention, especially for the Three Gorge Dam and the South-to-North water diversion project. People care whether there are enough water resources to support sustainable development. With the data of 180 weather stations, the hydrological trends and their spatial distribution were analysed in this basin in order to understand this issue better.

DATA AND METHODOLOGY

Data

Annual weather data at 180 stations from 1956 to 2005, including air temperature, precipitation, pan evaporation, wind speed and sunshine duration, were used in this study. Pan evaporation rate records were collected from the Climatic Data Center, National Meteorological Information Center, China Meteorological Administration, and other weather data released by the China Meteorological Data Sharing Service System (http://cdc.cma.gov.cn/).

Annual discharges from 1956 to 2000 at Datong hydrological station were regarded as the average runoff of the whole basin, which were used in the water balance analysis.

Trend analysis

Linear regression was used to identify trends of climatic and hydrological variables including pan evaporation, air temperature, sunshine duration, wind speed, precipitation, runoff and actual evaporation. The nonparametric Mann-Kendall test was applied for trend detection (Maidment *et al*, 1993) with a significance level (α) of 5%.

Water balance

In order to get the spatial average of hydrological variables, the kriging interpretation was used with the tool offered in ArcGIS.

Since the average of precipitation, potential evaporation and runoff can be measured or calculated for the whole basin, the actual evaporation can be simulated by the water balance if it is supposed that the water storage had no change. In reality, the water storage will change between different years. For the Yangtze River Basin, which has plenty of precipitation, the covering soil layer is thin and the groundwater storage is small, so the groundwater storage change can be ignored in the water balance analysis. Another concern is the impact of human activities as it is the observed surface runoff that is used in the water balance equation. The surface runoff would decrease and the evapotranspiration would increase if more water were drawn from rivers for irrigation. In the past 50 years, the water drawn from rivers has continually increased in China, so, as mentioned previously, the assumption that irrigation has not changed would not change the trend of the actual evaporation if it was estimated to have a decreasing trend.

RESULTS AND DISCUSSION

Evaporation paradox

Trends of annual average air temperature are shown in Fig. 1. The Yangtze River basin began warming in the past 50 years with global warming. Of the total 180 stations, air temperature at 151 stations increased and air temperature at 91 stations increased significantly. However, it was found that the air temperature decreased in the middle stream basin, although with no significance.

Trends of annual pan evaporation are shown in Fig. 2. It can be found that pan evaporation at 122 stations increased, among which 75 values increased significantly. In the middle stream basin, pan evaporation had different trends just like that of air temperature.

Therefore, we can conclude that evaporation paradox existed for the whole basin, since pan evaporation decreased at most stations with increasing air temperature. It is interesting for the

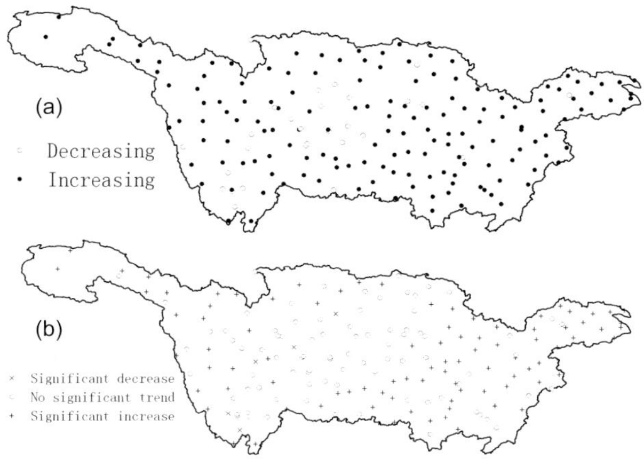

Fig. 1 Trends of annual average air temperature in Yangtze River basin in the past 50 years, (a) trends of air temperature, (b) significances of air temperature change.

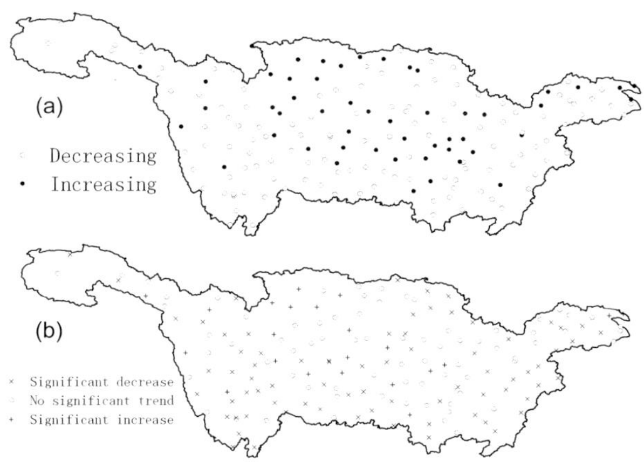

Fig. 2 Trends of annual pan evaporation in Yangtze River basin in the past 50 years, (a) trends of pan evaporation, (b) significances of pan evaporation change.

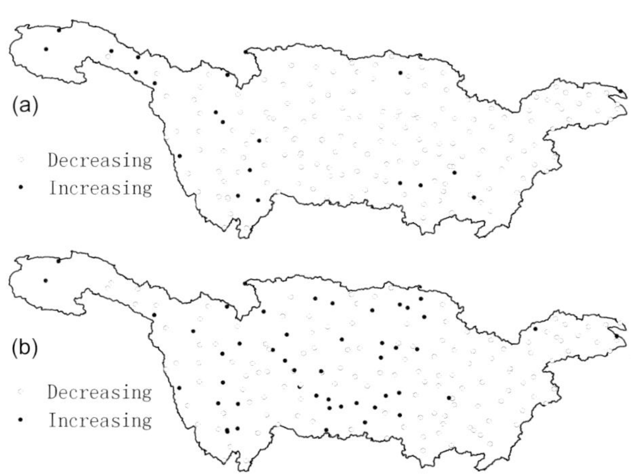

Fig. 3 Trends of sunshine duration and wind speed in Yangtze River basin in the past 50 years, (a) trends of sunshine duration, (b) trends of wind speed.

middle stream basin, where the pan evaporation increased at most stations with the air temperature decreasing.

Evaporation, including pan evaporation, potential evaporation and actual evapotranspiration, is influenced by solar radiation, air temperature, wind speed, vapour pressure deficit, etc. When using the standard Penman-Monteith function presented by Allen (1998), reference evapotranspiration can be calculated with radiation, wind speed, air temperature and vapour pressure deficit. As solar radiation observations were not available for this study, radiation was estimated based on sunshine duration according to Allen (1998).

Trends of sunshine duration and wind speed are shown in Fig. 3. From 1956 to 2005, sunshine duration data appeared as a decreasing trend. The decreasing trends existed at more than 2/3 of the stations for both pan evaporation and sunshine duration. The decreasing trend of sunshine duration may be caused by greater cloud coverage and increasing aerosol concentrations. The decrease in wind speed is another cause for the decrease in pan evaporation.

Hydrological trends

Precipitation from the weather stations showed that precipitation at most stations in the upstream basin and downstream basin increased and precipitation in most stations in middle stream decreased (in Fig. 4(a)), although these trends are not significant (in Fig. 4(b)). The decrease of precipitation in the middle stream can explain the increase of pan evaporation. Average

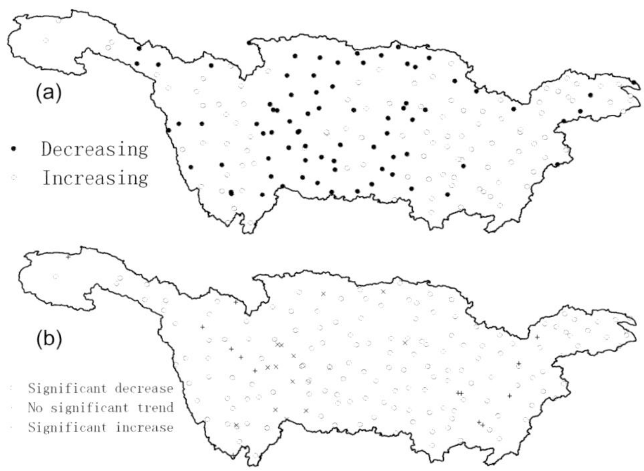

Fig. 4 Trends of annual precipitation in the Yangtze River basin in the past 50 years, (a) trends of precipitation, (b) significances of precipitation change.

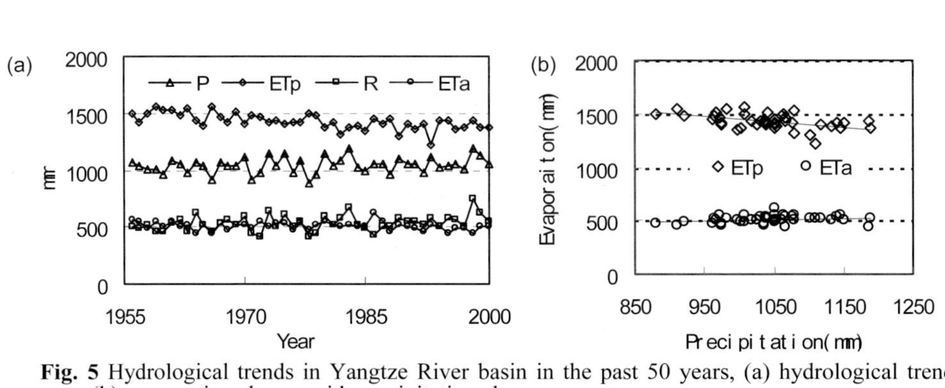

Fig. 5 Hydrological trends in Yangtze River basin in the past 50 years, (a) hydrological trends with year, (b) evaporation change with precipitation change.

precipitation is lowest in the middle-stream basin and highest in the upstream areas (in Fig. 4(c)). Based on the trends of precipitation interpretation, the increasing trends are higher in the downstream basin and the decreasing trends are higher in the middle stream basin (in Fig. 4(d)).

Annual discharges from 1956 to 2000 at Datong hydrological station were regarded as the average runoff of the whole basin. The spatial average of annual precipitation and annual pan evaporation were calculated with the kriging interpretation. Then the annual actual evaporation of the whole basin can be simulated by the water balance if the water storage was supposed to be constant.

For the whole Yangtze River basin, precipitation and runoff increased, pan evaporation decreased significantly, and actual evaporation decreased during the last 50 years, as shown in Fig. 5(a). Furthermore, it can be found that pan evaporation decreased and actual evaporation increased with the increase in precipitation (in Fig. 5(b)), which shows a complementary relationship between pan evaporation and actual evaporation.

With precipitation increasing, there is a complementary relationship between potential evaporation and actual evaporation. But the curve will change with the varied condition of energy. When the data in Yangtze River basin were divided into two groups, before 1985 and after 1986, we can find a different Budyko curve (in Fig. 6(a)). This phenomenon can be generalized into Fig. 6(b). When the precipitation increases from P1 to P2, the potential evaporation will decrease from ETp1 to ETp2, and the actual evaporation will decrease from ETa1 to ETa2. If the Budyko curve is fixed, the actual evaporation will increase. So we can conclude that the actual evaporation decreased because the energy, evaporation ability, decreased in the Yangtze River basin during the past 50 years.

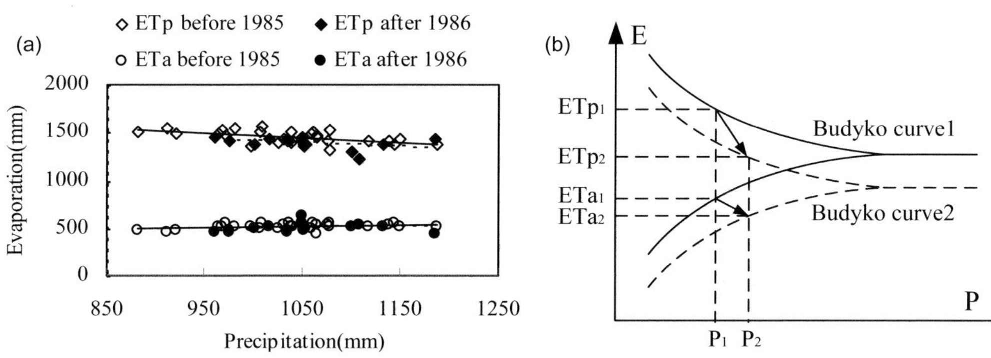

Fig. 6 Curve movement in Yellow River basin and Yangtze River basin, (a) evaporation change in different periods, (b) Budyko curve.

CONCLUSIONS

Using climate data from 180 stations in the Yangtze River basin, it was found that pan evaporation showed a decreasing trend as a whole during the past 50 years, with the increasing air temperature, which indicates the existence of an evaporation paradox in the this basin. Decreasing pan evaporation was caused by decreasing solar radiation and wind speed.

At the same time, actual evaporation estimated by water balance decreased with increasing precipitation. Based on the Budyko curve, it can be concluded that the decrease in actual evaporation was caused by the decrease in evaporation ability expressed by pan evaporation.

Acknowledgements This study is supported by the National Science Foundation of China grants 50721140161 and 50509011.

REFERENCES

Bronstert, A., Niehoff, D. *et al.* (2002) Effects of climate and land-use change on storm runoff generation: present knowledge and modelling capabilities *Hydrol. Processes* **2**, 509–529.

Brutsaert, W. & Parlange, M. B. (1998) Hydrologic cycle explains the evaporation paradox. *Nature* **396**, 29–30.

Chahine, T. M. (1992) The hydrological cycle and its influence on climate. *Nature* **239**, 373–381.

Chattopadhyay, N. & Hulme, M. (1997) Evaporation and potential evapotranspiration in India under conditions of recent and future climate change. *Agric. For. Met.* **87**(1), 55–73.

IPCC (eds.) (2007) *Climate Change 2007: The Physical Science Basis.* Contribution of Working Group I to the Fourth Assessment, Cambridge University Press, New York, USA.

Jackson, R. B., Carpenter, S. R. *et al.* (2001) Water in a changing world. *Ecol. Applications* **4**, 1027–1045.

Liu, B. H., Xu, M., Mark, H., Henderson, M. & Gong W. (2004) A spatial analysis of pan evaporation trends in China, 1955–2000. *J. Geophys. Res.* **109**, D15102, doi:10.1029/2004JD004511.

Liu, B. H., Xu, M., Henderson, M., *et al.* (2005). Observed trends of precipitation amount, frequency, and intensity in China, 1960-2000. *J. Geophy. Res.* 110 (D8): Art. no. D08103.

McCabe, G. J. & Wolock, D. M. (1997) Climate change and the detection of trends in annual runoff. *Climate Res.* **2**, 129–134.

Peterson, T. C., Golubev, V. S. & Groisman, P. Y. (1995) Evaporation losing its strength. *Nature* **377**, 687–688.

Roderick, M. L. & Farquhar, G. D. (2002) The cause of decreased pan evaporation over the past 50 years. *Science* **298**(15), 1410–1411.

Roderick, M. L. and Farquhar, G. D. (2004) Changes in Australian pan evaporation from 1970 to 2002. *Int. J. Climatol.* **24**, 1077–1090.

Thomas, A. (2000) Spatial and temporal characteristics of potential evapotranspiration trends over China. *Int. J. Climatol.* **20**, 381–396.

Yang, D., Li, C., Hu, H., Lei, Z., Yang, S., Kusuda, T., Koike, T. & Musiake, K. (2004) Analysis of water resources variability in the Yellow River of China during the last half century using historical data. *Water Resour. Res.* **40**, W06502, doi:10.1029/2003WR002763.

Yang, D., Sun, F., Liu, Z., Cong, Z. & Lei, Z. (2006) Interpreting the complementary relationship in non-humid environments based on the Budyko and Penman hypotheses. *Geophys. Res. Lett.* **33**, L18402, doi:10.1029/2006GL027657.

Yang, S. L., Gao, A., *et al.* (2005) Trends in annual discharge from the Yangtze River to the sea (1865–2004). *Hydrol. Sci. J.* **50**, 825–836.

Zhai, P. M., Sun, A., Ren, F. M., Liu, X., Gao, B. & Zhang, Q. (1999) Changes of climate extremes in China. *Climate Change* **42**, 203–218.

2 Hydrological Processes based on the New Observation Techniques

Daily process retrieval model for sensible heat flux using geostationary satellite data

CHANGSEN ZHAO[1,2,3], CHANGLEI SUN[4], YONGPING DANG[5], GUANGFENG LI[6], XUYONG LI[7] & ZHAOLIANG LI[3]

1 *Shandong University, Jinan, Shandong Province, 250013, China*
hzjohnson2003@163.com

2 *Department of Water Resources and Environment, Sun Yat-sen University, Guangzhou 510275, China*

3 *Institute of Geographical Sciences and Natural Resources Research, CAS, Beijing 100101, China*

4 *Shandong Design Institute of Water Resources, Jinan, Shandong Province, 250013, China*

5 *Shandong Survey Bureau of Hydrology and Water Resources, Jinan, Shandong Province, 250013, China*

6 *Dezhou Survey Bureau of Hydrology and Water Resources, Dezhou, Shandong Province, 253000, China*

7 *Yanggu Water Authority, Yanggu, Shandong Province, 252300, China*

Abstract Sensible heat flux is an important element in regional evapotranspiration estimation, which is of significant importance to the agriculture, hydrological cycle and regional climate, which greatly influence people's daily life. This paper presents a diurnal model to compute daily sensible heat flux processes on the basis of Fourier analysis of the daily surface temperature process. Its main advantages are in requiring no *in situ* ground meteorological data and having few unknowns to be calibrated. Calibration and validation of the model was conducted in a Spanish site of FLUXNET with data from the European geostationary satellite – Meteosat the Second Generation (MSG). Results show there is a higher correlation (R^2 = 0.97) and a better agreement (Nash-Sutcliffe efficiency coefficient = 0.97) between simulated and ground-measured sensible heat flux. Additionally, it has a root mean square error of 17.91 W m^{-2} in predicting sensible heat flux (H). After having obtained H, the energy balance equation can be employed to compute evapotranspiration.

Key words sensible heat flux; diurnal model; MSG; Fourier analysis

INTRODUCTION

Analysis of daily evapotranspiration (ET) is of significant importance to the agriculture, hydrological cycle and the regional climate, which greatly influence people's routine life. Retrieval of surface parameters, e.g. soil heat flux (G) and sensible heat flux (H), based upon the analysis of the daily variation of surface temperature has attracted many studies, but most of these studies need calculation of heat transfer resistances (Idso *et al.*, 1975, 1977; Seguia & Ltier, 1983; Soer, 1980; Carlson *et al.*, 1981; Price, 1982). Abdellaoui *et al.* (1982, 1986) and Raffy & Becker (1985) proposed methods for estimating the daily ET process without calculating resistances based on a Fourier analysis of the time variation of surface temperature. However, all of those studies required *in situ* ground meteorological data as inputs, and the many unknowns limited their further application.

The objective of this paper is to present a method for daily ET process retrieval based on the Fourier analysis of time variation of land surface temperature (LST). Its main advantages are requiring no *in situ* ground meteorological data as inputs, and having few unknowns to be calibrated.

METHODOLOGY

Sensible heat flux retrieval

We developed a daily sensible heat flux process retrieving model based on the data produced by a SVAT model – the SimSphere model (http://www.agry.purdue.edu/climate/dev/simsphere.asp). From analysis of observations of LST and H, we know they are period functions in daytime (i.e. from sunrise to sunset), and therefore sequences of LST and H in daytime can be fitted using the second-order Fourier series. Then the Fourier coefficients for LST and H (PLST and PH) were obtained. By analysing correlations between PLST and PH under the influence of ground albedo (GA), the models in Table 1 were developed.

So, one can first fit the Fourier coefficients P_{LST} based on the day LST process over time retrieved from MSG/SEVIRI. Then, P_H can be calculated using correlations between P_{LST} and P_H (Table 1). Finally, the daily H process can be retrieved using Fourier series.

Table 1 Correlations between P_{LST} and P_H influenced by ground albedo.

Parameter	Correlation between P_{LST} and P_H	Correlation coefficient, R^2	No.
GA	$P_{H,v1} = [a_{v1}(P_{LST,v1})^2 + b_{v1}P_{LST,v1} + c_{v1}]*\exp(GA)$	0.9638	(1)
	$P_{H,q1} = [a_{q1}(P_{LST,v1})^2 + b_{q1}P_{LST,v1} + c_{q1}]*\exp(GA)/\text{acos}(GA)$	0.9679	(2)
	$P_{H,v2} = [a_{v2}(P_{LST,v2})^2 + b_{v2}P_{LST,v2} + c_{v2}]*\exp(GA)$	0.9749	(3)
	$P_{H,q2} = [a_{q2}(P_{LST,v2})^2 + b_{q2}P_{LST,v2} + c_{q2}]*\exp(GA)/\text{acos}(GA)$	0.9782	(4)

a_{vi}, a_{qi}, b_{vi}, b_{qi}, c_{vi}, c_{qi} (i = 1,2) are coefficients that should be calibrated before these relationships are used.

LST retrieval

Wan & Dozier (1989, 1996) proposed a view-angle dependent split-window algorithm for retrieval of LST (equation (1)). This algorithm was applied by Jiang *et al.* (2006) to LST retrieval from MSG/SEVIRI data:

$$T_s = a_0 + (a_1 + a_2\frac{1-\varepsilon}{\varepsilon} + a_3\frac{\Delta\varepsilon}{\varepsilon^2})\frac{T_i + T_j}{2} + (a_4 + a_5\frac{1-\varepsilon}{\varepsilon} + a_6\frac{\Delta\varepsilon}{\varepsilon^2})\frac{T_i - T_j}{2} \tag{1}$$

where: $\varepsilon = \frac{\varepsilon_i + \varepsilon_j}{2}$, $\Delta\varepsilon = \varepsilon_i - \varepsilon_j$. T_i, T_j are the brightness temperature at top of atmosphere layer, measured through channels *i* and *j*, respectively, ε is the averaged emissivity, ε_i and ε_j are the emissivities in channels *i* and *j*, respectively, $\Delta\varepsilon$ is the emissivity difference, and a_0–a_6 are the unknown coefficients. Emissivity was retrieved according to Jiang *et al.* (2006).

In this paper, this algorithm was employed to retrieve LST from MSG/SEVIRI data.

Fourier series

From analysis of observations of LST and *H*, we know that they are period functions in the daytime (i.e. from sunrise to sunset), and therefore sequences of LST and *H* in the daytime can be fitted using Fourier Series as indicated in equation (2) (Xiong & Dai, 2006).

$$f(t) = a_0 + \sum_{k=1}^{\infty}[v_k \cos(k\omega t) + q_k \sin(k\omega t)] \tag{2}$$

where, ω ($=2\pi/t$) is the angular velocity for period function $f(t)$; a_0, v_k and q_k are the Fourier coefficients, *k* is the order number of harmonics.

$$a_0 = \frac{1}{\mathrm{T}}\int_0^{t_0+T} f(t)dt \tag{3}$$

$$v_k = \frac{2}{\mathrm{T}}\int_0^{t_0+T} f(t)\cos(k\omega t)dt \tag{4}$$

$$q_k = \frac{2}{\mathrm{T}}\int_0^{t_0+T} f(t)\sin(k\omega t)dt \tag{5}$$

Here, we used the second-order Fourier series (let $a_0 = 0$) to fit daily processes of LST and *H* over time.

DATA SOURCES

Spain is characterized by abundant sunshine and hence very many clear-sky days in this country are suitable for satellite data application, especially in August, the hottest month of the year.

Satellite data

Meteosat the Second Generation (MSG), a geostationary satellite, was jointly developed by the European Space Agency (ESA) and EUMETSAT. Its main payload, the Spinning Enhanced Visible and Infrared Imager (SEVIRI), can provides measurements of the Earth-disc (centred at 0 longitude and 0 latitude) every 15 minutes in 12 spectral channels ranging from visible to thermal infrared (0.35–14.40 μm) at fixed view angles. Its spatial resolution is 3 km for all channels except for the high resolution visible channel (Jiang *et al.*, 2006).

We ordered MSG images of August 2005 for model calibration and validation. Analysis of the cloud mask of MSG images indicated that 5 and 6 August were sunny days and so the images are the best to use for our model.

Ground data

FLUXNET is a global network of micrometeorological tower sites that use eddy covariance methods to measure the exchanges of carbon dioxide, water vapour, and energy between the terrestrial ecosystem and atmosphere. (http://www.fluxnet.ornl.gov). FLUXNET has eight sites in Spain, four of which provided us with 0.5 hour interval observations in the year 2005. These sites are Llano de los Juanes (LJu), Las Majadas del Tietar (LMa), El Saler-Sueca (ES2), El Saler (ES1). After analysis of the quality of satellite data coverage and ground-measured data for these sites, LJu was selected for our model calibration and validation.

LJu, a shrub-land plateau at 1600 m altitude in Sierra de Gádor (Almería, southeast Spain; 36°55′41.7″N; 2°45′1.7″W), is characterized by a semi-arid montane Mediterranean climate, with a mean annual temperature of 12°C and mean annual precipitation of 475 mm, falling mostly during autumn and winter, and a very dry season in summer. The vegetation is bio-diverse and somewhat sparse, with a predominance of two perennial species: *Festuca scariosa* (Lag.) Hackel (19.0%) and *Genista pumila* (Vierh) ssp. pumila (11.5%). The average height of the vegetation is 0.5 m. The soil parent material of the ecosystem is a dark dolomite with a layered distribution and a rock fragment content of 30–40% (Serrano-Ortiz *et al.*, 2007). Its surroundings (Fig. 1) show an even underlying condition, and therefore we have taken its observations as the pixel-averaged ones to calibrate and validate our model.

RESULTS AND DISCUSSION

Satellite data and ground data for LJu on 5 August 2005 were used to calibrate our model and the corresponding data for 6 August 2005 were employed to validate it.

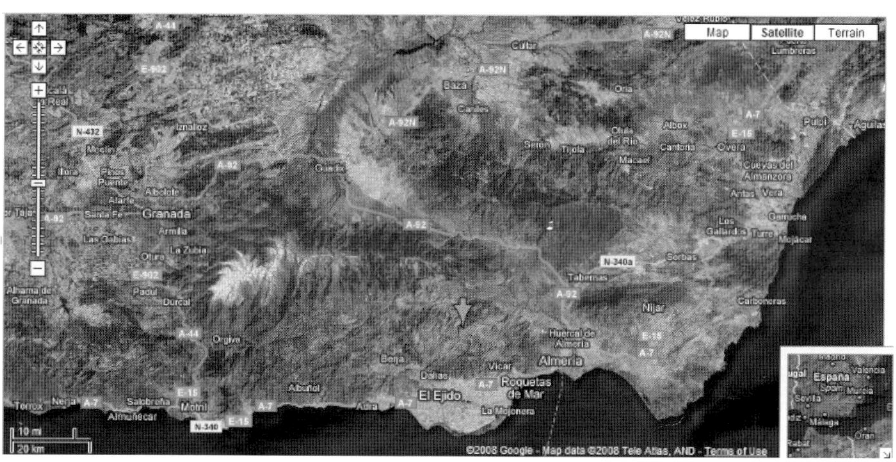

Fig. 1 LJu (arrow pointed) and its surroundings (from Google Earth).

Model calibration

The index of agreement (IA) (Quiring & Papakryiakou, 2003) and Nash-Sutcliffe efficiency coefficient (NSEC) (Nash & Sutcliffe, 1970) were employed to appraise the precision of calibration for the *GA* model in Table 1.

$$IA = 1.0 - \frac{\sum_{i=1}^{N}(O_i - P_i)^2}{\sum_{i=1}^{N}(|P_i - \bar{O}| + |O_i - \bar{O}|)^2} \tag{6}$$

$$NSEC = 1.0 - \frac{\sum_{i=1}^{N}(O_i - P_i)^2}{\sum_{i=1}^{N}(O_i - \bar{O})^2} \tag{7}$$

where, O_i is the observed value, P_i is the model predicted value, $\bar{O}$ is the averaged observation, N is the total number.

IA and *NSEC* are both between 0 and 1, the latter is stricter than the former. Only when all maximum and minimum values fit well, can *NSEC* have a high value.

The correlation coefficient (R^2), *IA* and *NSEC* are all calculated based on the modelled P_H (predicted P_H) and that was fitted through ground measurements of day H (measured P_H), as shown in Table 2.

The model coefficients and precision (Table 2) suggest there is a higher correlation between the predicted P_H and measured P_H ($R^2 \geq 0.80$) and a higher agreement between them ($IA > 0.90$) including a good agreement in the maximal and minimal points ($NSEC \geq 0.80$), as shown in Fig. 2.

Finally, we obtained the *GA* model as equation (8):

$$\begin{aligned} H(t) = {} & [-3.15(P_{LST,v1})^2 - 2.66(P_{LST,v1}) + 4094.44] * \exp(GA) * \cos(\omega t) \\ & + [0.15(P_{LST,v1})^2 - 40.69(P_{LST,v1}) + 1032.19] * \exp(GA) / a\cos(GA) * \sin(\omega t) \\ & + [-0.89(P_{LST,v2})^2 - 89.16(P_{LST,v2}) - 991.68] * \exp(GA) * \cos(2\omega t) \\ & + [-0.28(P_{LST,v2})^2 - 36.48(P_{LST,v2}) - 175.09] * \exp(GA) / a\cos(GA) * \sin(2\omega t) \end{aligned} \tag{8}$$

through which one can retrieve H process over time in a daytime period with Fourier coefficients ($P_{LST,v1}$, $P_{LST,v2}$) fitted from satellite retrieved daily LST process.

Model validation

We use data in LJu on 6 August 2005 to validation the *GA* model (equation (8)). Comparison between the predicted and measured H (Fig. 3) indicates that the *GA* model has a higher simulation precision; it has a higher R^2 (0.97) and there is good agreement between them (*IA* = 0.99, *NSEC* = 0.97). But, at some fluctuation points (for example 09:00, 11:00, 14:00, 14:30, 15:00), the *GA* model failed to agree with the measured data. This indicates that this model may not be suitable for simulation of a dramatically fluctuating flux process.

Table 2 Calibrated model coefficients and precision.

Fourier coeff. of H	a	b	c	Correlation coeff. R^2	IA	NSEC
$P_{H,v1}$	−3.15	−2.66	4094.44	0.80	0.94	0.80
$P_{H,q1}$	0.15	−40.69	1032.19	0.91	0.98	0.91
$P_{H,v2}$	−0.89	−89.16	−991.68	0.89	0.97	0.89
$P_{H,q2}$	−0.28	−36.48	−175.09	0.92	0.98	0.92

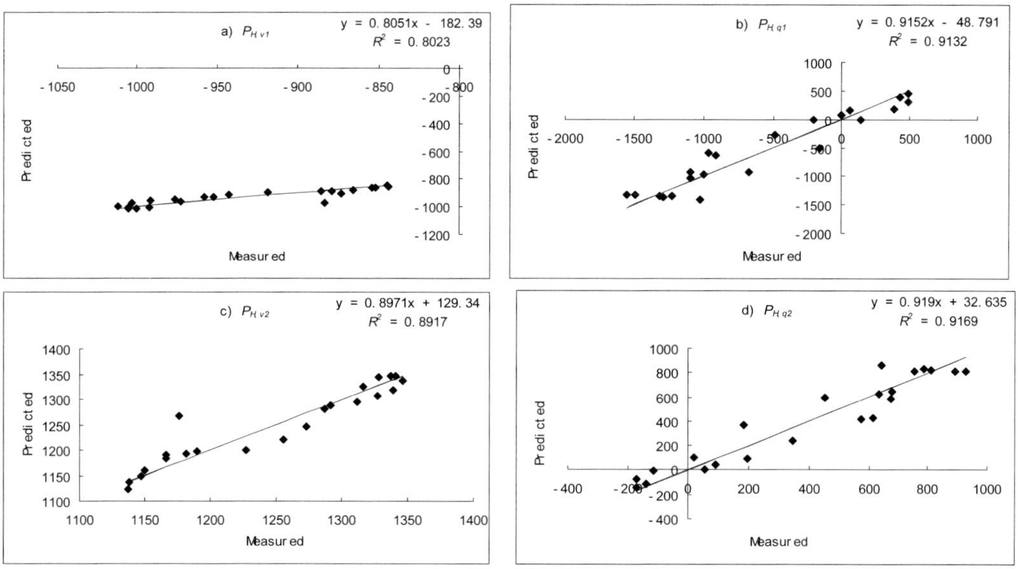

Fig. 2 Comparison between predicted and measured P_H in LJu on 5 August 2005, with the line being 1:1 line.

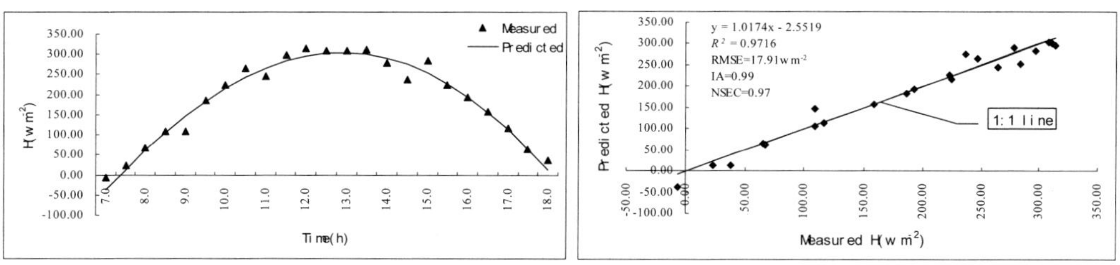

Fig. 3 Comparison between predicted and measured H for LJu on 6 August 2005.

Overall, the *GA* Model in Table 1 has a root mean square error (RMSE) of 17.91 W m^{-2} in predicting the sensible heat flux H in the LJu station of FLUXNET.

Having obtained a precise H, one can use the energy balance equation to retrieve the latent heat flux (LE) with net radiation data from MSG.

CONCLUSIONS

A daily sensible heat flux (H) process retrieval model is presented with support of data in various scenarios and using the second-order Fourier series. In this model (Table 1), four Fourier series coefficients of the sensible heat flux process (P_Hs) are retrieved through Fourier series coefficients of the land surface temperature process (P_{LST}s) under the influence of ground albedo (GA).

This model was calibrated and validated in Spain. The results imply that the model-retrieved sensible heat flux (H) has a high correlation (R^2 = 0.97) and a good agreement (Nash & Sutcliffe efficiency coefficient of 0.97) with the ground-measured sensible heat flux. Overall, it has a RMSE of 17.91 W m^{-2} in predicting H.

In this model, only the influence of ground albedo is considered. In practice, there are many factors impacting correlations between sensible heat flux and land surface temperature processes. From this perspective, there unavoidably exist some uncertainties in the GA model, to lessen which much more attention should be paid to other impacting factors in future. Also, in future, the model needs to be verified in more research areas for longer periods in order to overcome these uncertainties.

Acknowledgements This study was supported by Key Project of International Cooperation of the Natural Science Foundation of China (no. 40721140020), Key Project of the Natural Science Foundation of China (no. 50839005) and Project of the Research Initiative Foundation of Sun Yat-sen University for Young Teachers. Many acknowledgements to Penelope Serrano-Ortiz and Sergio Contreras from Spain. They provided valuable observations for our model.

REFERENCES

Abdellaoui, A., Becker, F., & Olory-Hechinger, E. & Raffy, M. (1982) Thermal inertia and soil moisture mapping. In: *Final Report of the Group Agromet Monitoring Project (GAMP)*, WG7 of Earsel, Chap. 6. ESA Contract no. 4660-81-D-IM-SC: 1-59.

Abdellaoui, A., Becker, F., & Olory-Hechinger, E. (1986) Use of Meteosat for mapping thermal inertia and evapotranspiration over a limited region of Mali. *J. Climate & Appl. Met.* **25**(11), 1489–1506.

Carlson, T. N., Dodd, J. K., Benjamin S. G. & Cooper, J. N. (1981) Satellite estimation of the surface energy balance moisture availability and thermal inertia. *J. Appl. Met.* **20**, 67–87.

Idso, S. B., Schmugge, T. J., Jackson, R. D. & Reginato, R. J. (1975) The utility of surface temperature measurements for the remote sensing of surface soil water status. *J. Geophys. Res.* **80**, 3044–3049.

Idso, S. B., Reginato, R. J. & Jackson, R. D. (1977) An equation for potential evaporation from soil, water and crop surfaces adaptable to use by remote sensing. *Geophys. Res. Lett.* **4**, 187–188.

Jiang, G. M. & Li, Z. L. (2006) Land surface temperature (LST) retrieval from MSG-SEVIRI data and comparisons with LST retrieved from AATSR and MODIS data. In: *Proceedings of the Second Recent Advances in Quantitative Remote Sensing Conference* (ed. by Jose A. Sobrino), 782–787. Publications de la Universitat de Valencia, Spain. ISBN: 84-370-6533-X; 978-84-370-6533-5.

Jiang, G. M., Li, Z. L. & Nerry, F. (2006) Land surface emissivity retrieval from combinged mid-infrared and thermal infrared data of MSG-SEVIRI. *Remote Sens. Environ.* **105**, 326–340.

Nash, J. E. & Sutcliffe, J. V. (1970) River flow forecasting through conceptual models. *J. Hydrol.* **10**, 282–290.

Price, J. C. (1982) Estimation of regional scale evaportranspiration through analysis of thermal infrared data. *IEEE Geos. Remote Sensing* **20**, 286–292.

Quiring, S. M. & Papakryiakou, T. N. (2003) An evaluation of agricultural drought indices for the Canadian prairies. *Agric. Forest Met.* **118**, 49–62.

Raffy, M. & Becker, F. (1985) An inverse problem occurring in remote sensing in the thermal infrared bands and its solutions. *J. Geophys. Res.* **90**, 5809–5819.

Seguin, B. & Itier, B. (1983) Using midday surface temperature to estimate daily evaporation from satellite thermal IR data. *Int. J. Remote Sens.* **4**(2), 371–383.

Serrano-Ortiz, P., Kowalski, A. S., Domingo, F., *et al.* (2007) Variations in daytime net carbon and water exchange in a montane shrubland ecosystem in southeast Spain. *Photosynthetica* **45**(1), 30–35.

Soer, G. J. R. (1980) Estimation of regional evapotranspiration and soil moisture conditions using remotely sensed crop surface temperature. *Remote Sens. Environ.* **9**, 27–45.

Wan, Z. & Dozier, J. (1989) Land-surface temperature measurement from space: physical principles and inverse modeling. *IEEE Trans. Geosci. Remote Sens.* **27**(3), 268–278.

Wan, Z. & Dozier, J. (1996) A generalized split-window algorithm for retrieving land-surface temperature from space. *IEEE Trans. Geosci. Remote Sens.* **34**(4), 892–905.

Xiong, D. H. & Dai, X. P. (2006) Realization of Fourier series approximate calculation and simulation. *Computer & Modernization* **5**, 15–17.

GIS assisted flash flood risk zonation for small ungauged basins in Urumchi, western China

PING XIE[1], GUANGCAI CHEN[2] & DE LI[3]

1 *State Key Laboratory of Water Resources and Hydropower Engineering Science, Wuhan University, Wuhan, Hubei Province 430072, China*
pxie@whu.edu.cn

2 *Water Resources Department of Changjiang River Scientific Research Institute, Wuhan 430010, China*

3 *Department of Hydropower and Hydrology planning, Guangxi Electric Power industry Investigation Design and Research Institute, Nanning 530023, China*

Abstract Flash floods create hazardous situations for people and extensive damage to property. It is critical to warn of flash flood conditions in a timely manner to minimize impacts, especially in arid ungauged areas like Urumchi. Of a total of 18 small basins, there are only six gauged with more than 30 years of hydrological data. In this paper, flash flood risk zonation was studied for flash flood risk warning and control. The critical rainfall method was used to make flash flood warnings. In order to carry out flash flood zoning, the modified rational formula method was adopted to calculate the design flash flood of ungauged basins first. The Genetic Algorithm approach was used to search for the parameters (n_1, n_2, u, m) in gauged basins. The geostatistical analysis module in ArcGIS was used to map storm parameters n_1/n_2. Regression analysis was used to obtain the empirical relationship between u and θ, and that between m and θ. Finally, in ungauged basins, by using the map to confirm n_1/n_2, and by using the empirical formula to get m and u, the design flash flood can be calculated with the rational formula. With the Manning formula and GIS, the risk map can be prepared. The predicted 24-h critical rainfall in Urumchi varies in the range 13–25 mm. The empirical relationship between u and θ and that between m and θ are: $m = 0.1486\ln\theta + 0.4952$; $u = -23.3425\ln(\ln\theta) + 43.7455$.

Key words Urumchi; flash flood; critical rainfall; rational formula method; GIS

INTRODUCTION

The Urumchi region is located in the central part of Xinjiang, China, next to northern slope of Tianshan Mountain and southern margin of Junggar basin. It is surrounded by mountains in three directions (east, west, south) and a vast flood plain to the north. The geological background of this region is very complicated and the landform types are various. The terrain is high in the south and low in the north. Moving from higher to lower, there are various topographic units such as the alpine zone characterized by ice, snow and periglacial action, a middle zone characterized by normal erosion and arid denudation, a lower zone characterized by arid and semi-arid denudation, including low mountain and hilly regions, mountain down-faulted basins, mountain piedmont alluvial-pluvial fans, and the Urumchi River piedmont plain.

Flash flooding triggered by rainstorms is one of the major natural disasters occurring in these mountainous areas where the vegetation coverage is low. Because flash floods often occur suddenly, with little time left for warning, and move fast and violently, they are a big threat to life and can cause severe damage to property and infrastructure (Montza & Gruntfest, 2002). Because flash flooding usually occurs in mountainous areas where it is difficult to observe and measure flows, there are few, or even, no observed data. It is a typical PUBs (Prediction in Ungauged Basins) problem.

Now, many important international organizations and plans, such as the International Human Program, World Climate Research Programme (WCRP), International Hydrology Program (IHP), International Geosphere-Biosphere Program (IGBP) have paid attention to the meteorology and hydrology workings of ungauged basin (Li *et al.*, 1996; IGPB, WCRD & IHDP, 2001; Xia *et al.*, 2002). Prediction in Ungauged Basins (PUB) was also one of the main topics of the International Association of Hydrological Science (IAHS) meetings held in Brazil (October, 2002) and Japan (March, 2003), and receives much attention from hydrologists in China. In order to solve the flood prediction problems in middle- or small-sized ungauged basin in the 1980s, some national projects were carried out in China. After analysing about 1500 basins, and 23 000 storm and flood events, many achievements were made (MWR, 1995). But the samples of storms and floods from the 1980s are now insufficient. Human activities have affected the underlying basin surface greatly, especially in the last 20 years. Those earlier achievements cannot fulfil the present requirements of flood calculation. So study of flash flood zonation in ungauged areas is of importance theoretically and has real practical significance. Rainfall zonation and flash flood calculation are the key parts of flash flood prevention programmes.

Nowadays there are two kinds of methods for flash flood warning: long-time prediction and real-time prediction. The long-time prediction method is mainly based on Florsid's theory. It is used to predict the generation periods and frequency of flash floods. It can be used to predict the trend of flash floods. The real-time prediction method usually adopts hydrometeorological models and runoff models. Real-time prediction can be classified into three kinds: first is the critical rainfall method; second is the rainfall analysis method; third is the manual observation and equipment observation method. The prediction models include statistical regression modelling and physical models. The technologies include statistics, grey system theory, ANN and 3S (GPS, GIS, RS) (Zhou *et al.*, 2001; Sun *et al.*, 2004). Because the critical rainfall method is easy and intuitive, it was adopted in this paper.

The purpose of hazard zoning of flash flooding is to estimate the inundated and impacted areas. The internationally widely-used method is the torrent classification and integrated index hazard-zone mapping method that was proposed by Aulitzky (1973), Wang (2000). The hazard-zone is classified as red, yellow and white based on the characteristics of the hazard and its grade. In this way, the flash flood can be predicted and warnings can be issued. Eldeen (1980) adopted hazard zone maps to judge the disaster type and its scale. The flood area is divided into four hazard zones based on the hazard grades. Hazard zone grading and mapping have played an important role in large-scale flash flood zonation. It has been widely used in disaster insurance, disaster prevention and disaster recourse (Zhou *et al.*, 2001; Sun *et al*, 2004).

The objective of this study is to: (1) identify the critical rainfall in the research area, including the gauged and ungauged areas; (2) calculate the flash flood runoff in ungauged basins; and (3) estimate the inundated area and map the hazard zone.

DATA

Urumchi Flash Flood Control Planning includes BanFangGou, HeiGou, etc., i.e. about 20 basins in total. There are only six basins with more than 30 years precipitation and runoff data; the others are ungauged basins. The basic properties of the basins, including basin area F, river length L and channel slope J, are listed in Table 1.

Table 1 The basic properties of basins in Urumchi region

Basin name	F (km^2)	L (km)	J (‰)	Data	Basin name	F (km^2)	L (km)	J (‰)	Data
BanFangGou	396	40.6	70.4	Yes	DongBaiYang	20.14	7.3	107	No
HeiGou	172	24	98	Yes	MoShiKe	19.7	9.3	124.4	No
SanGeShan	132	21	100	Yes	ZaiErDe	131.4	20	851	No
AKeShu	351	29.3	55.3	Yes	ShiRenZi	32.1	12.1	15.6	No
GaoYaZi	331	38.4	44	Yes	ShuiGouZi	15.4	7.5	79.8	No
ALaGou	1842	100	24.3	Yes	DaShiTou	40.9	15.1	19.1	No
SheYaoGou	11	8.7	57.5	No	WuShenCheng	234	20.4	42.6	No
BoErQinGou	72.8	13	973	No	MiaoErGou	15.4	7.5	79.8	No
KaLaGaYi	21	8	175	No	XiBaiYang	70	22.4	124.5	No
DongNanGou	67.1	14	56.2	No	HaXiongGou	112.9	16.5	822	No

METHODS AND CALCULATION RESULTS

Figure 1 charts the procedures used in this project.

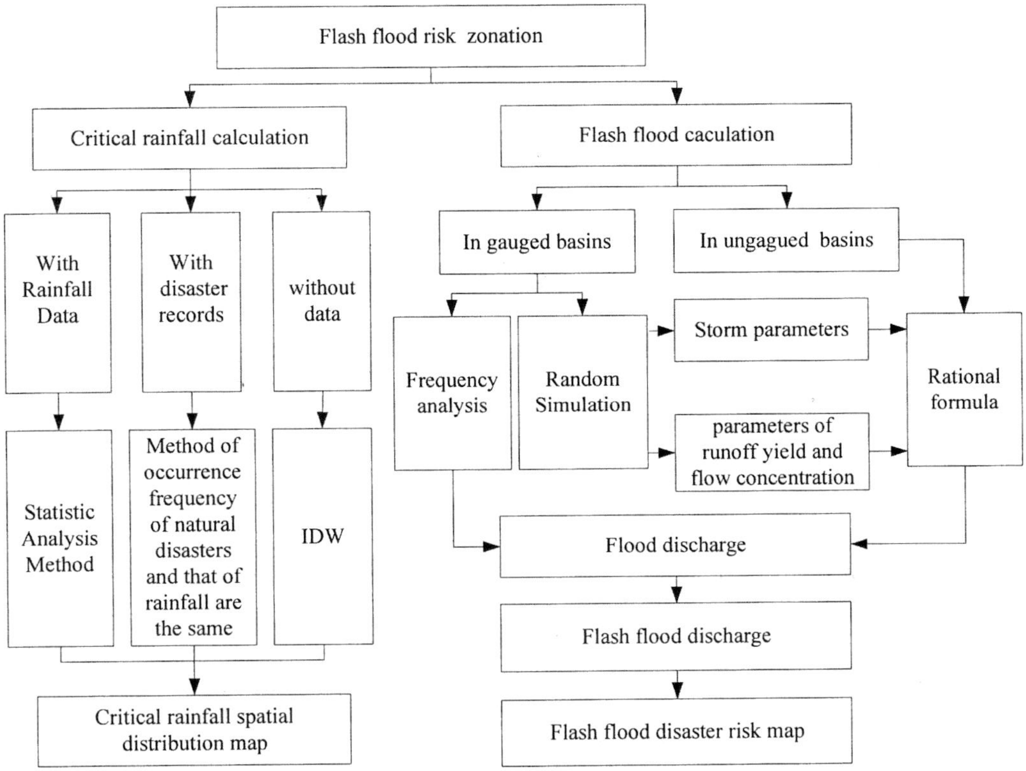

Fig. 1 Flash flood zoning flow chart.

Critical rainfall calculation

The critical rainfall method is used for flash flood warning. In a basin or area, when the amount of precipitation reaches or exceeds a certain magnitude or intensity, flash flooding will happen. Precipitation of such an amount or intensity is called the critical rainfall. There are three approaches to estimating the critical rainfall: (1) analysis of precipitation data directly in the basin with raingauges; (2) using in-basin disaster records; it is supposed that the occurrence frequencies of natural disasters and rainfall are the same, and the critical rainfall is validated by surveying the occurrence frequency of disasters; (3) the critical rainfall is estimated by spatial interpolation in basins having no data.

In basins with raingauges Collect the rainfall data and flash flood record, analyse the daily rainfall data with statistical methods to obtain the maximum daily rainfall and determine the rainfall amount that has caused flash floods. The critical rainfall is the minimum value of the maximum daily rainfall series. Daily rainfall data recorded by the meteorological department and hydrological department is rainfall in a fixed time period from 8:00 am to 20:00 pm. Generally it is lower than self-recording rainfall data. Therefore, the maximum daily rainfall from meteorological department and hydrological department records should further be multiplied with a coefficient K, usually $K = 1.1$, to obtain the critical rainfall.

In this study, daily rainfall data from 1958 to 2000 at four rainfall stations in Urumchi were collected, namely Urumchi, DaXiGou, XiaoQuZi, DaBanCheng. The daily rainfall data were analysed to get the maximum daily rainfall and determine the rainfall amount that has caused flash floods. The calculated 24-h critical rainfall results are as follows: Urumchi: 17.38 mm, DaXiGou: is 23.54 mm, XiaoQuZi: 24.97 mm, and DaBanCheng: 13.42 mm.

In basins with disaster records Through investigation to identify the occurrence n of flash flood disasters in the past, the flood occurrence frequency can be calculated as follows:

$$p = \frac{n}{N+1} \tag{1}$$

where p is flash flood frequency, n number of flash flood occurrences, and N is the number of years.

After p is calculated, then check the related maps of probable maximum storm to get the parameters for the annual maximum series of 24-h rainfall, namely the mean ($\overline{x}$), coefficient of variation (C_v) and skewness coefficient (C_s). The precipitation series are assumed to follow a Pearson III distribution. Use the three parameters ($\overline{x}$, C_v, C_s), and frequency model, the design point rainfall value R_p corresponding to frequency p can be obtained. In order to get the areal mean precipitation, check the map of "point to area converting coefficient" to get the converting coefficient α. Then the critical rainfall (CR) is calculated as follows: $CR = \alpha \times R_p$.

Historical investigation of flash floods has shown that there are 15 flash flood records in ALaGou basin during the 42 years from 1959 to 2000. So the calculated occurrence frequency of flash flood is 34.9%. Checking the maps of probable maximum storm in Xinjiang, we can get the storm parameters: $\overline{x}$ = 18.6 mm, C_v = 0.615, C_s = 3.5C_v. With the frequency P = 34.9%, the design 24-h point precipitation is 18.8 mm. In order to determine the areal mean precipitation, check the maps of probable maximum storm in Xinjiang to get the conversion coefficient: α = 0.876. So the design 24-h areal mean precipitation at P = 34.9% (critical rainfall) is 16.5 mm.

In basins without data When the critical rainfall values for the Urumchi, DaXiGou, XiaoQuZi, DaBanCheng and ALaGou station were obtained, the Inverse Distance Weighted interpolation method (IDW) was adopted to map the 24-h critical rainfall in Urumchi (Fig. 2). Critical rainfall in ungauged areas can be identified by checking the map. The values of 24-h critical rainfall of different basins in Urumchi are given in Table 2.

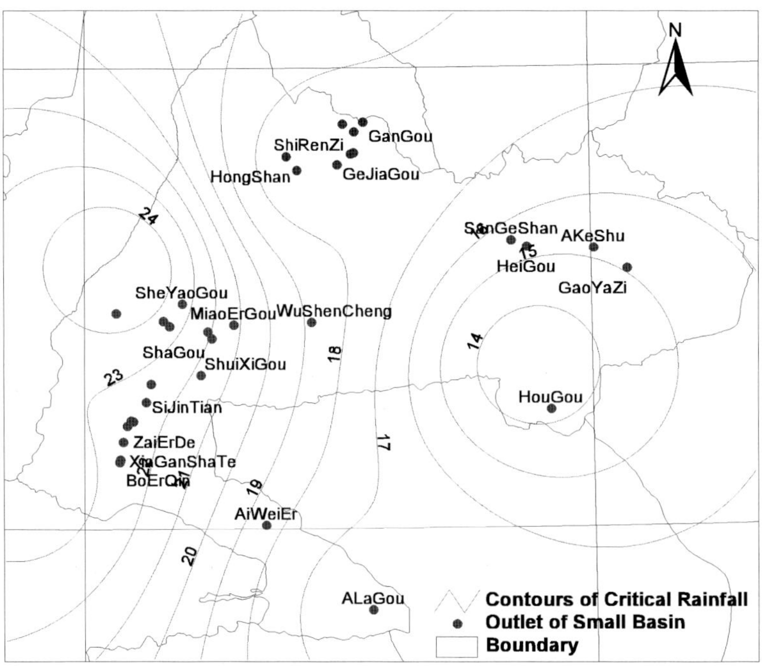

Fig. 2 Spatial interpolation of 24 h critical rainfall in Urumchi.

Table 2 24 h critical rainfall (CR_{24h}) of basins in Urumchi (unit: mm).

Basin name	CR_{24h}	Basin name	CR_{24h}	Basin name	CR_{24h}
BanFangGou	21.91	BoErQin	22.63	ShuiGouZi	17.63
HeiGou	15.33	KaLaGaYi	22.63	DaShiTou	17.54
SanGeShan	15.58	DongNanGou	22.61	WuShenCheng	18.82
AKeSu	15.22	DongBaiYang	23.39	MiaoErGou	21.21
ALaGou	16.5	MoSiKe	22.52	XiBaiYang	23.65
GaoYaZi	15.05	ZaiErDe	22.54	HaXiongGou	22.48
SheYaoGou	23.43	ShiRenZi	17.54		

Flash flood calculation in ungauged small basins

The purpose of hazard zoning of flash flooding is to estimate the flooded and impacted area, namely hazard zoning of flash flooding. It can be divided into two steps, namely runoff calculation and hazard regionalization.

Runoff calculation The most important task in hazard zoning of flash flooding is to estimate the submerged area. And the key issue in ungauged small basins is design flood calculation. The calculation methods include the rational method, empirical formula method, analogous watersheds method, historical flood investigation and analysis method, integrated unit hydrographs, and hydrological models. To some degree, the rational formula method (Chen *et al.*, 1983; Zhan *et al.*, 2000) can reflect the flood mechanism and its parameters can be empirically calibrated. The rational formula expresses the empirical relationship between runoff yield factors and runoff concentration factors. So the formula may be reasonable and practical to some extent.

Use the modified rational formula to calculate the design flood peak Q_{mp} as follows (Chen *et al.*, 1983; Zhan *et al.*, 2000):

$$Q_{mp} = \begin{cases} 0.278\left(\dfrac{S_p}{\tau^n} - \mu\right)F, & t_c \geq \tau \qquad (2) \\[2ex] 0.278\left(\dfrac{S_p t_c^{1-n} - \mu t_c}{\tau}\right)F, & t_c < \tau \qquad (3) \end{cases}$$

$$\tau = \frac{0.278L}{mJ^{1/3}Q_{mp}^{1/4}} \qquad (4)$$

where watershed characteristics parameters include: area F, main channel length L, main channel slope J; storm characteristics parameters: S_P is the average precipitation intensity over 1 h, n is the rainstorm attenuation index (in most areas n changes after t = 1 h, when t < 1 h, n is n_1; when $1 < t < 24$, n is n_2); runoff yield and concentration parameters: empirical yield loss parameter μ, empirical concentration parameter m; time parameters: yield time t_c, concentration time τ.

The procedure for the design flood peak Q_{mp} calculation involves the following steps:

(1) Determine the geometric characteristics (F, L and J) of the watershed from the topography map.
(2) Determine the 24-h design storm parameters ($\overline{x}$, C_v, C_s), parameter n (or n_1, n_2) in storm formula, and S_p.
(3) Determine the empirical yield loss parameter μ and empirical concentration parameter m based on their geographical distribution law.

Generally in an ungauged area, the geometric characteristics (F, L and J) of the watershed can be easily obtained, but the storm parameter n (n_1 or n_2), empirical yield loss parameter μ and empirical concentration parameter m can not be easily obtained. In order to continue with hazard zoning of flash flooding, the Genetic Algorithm method is used to search for the rainstorm parameters (n_1, n_2) and the flow concentration empirical parameters (u, m) in gauged areas. The relationship between parameters m and θ, and that between parameters u and θ is analysed to get the empirical formula. Then with the help of spatial extension and geostatic extension in ArcGIS, the spatial distribution of the storm parameter can be obtained. Finally, in ungauged basins, get n_1 and n_2 by using the map, and get m and u by using the empirical formula, so then the design flash flood can be calculated with the rational formula.

Flash flood discharge calculated in gauged area

After analysing the reliability, consistency and representativeness of the storm series and flood series in BanFangGou, HeiGou, SanGeShan, ALaGou, GaoYaZi and AKeShu, it was assumed that the series satisfied the flood calculation criteria and can be used to calculate the design storm and design flood. The P-III distribution curve was used to calculate the design storm and design flood. The statistic parameters and calculated results are listed in Table 3.

Table 3 The calculation results of design flash flood in gauged basins.

Basin name	Annual maximum 24-h precipitation parameter			Annual maximum flood peak discharge parameter			Design flood peak (m^3/s)	
	$\overline{x}$ (mm)	C_v	C_s	$\overline{x}$ (m^3/s)	C_v	C_s	P=1%	P=10%
BanFangGou	30.00	0.60	2.10	25.0	1.40	6.30	189.7	43.6
HeiGou	25.79	0.53	1.86	24.9	1.05	4.20	140.4	50.0
SanGeShan	26.06	0.53	1.86	32.2	1.10	4.40	190.3	64.8
AKeSu	26.32	0.52	1.84	55.1	0.83	3.32	245.6	106.7
GaoYaZi	25.63	0.53	1.87	58.43	0.70	2.53	216.0	109.3
AlaGou	18.58	0.61	2.15	92.5	1.20	4.80	596.4	185.4

Spatial distribution of the yield and concentration parameters

(1) Design storm and design flood random samples generated in gauged area

Because the statistical parameters of both the maximum 24-h storms and the annual maximum flood peak discharges in six gauged basins are known (Table 3), their P-III population distributions are known. It is assumed that p obeys a uniform distribution, so it was just necessary to simulate random series p in [0, 1]. Then with the P-III distribution, the random storm samples and design flood samples were obtained.

(2) Parameters of rational formula

Generate 1000 groups of random design storm and design flood samples for each basin by the linear congruence method (Xu, 1995). With the rational formulae (2), (3), (4) and Genetic Algorithm, the optimal parameters were found using following objective function:

$$\min f(x_k, p_i) = \sum_{i=1}^{N} \frac{|Q_0(x_k, p_i) - Q(x_k, p_i)|}{Q_0(x_k, p_i)} \tag{6}$$

where $Q_0(x_k, p_i)$ is the randomly generated design flash flood at some frequency p_i (direct method); $Q(x_k, p_i)$ the flash flood calculated by the rational formula method at frequency p_i (indirect method); p_i is the random simulation frequency; x_k is the parameters (n_1, n_2, μ, m) to be optimized; N is simulation times; k is the number of parameters needing optimisation.

The basin characteristic parameter θ can be calculated as follows:

$$\theta = L / J^{1/3} / F^{1/4} \tag{7}$$

Generate 1000 groups of design storm and design flash floods again to validate the optimized parameters in the flash flood calculation.

The results are given in Table 4, which shows that the optimized parameters demonstrate a higher efficiency, which means that the rational formula is suitable in Urumchi.

Table 4 The parameter calibration and verification of rational formula in gauged basins.

Basin name	n_1	n_2	μ	θ	m	R_c (%)	R_v (%)
BanFangGou	0.57	0.62	14.51	22.04	1.03	91	82
HeiGou	0.47	0.71	21	14.37	0.91	97	94.5
SanGeShan	0.46	0.76	24.99	13.35	0.86	97.1	96.7
AKeSu	0.56	0.72	17.5	17.77	0.85	90	90
GaoYaZi	0.55	0.76	14.80	25.44	1.0	79	79
ALaGou	0.41	0.8	13.97	52.70	1.06	97	97

R_c is the calibration efficiency coefficient; R_v is the validation efficiency coefficient, the definition of R_c and R_v can be found in Kachroo (1992).R_c is the calibration efficiency coefficient; R_v is the validation efficiency coefficient, the definition of R_c and R_v can be found in Kachroo (1992).

(3) Identification of the spatial distribution of yield and concentration parameters

With linear regression analysis, the relationship between the yield parameter m and basin characteristic θ, and the relationship between empirical yield loss parameter μ and characteristic θ are established as follows:

$$m = 0.1486 \ln\theta + 0.4952 \qquad R_m = 0.827 \tag{8}$$

$$\mu = -23.3425\ln(\ln\theta) + 43.7455 \qquad R_\mu = 0.825 \tag{9}$$

where R_m is the correlation coefficient of the regression equation (8), R_μ is the correlation coefficient of regression equation (9). Setting the significance level α at 0.05, the correlation test shows that m and θ, μ and θ are correlated well in Urumchi area.

(a)

(b)

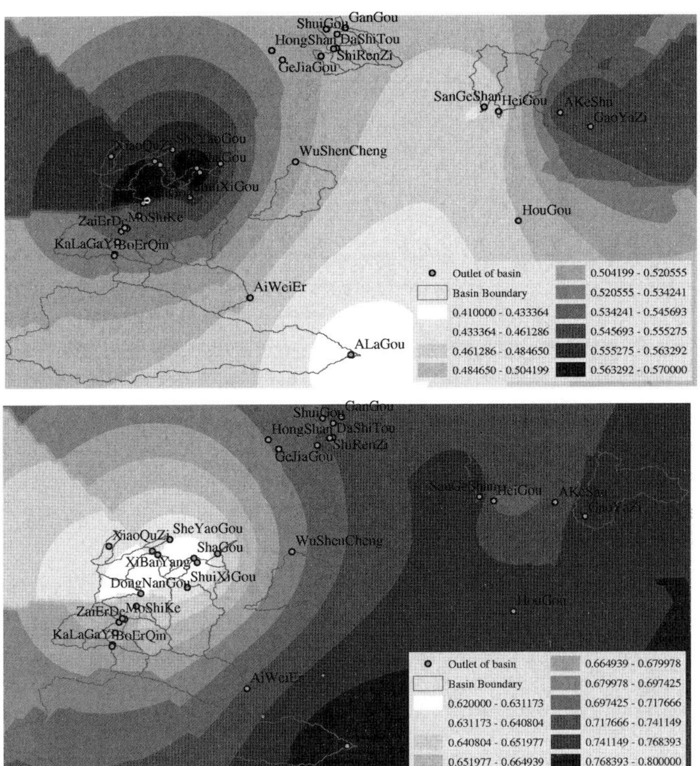

Fig. 3 (a) The spatial distribution map of n_1. (b) The spatial distribution map of n_2.

Spatial distribution of storm parameters

In the geostatic module of ArcGIS (McCoy & Johnston, 2001), at least 10 samples are needed. But there are only six samples in the Urumchi area. In order to satisfy the kriging interpolation requirement in ArcGIS, a spatial interpolation module (McCoy & Johnston, 2001) was first used to generate the samples. Then with ordinary kriging, spatial distribution maps of storm parameters were obtained as in Fig 3(a) and (b). The error variance of the kriging estimates is near 0, which shows a good interpolation precision.

Full test of calculation accuracy for rational formula method in gauged basins

In order to test the calculation accuracy of the rational formula method in gauged basins, first the design storm and the design flash flood are calculated by frequency analysis. The results can be seen as the "direct method" calculation results. Then calculating the basin characteristic parameter θ, with equations (8) and (9), m and μ can be obtained. Check the storm parameters distribution maps in Figs 2 and 3 to obtain storm parameters n_1/n_2. Use the rational formula method to calculate the flash flood peak discharge. The results can be seen as the "indirect method" calculation results. Compare the "direct" and "indirect method" calculation results (Table 5). The results show that flood peak discharge calculated by the "direct method" and that by the "indirect method" are very close. The largest relative error is less than 20%. It shows that the rational formula method has high accuracy and can be used to calculate the flash flood in ungauged areas.

Results of flash flood calculation in ungauged basins

In ungauged basins, check the maps of probable maximum precipitation to get the storm parameters $\bar{x}$, C_v, C_s and "point to area" converting coefficient β. Calculate the design area

Table 5 The whole test of calculation accuracy for rational formula method in gauged basins.

Basin name	Design peak flow, direct method (m^3/s)		Design peak flow indirect method (m^3/s)		Relative error (%)	
	P=1%	P=10%	P=1%	P=10%	P=1%	P=10%
BanFangGou	189.7	43.6	173.6	49.2	8.49	12.84
HeiGou	140.4	50	151.6	56.6	7.98	13.20
SanGeShan	190.3	64.8	178.5	70.4	6.20	8.64
AKeSu	245.6	106.7	260.8	99.3	6.19	6.94
GaoYaZi	216	109.3	234.8	93	8.70	14.91
AlaGou	596.4	185.4	592	218.8	0.74	18.02

Table 6 The calculation of design flash flood in ungauged basins.

Basin Name	θ	m	u	Storm parameters		Design peak flow (m^3/s)	
				n_1	n_2	P=1%	P=10%
SheYaoGou	12.38	0.869	22.207	0.562	0.632	8.5	2.5
BoErQin	4.49	0.718	34.252	0.529	0.667	63	19.5
KaLaGaYi	6.68	0.777	28.774	0.53	0.666	38.4	11.9
DongNanGou	12.77	0.874	21.921	0.566	0.626	49.9	14.3
DongBaiYang	7.26	0.790	27.772	0.56	0.632	26.2	7.6
MoSiKe	8.84	0.819	25.562	0.539	0.655	26	7.9
ZaiErDe	6.23	0.767	29.647	0.538	0.657	84.5	25.7
ShiRenZi	20.34	0.943	18.003	0.503	0.719	39.2	15
ShuiGouZi	8.79	0.818	25.623	0.503	0.719	36.2	15.1
DaShiTou	22.34	0.957	17.288	0.506	0.713	38.4	15.8
WuShenCheng	14.93	0.897	20.531	0.509	0.688	164.2	55.3
MianErGou	8.79	0.818	25.623	0.556	0.636	17.1	5
XiBaiYang	7.2	0.789	27.870	0.559	0.634	31.4	9.1
HaXiongGou	5.4	0.746	31.547	0.539	0.655	77.2	23.4

precipitation and S_p. Use the map (Fig. 3(a) and (b)) to validate n_1/n_2, use the empirical formula to get m and u, then the design flash flood can be calculated with the rational formula. The results are listed in Table 6.

Hazard zoning Hazard zoning is mainly used to determine the submerged area. With the river cross-section map and hydraulic formula, the water level can be calculated. Then with the topographic map, the submerged area can be identified.

Natural river discharge Q can be express with Manning formula (Wang *et al.*, 2000):

$$Q = A \cdot V_c \tag{10}$$

where Q is discharge capacity (m^3/s), A is the cross-section area (m^2), V_c is average flow velocity of the cross section (m/s), V_c is calculated as follows:

$$V_c = (1/an) \cdot R^{2/3} \cdot J^m \tag{11}$$

$$R = \frac{A}{\chi} \tag{12}$$

$$A = A_1 + A_2 + \ldots + A_i \tag{13}$$

$$\chi = \chi_1 + \chi_2 + \ldots + \chi_i \tag{14}$$

where n is Manning's roughness coefficient; m is an index, m is 1/10 when flash flood; α is the resistance coefficient, $\alpha = (\varphi \cdot r_s + 1)^{1/2}$, φ is correction coefficient, $\varphi = (r_c - r_w)/(r_s - r_c)$; r_c is the flash flood unit weight (t/m^3), 1.1; r_i is sand unit weight (t/m^3), 2.56; r_w is the water unit weight, 1.0 (t/m^3); R is hydraulic radius of the cross-section; χ is wetted perimeter; I is the number of observation points; J is the hydraulic slope and generally is the channel longitudinal slope.

After calculating Q, a trial method is used to estimate the water level of the cross-section. The procedure is as follows: First assume a water level h, calculate Q_c with equation (10); then, compare Q_c with the Q, if the error between Q_c and Q can be accepted then stop, else change the water level h, repeat the calculation and comparison until the error between Q and Q_c is acceptable.

The trial method was used to estimate the water level of the cross-section. The results are listed in Table 7. Using the topographic map, the submerged area was obtained (Fig. 4). From Fig. 4 we found that the most hazardous areas are mainly concentrated around the river and the flat plain which are located in the lower river area.

Table 7 Water level calculation results (unit = m).

River name	Calculated water level		River name	Calculated water level	
	P=1%	P=10%		P=1%	P=10%
SheYaoGou	8.55	8.51	BanFangGou	11.01	10.13
BoErQin	12.66	11.86	ZaiErDe	11.08	10.53
KaLaGaYi	10.48	10.08	HeiGou	10.56	9.95
DongNanGou	9.58	9.34	SanGeShan	11.02	10.06
XiBaiYang	10.46	10.18	AKeSu	10.04	9.37
DongBaiYang	8.63	8.35	GaoYaZi	8.99	8.75
HaXiongGou	9.37	8.89	ShiRenZi	9.54	8.74
MoSiKe	9.34	8.98	DaShiTou	9.6	9.19

CONCLUSION

In this paper, we calculated the critical rainfall in gauged and ungauged drainage basins in Urumchi. The IDW method was used to map the critical rainfall and derive certain critical values

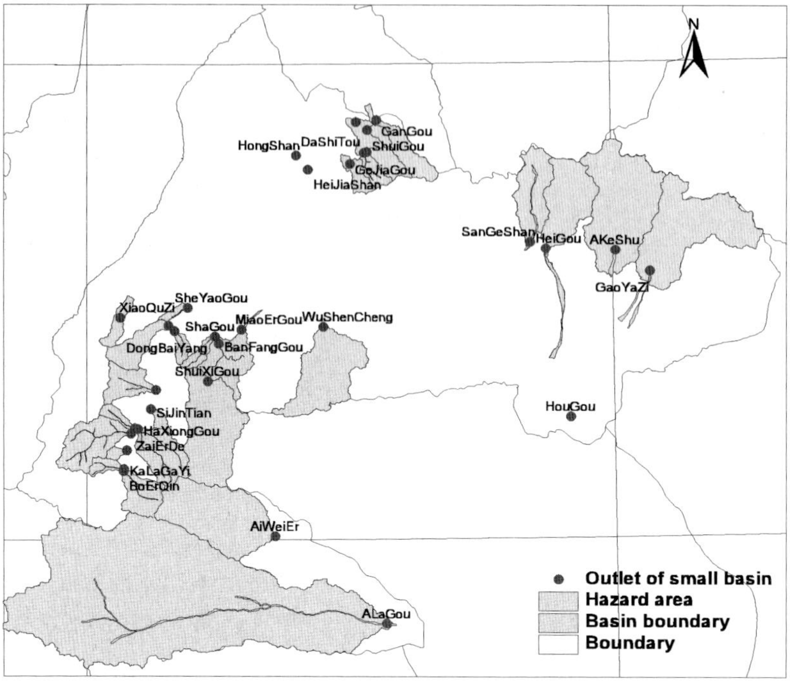

Fig. 5 Flood hazard areas in Urumchi.

of rainfall in small ungauged drainage basins. We found the predicted 24-h critical rainfall in Urumchi varies in a range 13–25 mm, which means that flash floods may easily occur if the rain is heavy.

The rational formula method was used to calculate flash flooding in the Urumchi area. We found the relationships between parameters m and θ, and between parameters u and θ in Urumchi are as follows: $m = 0.1486\ln\theta + 0.4952$, $u = -23.3425\ln(\ln\theta) + 43.7455$, respectively.

The flood disaster risk maps of all ungauged basins were obtained finally. The hazard areas are mainly concentrated around the river and the flat plain which is located in the lower river area. More attention should be paid to this area in Urumchi Flash Flood Control Planning.

Acknowledgements This research was supported by National Natural Science Foundation of China (no. 50979075, no. 50839005, no. 50579052).

REFERENCES

Chen Jiaqi & Gongsu Zhang (1983) *Storm Flood Calculation of Small Basin*. WaterPower Press, Beijing, China.

Eldeen, M. T. (1980) Predisaster physical planning: interpretation of disaster risk analysis into physical planning – a case study in Tunisia, *Disasters* **4**(2), 211–222.

IGPB, WCRD & IHDP (2001) Abstract volume of *Challenge of a Changing Earth*, 10–13. Global Change Opening Science Conference, Amsterdam, The Netherlands.

McCoy, J. & Johnston, K. (2001) *Using ArcGIS Spatial Analyst*. ESRI Press, USA.

Kachroo, R. K. (1992) River flow forecasting. Part 1: A discussion of the principles. *J. Hydrol.* **133**, 1–15.

Ma Dongtao, Cui Peng, Yang Kun, *et al.* (2003) An approach to debris flow hazards along the Xinjing-Tibetan Highway Line in Xinjiang. *Arid Land Geography* **2**(4), 349–354.

Montza, B. E. & Gruntfest, E. (2002) Flash flood mitigation: recommendations for research and applications. *Environmental Hazards* **4**, 15–22.

Sun Houcai, Shayun & Huang Zhipeng (2004) Review of present situation in studying mountain torrent disaster. *J. Yangtze River Scientific Research Institute* **21**(6), 77–80.

Wang Lixian & Shao Song-dong (1998) Analysis of characters of debris flow. *Int. J. Sediment Res.* **12**(4), 25–31.

Wei Jinglian, Zhao bo & Dong Guizhi (1994) Analysis of characters of debris-flow, rainfall and first study on forecast of rainfall in Beijing Maintain Area. *Chinese J. Geological Hazard and Control* **10**(s1), 45–52.

Xia Jun & Tan Ge (2002) Hydrological science towards global change: progress and challenge. *Resources Science* **3**, 1–7.

Xu Shiliang (1995) *FORTRAN Arithmetic Program Library In Common Use*. Tsinghua University Press, Beijing, China.

Zhan Daojiang & Ye Shouze (2000) *Project Application of Hydrology.* China WaterPower Press, Beijing, China.

Zhou Jinxing, WangLixian, Xie Baoyuan, *et al.* (2001) A review on the technique of forecasting and alarming flash flood and debris flow disaster. *J. Mountain Science* **19**(6), 527–532.

Study of the energy and water cycle over the heterogeneous landscape of the northern Tibetan Plateau

YAOMING MA[1], WEIQIANG MA[2], YONGJIE WANG[1], LEI ZHONG[1], MAOSHAN LI[2] & HIROHIKO ISHIKAWA[3]

1 *Key Laboratory of Tibetan Environment Changes and Land Surface Processes, Institute of Tibetan Plateau Research, Chinese Academy of Sciences, Beijing 100085, China*
ymma@itpcas.ac.cn

2 *Cold and Arid Regions Environmental and Engineering Research Institute, Chinese Academy of Sciences, Lanzhou 730000, China*

3 *Disaster Prevention Research Institute, Kyoto University, Kyoto 611-0011, Japan*

Abstract The exchange of heat between the land surface and atmosphere over the Tibetan Plateau area plays an important role in the Asian monsoon system, which in turn is a major component of both the energy and water cycles of the global climate system. Its study is also regarded as the main task in the CEOP (Coordinated Enhanced Observing Period) Asia-Australia Monsoon Project (CAMP) on the Tibetan Plateau (CAMP/Tibet, 2001–2005). Firstly, the field experiments of the CAMP/Tibet are introduced and some results on the local energy partitioning (the diurnal variations and inter-monthly variations of radiation energy budget, land surface energy budget and land surface heating field) are presented in this study. The study of the regional distribution of land surface heat fluxes is of paramount importance over the heterogeneous landscape of the Tibetan Plateau and is also one of the main scientific objectives of the CAMP/Tibet. Therefore, the regional distributions and their inter-monthly variations of surface heat fluxes (net radiation flux, soil heat flux, sensible heat flux and latent heat flux) are also presented here by combining five Landsat-7 ETM images and one NOAA/AVHRR image with the field observations. The derived results were validated by using the "ground truth", and it shows that the regional distributions and their inter-monthly variations of land surface heat fluxes derived using the method in this study are reasonable. Further improvement of the method and its field of application are also discussed.

Key words observation platform; satellite data; land surface heat fluxes; heterogeneous landscape; northern Tibetan Plateau

INTRODUCTION

As the most prominent and complicated terrain of the globe, the Tibetan Plateau (Ye, 1981; Ye & Wu, 1998), with an average elevation of more than 4000 m above mean sea level makes up approximately one fourth of the land area of China. Long-term research on the Tibetan Plateau has shown that the giant prominence exerts thermal effects on the atmosphere, thus greatly influencing circulations over China, Asia and even the globe (Ye, 1981; Ye & Wu, 1998; Ma & Tsukamoto, 2002). Due to its topographic character, the plateau surface absorbs a large amount of solar radiation energy (much of which is redistributed by cryospheric processes), and undergoes dramatic seasonal changes of surface heat and water fluxes. The lack of quantitative understanding of interactions between the land surface and atmosphere makes it difficult to understand the complete energy and water cycles over the Tibetan Plateau and their effects on the Asian monsoon system with numerical models. Therefore, the number of land surface process studies over the Tibetan Plateau has increased in recent years. But the previous experiments were only carried out over short periods and the observational items were limited, and the previous investigations were only made in summer periods and at a few points or at the local level (e.g. Ye & Wu, 1998; Ma & Tsukamoto, 2002).

The intensive observation period (IOP) and long-term observation of the GAME/Tibet (GEWEX: Global Energy and Water cycle Experiment) Asian Monsoon Experiment on the Tibetan Plateau, 1996–2000; Ma & Tsukamoto, 2002) and the CAMP/Tibet (CEOP: Coordinated Enhanced Observing Period) Asia-Australia Monsoon Project (CAMP) on the Tibetan Plateau, 2001–2005) have been successfully completed in the past 12 years. A large amount of data has been collected, which form the best data set so far for the study of energy and water cycles over the Tibetan Plateau. It gives us a chance to investigate the land surface heat fluxes over the Tibetan Plateau in detail.

The purpose of this study is to analyse the characteristics of local and regional partitioning of the surface energy by using the field observation data, Landsat-7 ETM data and NOAA-14/AVHRR data collected during the GAME/Tibet and the CAMP/Tibet.

DESCRIPTION OF EXPERIMENTS

The objectives of CAMP/Tibet are: (1) quantitative understanding of an entire seasonal hydro-meteorological cycle, including winter processes, by solving surface energy "imbalance" problems in the Tibetan Plateau; (2) observation of the local circulation and evaluation of its impact on the plateau-scale water and energy cycle; and (3) establishment of quantitative observational methods for the entire water and energy cycle between the land surface and atmosphere by using satellites.

To achieve the scientific objectives of the CAMP/Tibet, a meso-scale observation network (150 × 250 km, 91°–92.5°E, 30.7°–33.3°N) was implemented in the central plateau (Fig. 1). It included: (1) a basic observation station (BJ); a flux measuring tower (20 m), a Sky Radiometer, a LIDAR system, a Wind Profiler and RASS, a radiosonde system, and four automatic weather stations (AWSs) were set up at this station. (2) AWS networks; six AWS (D105, D110, NPAM–MS3478, BJ, MS3608 and ANNI) stations were deployed in this area. (3) Soil moisture and soil temperature (SMTMS) networks; seven SMTMS sites (D105, D110, Amdo-Anduo, BJ, MS3608, ANNI and MS3637) were deployed. (4) Two deep soil temperature measurement equipment sets were put in D105 and NaquDS (near the BJ basic station). (5) Amdo (Anduo) PBL station (91°37′30″E, 32°14′28″N); a PBL tower (17 m, wind speed, wind direction, air temperature and humidity at three levels) and a radiation observation system were already installed and operated continuously for 12 years from 1997, and will be continued for another five years until the end of 2010. The enhanced automated observing period (EAOP) and IOP of CAMP/Tibet started from 1 October 2002 and were stopped at the end of September 2004.

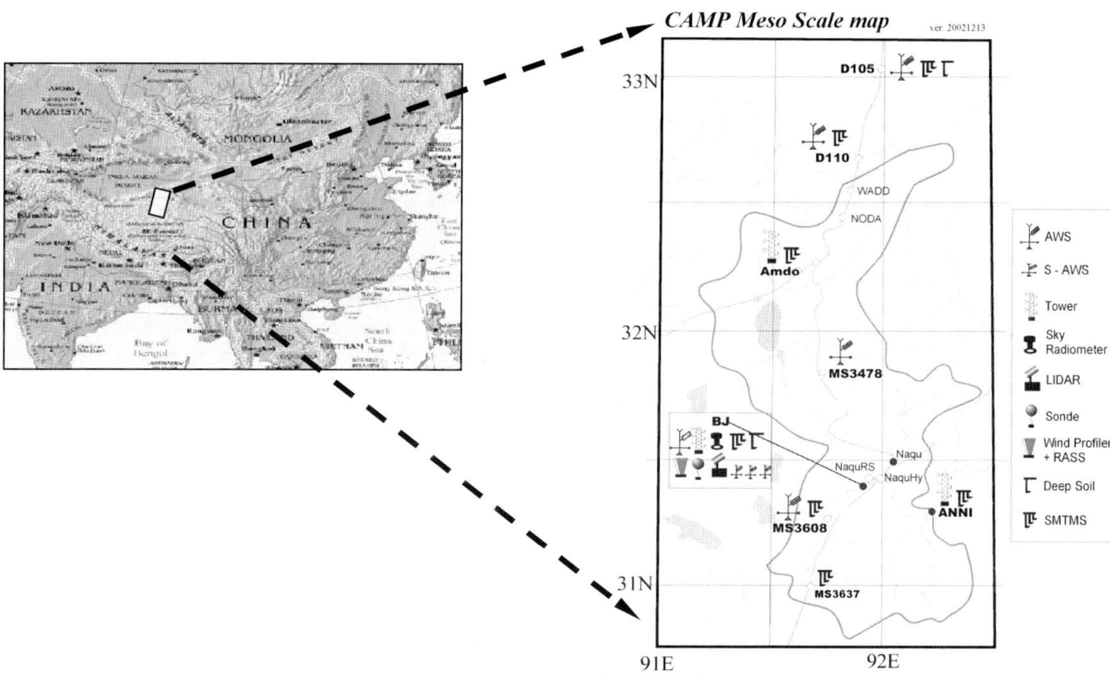

Fig. 1 The geographic map and the sites layout during the CAMP/Tibet.

IN SITU DATA ANALYSIS

Diurnal and inter-monthly variation of local radiation fluxes land surface heat fluxes

The field data observed at three AWS stations (D105, NPAM-MS3478 and BJ) of the CAMP/Tibet under clear day conditions will be used in this section.

Figures 2 and 3 show the diurnal variation and inter-monthly variations of each component of the surface radiation budget (downward shortwave radiation flux (*SD*), downward longwave radiation flux(*LD*), upward longwave radiation flux(*LU*) and upward shortwave radiation flux (*SU*)), net radiation flux, sensible heat flux H, latent heat flux λE and soil heat flux G_0 at stations D105, BJ and NPAM-MS3478 of the CAMP/Tibet. One clear day's data was selected in each month, and the point of the diurnal cycle for one month stops and the next month starts just at the centre between the two days' curves in Figs 2 and 3. The sensible heat flux H, latent heat flux λE and soil heat flux G_0 at the surface were determined by the combination method (Hu & Qi, 1991), which is:

$$H_0 = \rho C_p k^2 z_A{}^2 \frac{\partial u}{\partial z}\frac{\partial \theta}{\partial z} \approx \rho C_p k^2 z_A{}^2 \frac{\Delta u}{\Delta z}\frac{\Delta \theta}{\Delta z} \quad (1)$$

$$\lambda E_0 = \rho \lambda k^2 z_A{}^2 \frac{\partial u}{\partial z}\frac{\partial q}{\partial z} \approx \rho \lambda k^2 z_A{}^2 \frac{\Delta u}{\Delta z}\frac{\Delta q}{\Delta z} \quad (2)$$

$$z_A = \sqrt{z_i z_{i+1}} \quad (3)$$

$$G_0 = G_z + \int_0^z C_s \frac{\partial T}{\partial t} dz \approx G_z + C_s \frac{\Delta T}{\Delta t} z \quad (4)$$

$$F = \frac{R_n - G_0}{H_0 + \lambda E_0} \quad (5)$$

$$H = H_0 F \quad (6)$$

$$\lambda E = \lambda E_0 F \quad (7)$$

where H_0 and λE_0 are the sensible and latent heat fluxes, which are uncorrected by the stratification. λ is latent heat of vaporization, z_A is geometric mean value between z_i and z_{i+1}, C_s is the volumetric heat capacity ($C_s = 1.18 \times 10^6$ J m^{-3} K^{-1} in the study area over the Tibetan Plateau (Tanaka *et al.*, 2001)), T is soil temperature at the observed depth z or surface temperature, t is time, G_z is the observed soil heat flux below z from the surface (z = 10 cm was used here), and k is the Von-Karman constant (k = 0.4), and the potential temperature (θ) and specific humidity (q) are:

$$\theta = T_a \left(\frac{1000}{p}\right)^{0.286} \quad (8)$$

$$q = \frac{0.622e}{p - 0.378e} \quad (9)$$

where T_a is air temperature, p is pressure, e ($=E \times RH$) is vapour pressure, E is saturation water vapour pressure, and RH is relative humidity.

It can be seen that: (1) The inter-monthly variation of the downward shortwave radiation flux is very obvious. The summer values are much larger than those in winter; the minimum value is around December. Although the upward shortwave radiation had different values in different months, their inter-monthly variations were not clear. (2) The downward shortwave radiation (*SD*) reaches about 1200 W/m^2 or more at local noon on fine days in summer. The transmission rate of incoming solar radiation from the top of the atmosphere was estimated as about 85% in cloudless conditions. The values are about 10–15% greater than those observed at the typical sea level

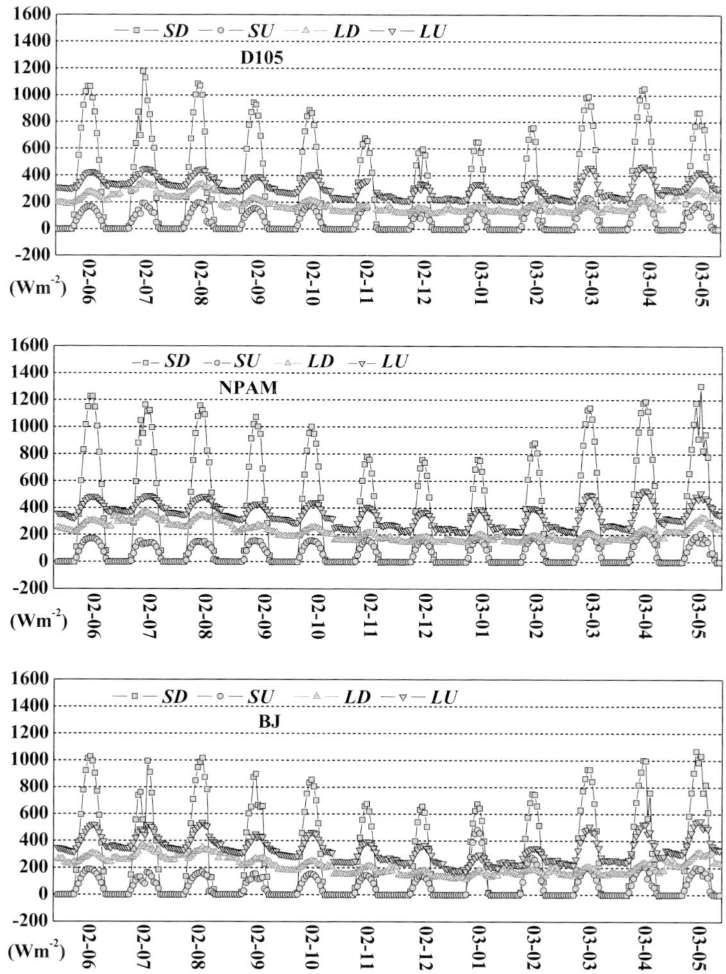

Fig. 2 The inter-monthly variations of radiation flux at the CAMP/Tibet stations. *SD*: downward short wave radiation flux; *LD*: downward long wave radiation flux; *LU*: upward long wave radiation flux; and *SU*: upward short wave radiation flux.

station. This is due to the high altitude of the sites (the elevation of D105 is 5039 m and NPAM–MS3478 is at 4620 m), thus a shallower atmospheric layer between the top of the atmosphere and the ground surface. (3) The inter-monthly variation of the upward longwave radiation flux (*LU*) was obvious. The summer values were larger than those in winter, and it reaches the minimum value around December. This means that in summer the surface temperature is higher than in winter. Although the values of the downward longwave radiation flux in summer are a little higher than those in other months, their inter-monthly variations were not clear. (4) The diurnal variations of downward shortwave radiation, upward shortwave radiation and upward longwave radiation were very obvious, but the diurnal variation of downward longwave radiation was not clear, especially during the pre-monsoon period (October 2002–April 2003). (5) The diurnal variation and inter-monthly variation of the net radiation flux was very obvious. The summer values were much larger than those in winter, and it reached the minimum value around December. (The net radiation flux reached the minimum value in January 2003 at BJ station because snow-ice cover was present at the station and the upward shortwave radiation was

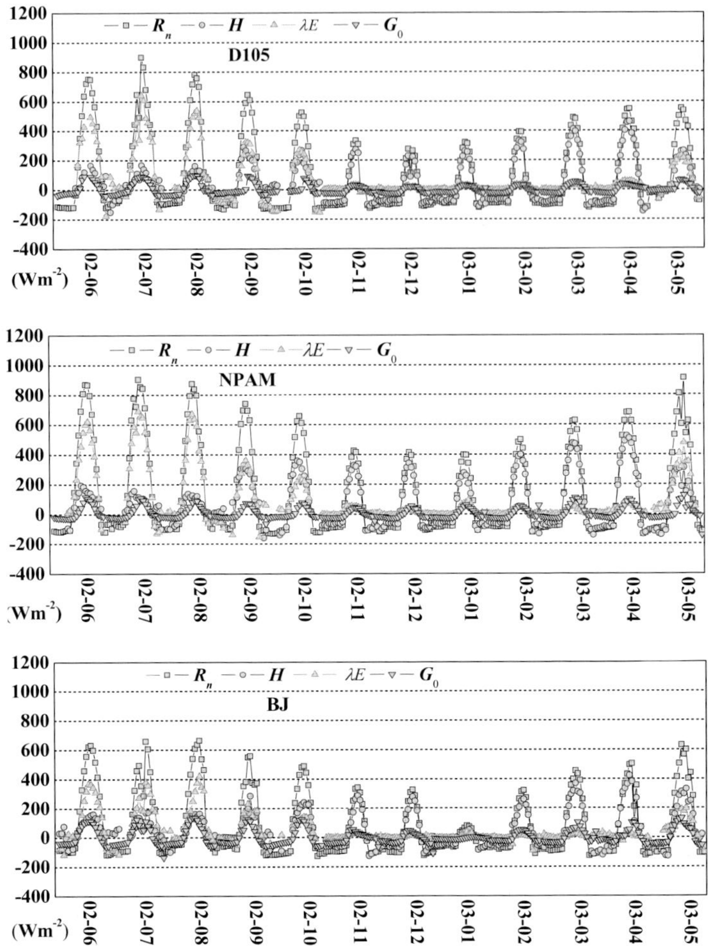

Fig. 3 The diurnal variation and inter-monthly variations of land surface heat fluxes at the CAMP/Tibet stations.

much larger over it, see Fig. 3). (6) Sensible heat and latent heat fluxes play different roles in the partition of net radiation flux in different months in the Tibetan Plateau. In other words, sensible heat flux plays the main role in winter and latent heat flux plays the main role in summer and autumn.

Land surface heating field

The characteristic of the atmospheric heating from the land surface of the Tibetan Plateau has very important effects on the weather and climate processes over the Tibetan Plateau, eastern Asia and even the northern part of the globe. Therefore, the study of the surface heating field over the Tibetan Plateau area has been a major research topic in recent years. We define the surface heating intensity here as:

$$R_n - G_0 = H + \lambda E \qquad (10)$$

It means that the land surface is the heat source to the atmosphere when $R_n - G_0 > 0$, otherwise, the land surface is a heat sink to the atmosphere when $R_n - G_0 < 0$. Figure 4 shows the daily

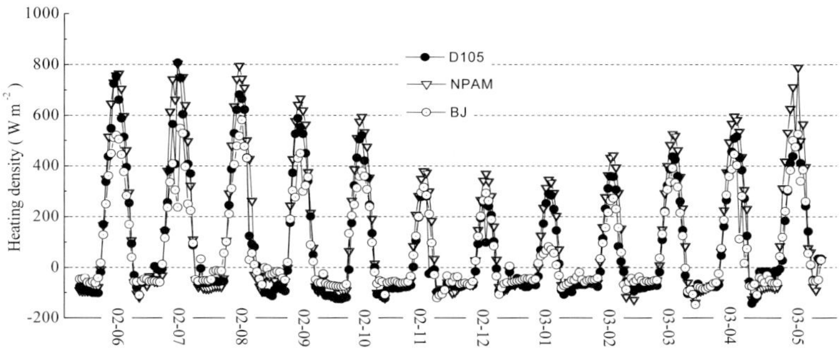

Fig. 4 The diurnal variation and inter-monthly variations of land surface heating density at the CAMP/Tibet stations.

diurnal variations of the surface heating intensity over the northern Tibetan Plateau area for each of the 12 months. It can be seen that: (1) the land surface is a very strong heat source in the day time, and it is a weak surface heat sink at night, but the daily mean is of a strong surface heat source. (2) The diurnal variation of the intensity of surface heating field is clear. It means that when sunrise is around 08:00 (Beijing Standard Time, BST) in the morning, the intensity of the surface heating field increases with the increasing of solar altitude angle, and the maximum value occurs at about 13:00 BST, then it decreases with the decreasing of solar altitude angle. It becomes a surface heat sink at about 20:00 BST at nightfall. (3) The surface heating densities in summer are much larger than those in winter, and it reaches the minimum value around January. In other words, the surface of the Tibetan Plateau is a strong heat source in summer and a weak heat source in winter.

THE DISTRIBUTION OF REGIONAL LAND SURFACE HEAT FLUXES

The study on the regional surface energy partitioning and its inter-monthly variation is of paramount importance over the heterogeneous landscape of the Tibetan Plateau and it is also one of the main scientific objectives of the GAME/Tibet and CAMP/Tibet programmes. Therefore, here we will try to derive the regional land surface heat fluxes by using the Landsat-7 ETM data, NOAA-14/AVHRR data and the field observational data. Here, "Regional" land surface heat fluxes are not "aggregated" fluxes (Batchvarova *et al.*, 2001), but surface flux fields or surface heat fluxes on each pixel of Landsat-7 ETM images and NOAA-14/AVHRR images.

The methodology proposed to determine regional land surface heat fluxes over the Tibetan Plateau was to use Landsat-7 ETM data, NOAA-14/AVHRR data and field observations (Ma & Tsukamoto, 2002; Ma *et al.*, 2002a,b). The surface reflectance for short-wave radiation (r_0) and land surface temperature (T_{sfc}), are retrieved from Landsat-7 ETM data and NOAA-14/AVHRR data with the atmospheric correction by using the radiative transfer model MODTRAN (Berk *et al.*, 1989), a four-stream radiative transfer assumption (Verhoef, 1997) and aerological observation data. The radiative transfer model also computes the downward short- and long-wave radiation at the surface. With these results the surface net radiation flux (R_n) is determined. The surface soil heat flux (G_0) is estimated from R_n, T_{sfc}, r_0 and *MSAVI* (Modified Soil Adjusted Vegetation Index, Qi *et al.*, 1994), which is also derived from Landsat-7 ETM data or NOAA-14/AVHRR data. Sensible heat flux (H) is estimated from T_{sfc}, surface and aerological data with the aid of the so-called "blending height" approach (Mason, 1988), and the regional latent heat flux (λE) can be derived as the residual of the energy budget theorem for the land surface.

Results analysis

It is better to select satellite data for clear days to study the distribution and inter-monthly variation of energy budget components. Unfortunately, it is difficult to select this kind of satellite data over the Tibetan Plateau area because of the presence of strong convective clouds when the satellite observations took place. Only five Landsat-7 ETM images (9 June 2002, 24 July 2002, 28 August 2002, 2 December 2002 and 24 March 2003) were suitable during the whole EAOP of the CAMP/Tibet, and one NOAA/AVHRR image (12 June 1998) was selected as a case to compare with the Landsat-7 ETM (9 June 2002).

Using the methodology shown in the previous sections, the regional distributions of land surface variables (surface reflectance and surface temperature), vegetation variables (*MSAVI*, vegetation coverage P_v and leaf area index *LAI*) and land surface heat fluxes were derived. The land surface heat fluxes derived from the satellite data were also compared with the field measurements at BJ station and ANNI station (see Table 2). The field observation data, which was used for validation here, was just measured at the time of satellite over pass, ±10 minutes. The absolute percent difference (*APD*) was computed as a quantitative measure of the difference between the derived results at point *i* ($H_{\text{derived}(i)}$) and measured value at point *i* ($H_{\text{measured}(i)}$) of one scene as:

$$APD = \frac{\left| H_{\text{derived}(i)} - H_{\text{measured}(i)} \right|}{H_{\text{measured}(i)}} \tag{11}$$

The results show that: (1) The derived surface variables (land surface reflectance and surface temperature), vegetation variables (*MSAVI*, vegetation coverage P_v and leaf area index *LAI*) and surface heat fluxes (net radiation flux R_n, soil heat flux G_0, sensible heat flux H and latent heat flux λE) in five different months over the study area are in good accordance with the land surface status. These parameters show a wide range due to the strong contrast of surface features (Fig. 1 and Table 1). Surface reflectance, surface temperature and sensible heat flux around the lake in the distribution maps are much higher in summer (June, July and August), and at the same time, net radiation flux, soil heat flux and latent heat flux are lower in the area. The reason is that much of the area around the lake area is desertification grassland. (2) The derived pixel value and average values of surface temperature, net radiation flux, soil heat flux and latent heat flux in June, July and August are higher than that in December and March. It means that there is much greater evaporation in summer than in winter (December) and spring (March) in the northern Tibetan Plateau area. And it also illustrates out that the heating density ($R_n - G_0 = H + \lambda E$) in summer is much higher than in winter and spring in the northern Tibetan Plateau area, which is same as the results shown in Fig. 4. But, the sensible heat flux is the main part of the net radiation flux in December and March. (3) During the experimental periods, the derived net radiation flux was larger than that in the HEIFE area (Ma *et al.*, 2002b) due to the high altitude (the higher value of downward shortwave radiation) and land surface coverage of grassy marshland (the lower value of the upward longwave radiation) in this area. For example, the regional average value of net radiation flux was 470 W m^{-2} over the HEIFE area on 9 July 1991, and it was 725 W m^{-2} over the CAMP/Tibet area on 24 July 2001. (4) Although the resolution of Landsat-7 ETM is different to the NOAA/AVHRR, the results derived here are also comparable to those derived from NOAA/

Table 1 The distribution ranges of land surface variables, vegetation variables and land surface heat fluxes over the CAMP/Tibet area.

	June	July	August	December	March
MSAVI	0.00~0.372	0.000~0.821	0.000~0.894	0.000~0.394	0.000~0.429
P_v	0.00~0.86	0.00~0.86	0.00~1.00	0.00~0.86	0.00~0.80
LAI	0.00~2.57	0.00~2.00	0.00~3.00	0.00~2.80	0.00~2.58
r_0	0.010~0.490	0.010~0.540	0.01~0.540	0.110~0.550	0.001~0.050
T_{sfc} (°C)	0~52	0~57	0~48	–59~18	–20~40
R_n (W/m^2)	164~774	125~900	151~900	22~550	53~660
G_0 (W/m^2)	25~221	0~270	0~245	0~150	0~170
H (W/m^2)	0~450	0~450	0~455	0~347	0~550
λE (W/m^2)	0~550	0~630	0~640	0~155	0~490

AVHRR data from around the relatively homogeneous sites of the CAMP/Tibet area. For example, the derived results for net radiation flux, soil heat flux, sensible heat flux and latent heat flux from Landsat-7 ETM data at the BJ site in June are 562, 105, 163 and 294 W m^{-2}, respectively, and at the same time, the derived values from NOAA/AVHRR data are 578, 108, 178 and 292 W m^{-2}, respectively. But this should be validated over the heterogeneous sites (e.g. ANNI site). Unfortunately, we have no measurements at the ANNI site during the GAME/Tibet period. (5) The surface reflectance and surface temperature derived in this research are in good accordance with the field measurements with APD less than 9.0% (Table 2). (6) The derived net radiation fluxes over the study area are very close to the field measurement with APD less than 8.0% (Table 2). (7) The regional soil heat flux derived from equation (15) and the relationship between soil heat flux and net radiation flux (Ma *et al.*, 2002a) is suitable for the heterogeneous land surface of the CAMP/Tibet area, and the APD is less than 8.0% at validation sites because the relationship itself was derived from the same area (Table 2). Finally, (8) the derived regional sensible heat flux and latent heat flux (APD less than 9.5% at the validation site) in the CAMP/Tibet area, is in good agreement with field measurements (Table 2). It is concluded that the proposed parameterization methodology for sensible heat flux and latent heat flux is reasonable, and may be used over the northern Tibetan Plateau area.

Table 2 Comparison of the derived surface reflectance and surface temperature (Cal.) *versus* those measured values (Meas.) at the CAMP/Tibet site with APD.

r_0 (-)	June		July		August		December		March	
	BJ	ANNI	BJ	ANNI	BJ	ANNI	BJ	ANNI	BJ	ANNI
Cal.	0.159	0.156	0.196	0.192	0.165	0.182	0.219	0.222	0.220	0.211
Meas.	0.164	0.161	0.180	0.179	0.174	0.186	0.234	0.239	0.238	0.228
APD	3.0%	3.1%	8.9%	7.3%	5.5%	2.2%	6.4%	7.1%	7.6%	7.5%
T_{sfc}	June		July		August		December		March	
(^{0}C)	BJ	ANNI	BJ	ANNI	BJ	ANNI	BJ	ANNI	BJ	ANNI
Cal.	28.0	31.0	28.0	26.0	17.0	16.0	-26.0	-25.0	6.0	5.0
Meas.	26.5	28.8	28.7	28.2	18.5	17.0	-27.8	-23.6	5.6	4.8
APD	5.7%	7.6%	2.4%	7.8%	8.1%	5.9%	6.5%	5.9%	7.1%	4.2%
R_n	June		July		August		December		March	
(W m^{-2})	BJ	ANNI	BJ	ANNI	BJ	ANNI	BJ	ANNI	BJ	ANNI
Cal.	562	565	644	667	540	684	380	403	413	548
Meas.	540	542	624	633	506	662	357	385	385	520
APD	4.1%	4.2%	3.2%	5.4%	6.7%	3.3%	6.4%	4.7%	7.3%	5.4%
G_0	June		July		August		December		March	
(W m^{-2})	BJ	ANNI	BJ	ANNI	BJ	ANNI	BJ	ANNI	BJ	ANNI
Cal.	105	104	150	147	154	152	74	73	85	83
Meas.	98	99	140	142	147	142	69	68	79	78
APD	7.1%	5.1%	7.1%	3.5%	4.8%	7.0%	7.2%	7.4%	7.6%	6.4%
H	June		July		August		December		March	
(W m^{-2})	BJ	ANNI	BJ	ANNI	BJ	ANNI	BJ	ANNI	BJ	ANNI
Cal.	163	152	247	236	191	205	239	310	247	364
Meas.	157	158	255	248	216	211	234	326	257	345
APD	3.8%	3.8%	3.1%	3.2%	6.9%	2.8%	2.1%	4.9%	3.9%	5.5%
λE	June		July		August		December		March	
(W m^{-2})	BJ	ANNI	BJ	ANNI	BJ	ANNI	BJ	ANNI	BJ	ANNI
Cal.	294	309	247	284	195	327	67	20	81	101
Meas.	284	283	226	265	190	350	62	19	76	99
APD	3.5%	9.2%	9.3%	7.2%	2.6%	6.6%	8.1%	5.3%	6.6%	2.0%

CONCLUDING REMARKS

In this paper, the field experiments of the GAME/Tibet and the CAMP/Tibet are introduced and some first results on the local energy partitioning (the diurnal variations and inter-monthly variations of radiation energy budget, land surface energy budget and land surface heating field) are obtained. The regional distributions and inter-monthly variations of land surface heat fluxes (net radiation, soil heat flux and sensible heat flux) over the heterogeneous area of the CAMP/Tibet and GAME/Tibet are also derived by using Landsat-7 ETM data, NOAA-14/AVHRR data and field observations. The results are in good agreement with the field observations. In other words, the regional land surface heat fluxes over the heterogeneous landscape can be determined by using the satellite remote sensing and the atmospheric boundary layer observations.

All the results in this paper were only obtained from the meso-scale area over the Tibetan Plateau. In order to up-scale the land surface heat fluxes to all of the heterogeneous landscapes of the Tibetan Plateau area, more field observations and more satellite data should be used in the future. The proposed parameterization method should also be improved to determine the regional land surface heat flux over whole Tibetan Plateau area.

Acknowledgements This work was done under the auspices of the National Natural Science Foundation of China (40825015, 40810059006 and 40705004), Innovation Projects of the Chinese Academy of Sciences (KZCX2-YW-Q11-01), EU-FP7 project "CEOP-AEGIS" (212921). The CAMP/Tibet project was founded by the Japan Science and Technology Corporation (JST). The authors thank all the scientists in the CAMP/Tibet group from China, Japan and Korea for their hard work in the collection of field data.

REFERENCES

Batchvarova, E., Gryning, S. E. & Hasager, C. B. (2001) Regional fluxes of momentum and heat over a sub-artic landscape during late winter. *Bound.-Layer Met.* **99**, 489–507.

Berk, A., Bernstein, L. S. & Robertson D. C. (1989) MODTRAN, A moderate resolution model for LOTRAN 7. GL-TR-89-0122.

Hu, Y.& Qi, Y. (1993) The combinatory method to determine the turbulent fluxes and the universal functions in the surface layer. *ACTA Meteor. Sinica* **7**, 101–109.

Ma, Y. & Tsukamoto, O. (2002) *Combining Satellite Remote Sensing with Field Observations for Land Surface Heat Fluxes Over Inhomogeneous Landscape*. China Meteorological Press, Beijing, China.

Ma, Y., Su, Z., Li, Z.-L., Koike, T. & Menenti, M. (2002a) Determination of regional net radiation and soil heat flux densities over heterogeneous landscape of the Tibetan Plateau. *Hydrol. Processes* **16**(15), 2963–2971.

Ma, Y., Tsukamoto, O., Ishikawa, H., Su, Z., Menenti, M., Wang, J. & Wen, J. (2002b) Determination of regional land surface heat flux densities over heterogeneous landscape of HEIFE integrating satellite remote sensing with field observations. *J. Met. Soc. of Japan* **80**(3), 485–501.

Mason, P. (1988) The formation of areally-averaged roughness lengths. *Quart. J. Royal Met. Soc.* **114**, 399–420.

Qi, J. G., Chehbouni, A., Huete, A. R., Kerr, Y. H. & Sorooshian, S. (1994) A modified soil adjusted vegetation index. *Remote Sens. Environ.***48**, 119–126.

Tanaka, K., Ishikawa, H., Tamagawa, I. & Ma, Y. (2001) Surface energy budget at Amdo on Tibetan Plateau using GAME/Tibet IOP'98 data. *J. Met. Soc. Japan* **79**(1B), 505–517.

Verhoef, W. (1997) Theory of radiative transfer models applied in optical remote sensing of vegetation canopies. PhD Thesis, Remote Sensing Department of National Aerospace Laboratory, The Netherlands.

Ye, D. (1981) Some characteristics of the summer circulation over the Qinghai-Xizang (Tibet) Plateau and its neighborhood. *Bull. Am. Met. Soc.* **62**, 14–19.

Ye, D. & Wu, G. (1998) The role of the heat source of the Tibetan Plateau in the general circulation. *Met. Atmos. Phys.* **67**, 181–198.

Quantitative remote sensing of information indices for eco-water (layer): a case study at Maoergai area in the upper Minjiang River

WU-NIAN YANG, JI JIAN, YU-XIA LI, LI PENG, XIN-NAN WAN, HAN-HU LIU, HUAI-YONG SHAO, XIAO-AI DAI, TAO ZENG & XUE-MING WU

State Key Laboratory of Geo-hazard Prevention and Geo-environment Protection /Key Lab of Earth Exploration & Information Techniques of Ministry of Education/ Institute of RS & GIS, Chengdu University of Technology, Chengdu, Sichuan, 610059, China

pengli@cdut.edu.cn

Abstract Eco-water is defined as the water body closely related to the ground vegetation, including water conserved in the vegetation layer, the vegetation humus layer and the root soil layer. It is capable of precipitation interception and rivers or groundwater supplementation. This paper applies quantitative remote sensing method to experiments performed in the Maoergai area in the upper Minjiang River in China to examine properties, functions, spatial distribution patterns and transformation rules of the eco-water (layer). Based on the features of the eco-water and their reflection in the remotely sensed data, the information indices for the eco-water (layer) were identified, including the Modulus of Eco-water Conservation (MEC), the Modulus of Eco-water Runoff (MER), the Vegetation Moisture Content (VMC), the Soil Moisture Saturation (SMS) and the Soil Moisture Content (SMC). Limited by time, cost and available data, this paper mainly presents the methods for remote sensing information extraction, inversion model construction and quantitative estimation for the eco-water (layer). The consistency between the inverted resource quantity of the eco-water layer and the measured data suggest a valid vegetation-centred quantitative approach to study the eco-water (layer).

Key words eco-water (layer); SMC; SMS; VMC; MEC; quantitative remote sensing

INTRODUCTION

In the research field of the hydrological cycle, the Biospheric Aspects of the Hydrological Cycle (BAHC) was launched as a core project of the International Geosphere-Biosphere Programme (IGBP) (Hutjes *et al.*, 1998; Xia & Zuo, 2006), by ecologists and hydrologists in 1991. This project has addressed biospheric effects on the hydrological cycle to explore the interaction processes between vegetation and the physical processes of the hydrological cycle, which has clarified that from the macro-perspective, research of the hydrological environment and resources should not be confined to the water body itself, but should go further into its correlated factors such as vegetation, ground, and especially the water-body-related vegetation, i.e. the eco-water (layer) (Wan *et al.*, 2004; Jian, 2006; Li, 2007). Most current BAHC implementations have been confined to research on the hydrological effects of vegetation. Few studies have examined the eco-water (layer) as an entity of the hydrological cycle. Also, few studies have introduced modern remote sensing methods to high precision quantitative research (Bi & Zhong, 2002; Wan *et al.*, 2004; Jian, 2006; Li, 2007).

Based on the analysis of our previous experiments, the eco-water is defined as the water body closely related to the ground vegetation; the eco-water layer is defined as a vegetation-centred layer formed by the transformation of atmospheric precipitation conserved in the vegetation layer, the vegetation humus layer and the root soil layer. This water is closely related to the ground vegetation and its existing form and movement pattern differ from the groundwater and underground water. Its conserved water quantity provides the source for rivers or groundwater supplementation. The relation pattern between surface runoff and sparse/flush vegetation shown in Fig. 1 demonstrates the positive correlation between vegetation cover degree and the formation of the eco-water.

The eco-water essentially consists of three parts: (a) water intercepted from different forms of precipitation and conserved in vegetation layers (including leaves, roots and vegetation humus layers); (b) water conserved in the host of vegetation, that is, the root soil aeration zone; (c) the biologic

Fig. 1 Relation pattern between vegetation and eco-water: (a) flush vegetation, (b) sparse vegetation.

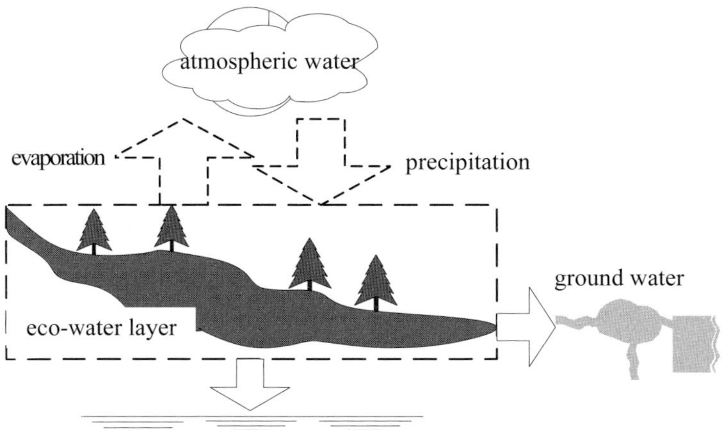

Fig. 2 Roles of eco-water layer in hydrological cycle (Wan, 2004).

water held by plants themselves in multi-forms. Here, the eco-water is not involved directly in the surface or underground runoff. Instead, as an important sort of supplementation for them, it influences the movement of surface water and groundwater in a particular way, i.e. as "biological water" movement. The process of the water cycle suggests that the vegetation-centred eco-water serves as a special transfer zone which determines the proportioning of precipitation to surface runoff or groundwater runoff. It can be treated as a filtering layer, the so called eco-water layer. The role of the eco-water layer is shown in Fig. 2. In primitive forests, water intercepted by the eco-water layer forms up to 20–30% of the total atmospheric precipitation, or even 40% sometimes. Therefore, its impact on the hydrological cycle should not be underestimated. Its quantity has direct consequences on the distribution of various types of water resource quantity and their corresponding retention time (Mo, 1998; Pei & Yu, 2001; Wang, 2001; Wu & Dong, 2001; Wang *et al.*, 2001; Dong, 2003; Feng *et al.*, 2003; Huang *et al.*, 2003; Jiang *et al.*, 2003; Peng *et al.*, 2003; Tian & Kang, 2003; Xia *et al.*, 2003).

The eco-water (layer) is capable of conserving partial precipitation. Moreover, it has the capability of microclimate environment adjustment through intercepting and cushioning rainstorms to prevent floods or flood peaks, and supplementing surface water and groundwater to avoid lands and rivers drying up in drought seasons, and maintaining the survival of vegetation. In the areas where the eco-water is well developed, the water resource appears to be rich and the ecological environment appears to be well-functioning. Otherwise the ecological environment appears to be poorly functioning. As a result, the eco-water (layer) plays an important role in the hydrological cycle, supporting the sustainable use of water resources (Entekhabi *et al.*, 1994; Kang & Liu, 1994; Lei & Yang, 1998; Ma & Ye, 1998; Sun *et al.*, 1998; Liu & Sun, 1999; Canadell & Noble, 2001; Wan *et al.*, 2004).

Because of the scale requirements of the eco-water conserved in the eco-water layer arising from measuring difficulties, this paper introduces the quantitative remote sensing method for the examination of the eco-water. Taking advantage of the combination of classical mathematical, physical theories and remote sensing practices, the basic idea of quantitative remote sensing (or quantitative study of remote sensing) is to build quantitative inversion models according to the interactions between ground objects and spectral features. The various properties of the eco-water have different information presentation according to different wavelengths. Thus, based on the radiation theory, physical features of the vegetation/soil layer and the remote sensing imaging mechanism, the remotely sensed data in a wide range of wavelength, from visible light, near infrared, thermal infrared to microwave can be used to study the resource quantity of eco-water by correlation analysis between its spectral properties, reflectance or brightness of ground objects and its environmental factors.

This paper carries out studies on the properties, functions, spatial resource distributions and the transformation rules of the eco-water layer by remote sensing techniques together with principles from ecology, botany, hydrogeology, forest hydrology and mechanism of remote sensing information. The information index system presented here mainly includes the Modulus of Eco-water Conservation (MEC), the Modulus of Eco-water Runoff (MER), the Vegetation Moisture Content (VMC), the Soil Moisture Saturation (SMS) and the Soil Moisture Content (SMC) (Yang *et al.*, 2008). The inversion models for SMC, SMS, VMC and MEC are presented here. The purpose of this study is to suggest a vegetation-centred approach to model and estimate the eco-water, which may provide quantitative data for flood and drought prediction in gauged and ungauged basins.

METHODS

Study area

The Maoergai area (from north 32°20′–32°40′ to east 103°00′–103°30′) is located at the upper Minjiang River in north Sichuan province in the southwest of China. This area represents a variety

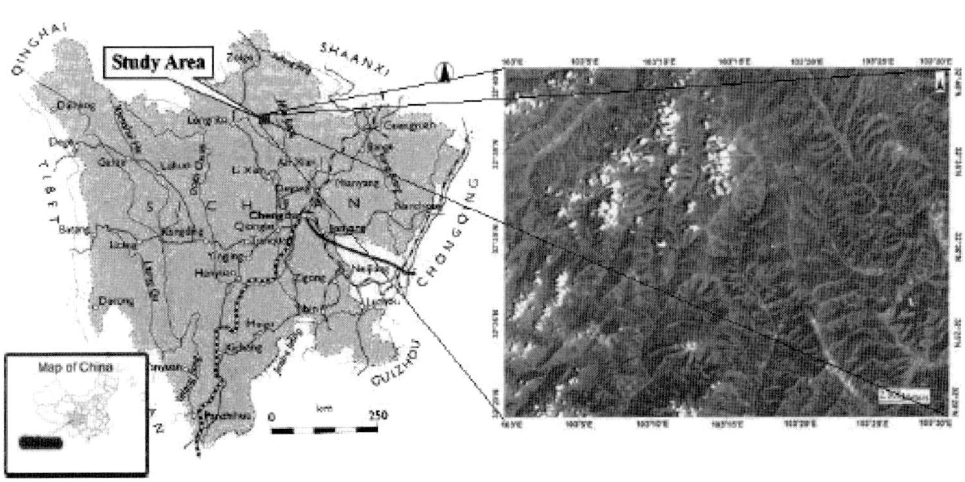

Fig. 3 Location of the study area.

of different types of vegetation, soil and terrain. The forest area reached up to 1.20 million hm^2, about 50% of the whole area 600 years ago. However, the over deforestation had reduced the forest area to 0.74 million hm^2, about 32% of the whole area by 1950. Currently, the Maoergai area is transforming from an uncultured region to an area with expanding human activities, which brings increasing threats to its ecosystem. Figure 3 shows the location of the study area.

Feasibility analysis for remotely studying eco-water layer

The positive correlation between the Vegetation Index (VI) and the vegetation moisture content suggests the possibility of estimating indirectly the eco-water. Among the VIs, generally, the Normalized Difference Vegetation Index (NDVI) can reflect vegetation status that is related to vegetation evaporation and soil moisture content. Cihlar *et al.* (1991) have applied NDVI data of every 15 days in crop growth season and meteorologic data to soil moisture models to invert roots moisture in different depths, together with ecological and soil information. Also, their research was performed on a GIS platform to discover the relation between ecological variables and NDVI (from NOAA/AVHRR), where the possibility for evaporation estimation using NDVI was discussed. Their research has revealed the close correlation between NDVIs of different vegetation-soil combination and their Potential Evaporation (PE) curve. The correlation coefficient of Actual Evaporation (AE) and NDVIs was 0.77; the correlation coefficient of PE and NDVIs in the first 15 days was 0.86; the correlation coefficient of total evaporation and total NDVIs in whole season was 0.96. Therefore the NDVI and PE could be used to estimate 15-day AE with 10–15% errors. The result has demonstrated that the NDVI curve is capable of revealing temporal information of vegetation besides estimating AE in acceptable precision.

GVI can reflect soil brightness, greenness and wetness which are related to soil moisture and vegetation types. PVI can filter out soil background influence and is less sensitive to atmospheric effect than other VIs. DVI, RVI and MVI are related to Leaf Area Index (LAI), reflecting vegetation cover degree. Some researchers have studied the relationship between VI and soil moisture in arid and semi-arid areas. They discovered a strong empirical relationship between VI and the measured soil moisture, and a relationship between soil water stress and VI. It is concluded that vegetation moisture content and soil moisture content can be estimated indirectly from VI.

Method outline

As shown in Fig. 4, the method outline consists of following four steps:

(a) pretreatment for remote sensing images: geometric correction and radiometric correction (including radiometric scaling, fluctuation correction and atmospheric correction) to eliminate errors from position variance surface fluctuation, atmospheric scattering and atmospheric absorbing,
(b) bio-physical parameter extraction: VI, soil brightness, soil wetness, soil greenness, surface temperature, etc.,
(c) models building/selection and information indices inversion,
(d) validation: model validation against data acquired from the project of Primary Comprehensive Remote Sensing Survey for the Ecological Environment in 2000.

RESULTS

Inversion model for SMC

Based on the current quantitative remote sensing model to estimate soil moisture, a Ts/NDVI slope method was selected to retrieve soil moisture from the remotely sensed data at the study area. Figure 5 shows the inverted result of SMC.

Inversion model for SMS

A remote sensing inversion model of soil moisture saturation was built based on the analysis of the relationship between the tone on the remote sensing image, the moisture and the soil brightness obtained by K-T transformation upon the remotely sensed data. The soil moisture saturation of the study area was then calculated by this model. Figure 6 shows the soil moisture saturation index retrieved from the ETM image.

Inversion model for VMC

Remote sensing inversion model of vegetation moisture content was built based on the analysis of the relationship between the tone on the remote sensing image, the moisture and the greenness obtained by K-T transformation upon the remotely sensed data. The vegetation moisture of the study area was then estimated by this model. Figure 7 shows the vegetation moisture content retrieved from the ETM image.

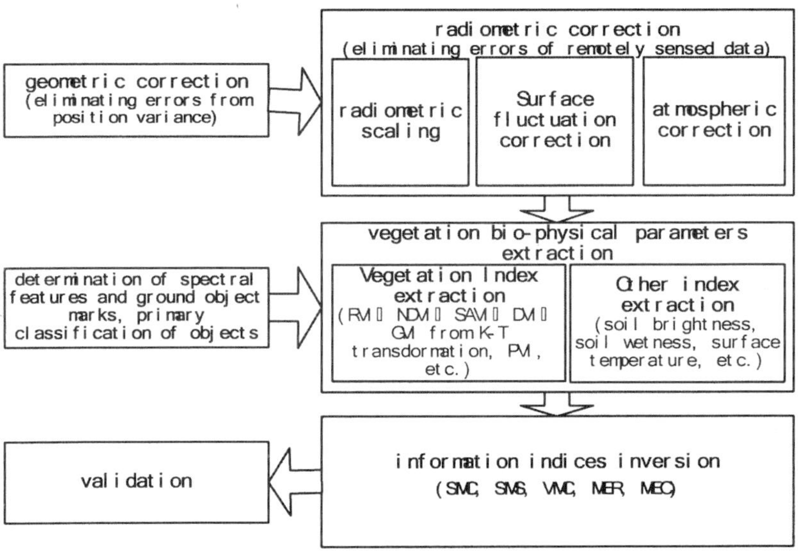

Fig. 4 Method outline.

Fig. 5 Soil moisture content retrieved from ETM.

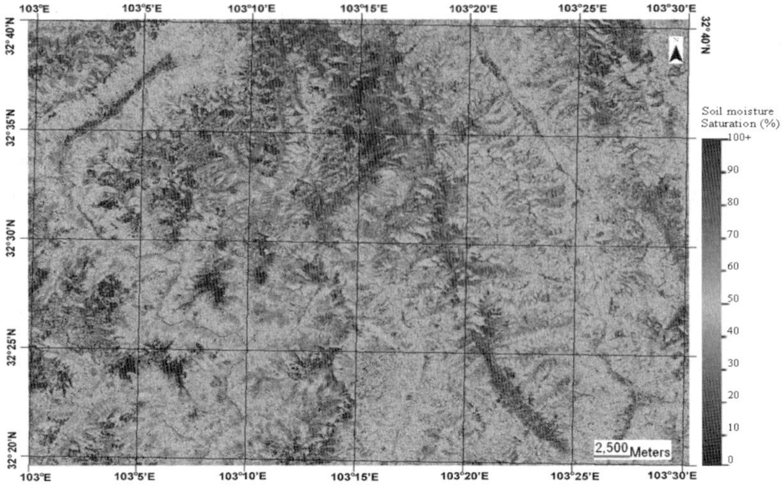

Fig. 6 Soil moisture saturation index distribution retrieved from ETM.

Fig. 7 Vegetation moisture content index distribution retrieved from ETM.

Fig. 8 MEC distribution estimation.

MEC model and Eco-water resource quantity estimation

The definitions of the eco-water and the MEC indicate that the MEC, which is related to the parameters mainly including the VMC, the SMC, the Vegetation Canopy Interception Content (VCIC) and the Topographic Measures (Mtopo), characterizes the quantity of eco-water conserved per unit area of the eco-water layer. Thus it is possible to estimate the total water quantity of the eco-water layer in the study area by constructing a remote sensing inversion model for the MEC. Figure 8 shows the MEC distribution calculated from the VMC, the SMC, the VCIC and the Mtopo.

DISCUSSION AND CONCLUSIONS

The eco-water develops its distribution under the consequences of the precipitation, the water system, the vegetation type, the soil type and the basic eco-environmental factors. Remotely sensed data of the study area in 2000 were used to validate the models (Jian, 2006; Li, 2007). This paper presents only the inverted results and left the detailed validation to two other papers.

Soil type Generally, different soils have different soil moisture saturations: sub-alpine meadow soil with comparatively low soil moisture saturation but brown soil with comparatively high soil moisture saturation. In some cases, however, the same kind of soil at different locations has different soil moisture saturations due to its different water potential energy. The soil type presents positive correlation with vegetation moisture content, which can be identified more obviously in summer when water conserved in the soil supports higher evaporation than in other seasons.

Vegetation distribution Vegetation distribution has a high correlation with soil moisture saturation. The distribution of soil moisture saturation conforms to the vegetation distribution: the forest with comparatively low soil moisture saturation, but crops with comparatively high soil moisture saturation since different types of vegetation have different capabilities of water conversion. The distribution of vegetation moisture content conforms to the vegetation distribution: different types of vegetation with different vegetation moisture content due to different types of vegetation having different evaporation abilities.

Temporal effects Time is a factor that should be included since vegetation status varies from time to time.

Based on the analysis above, we can conclude that the vegetation type, the vegetation distribution and the soil type have high correlation with the eco-water.

As a conclusion, the eco-water maintains the normal functions of the ecological environment, but it is difficult to quantify by ordinary measurement methods. The proposed vegetation-centred quantitative remote sensing method suggests a valid approach to study the eco-water (layer). This paper has no consideration for some uncertain factors such as wind speed and rainfall that may have relation to the eco-water layer. Further works are necessary to handle these factors to acquire more accurate results in later studies.

Acknowledgements Wu-Nian Yang was sponsored by the Open Research Foundation Project of Scientific & Technical Hazard Alleviation and Home Rebuilding of the State key Laboratory of Geo-hazard Prevention and Geo-environment Protection (ZYKT2009-02) and the Key Fundamental Research Foundation Project of the Educational Department of Sichuan Province (07ZA007).

REFERENCES

Bi, H. X. & Zhong, B. L. (2002) Review on integrating of RS and GIS with hydrology. *J. Soil & Water Conserv.* **12**(2), 45–49.

Canadell, J. &Noble, I. (2001) Challenges of a changing Earth. *Trends in Ecology & Evolution* **16**(12), 664–666.

Cihlar, J., St.-Laurent, L. *et al.* (1991) Relation between the normalized difference vegetation index and ecological variables. *Remote Sens. Environ.* **35**(2–3), 279–298.

Dong, Z. R. (2003) Theoretical framework for eco-hydraulics. *J. Hydraul. Engng* **1**, 1–6.

Entekhabi, D., Nakamurai, H. & Njoku, E. G. (1994) Solving the inverse problem for soil moisture and temperature profiles by sequential assimilation of multi-frequency remotely sensed observations. *IEEE Trans. Geosci. Remote Sens.* **3**(32), 438–448.

Feng, H. L., Xia, J. & Zhan, C. S. (2003) Advances on ecological and environmental water requirement research. *Progr. Geogr.* **22**(6), 591–598.

Gao, B. (1996) NDWIA normalized difference water index for remote sensing of vegetation liquid water from space. *Remote Sens. Environ.* **58**(3), 257–266.

Gao, Y. C. & Wang, C. Y. (2000) Biospheric aspects of hydrological cycle: BAHC plan and its research progress. *Progr. Geogr.* **19**(2), 97–102.

Huang, Y. L., Fu, B. J. & Chen, L. D. (2003) Advances in ecohydrological process research. *Acta Ecol. Sinica* **23**(3), 580–587.

Hutjes, R. W. A., Kabat, P., Running, S. W., Shuttleworth, W. J., Field, C. B., Bass, B., da Silva Dias, M. A. F., Avissar, R., Becker, A., Claussen, M., Dolman, A. J., Feddes, R. A., Fosberg, M., Fukushima, Y., Gash, J. H. C., Guenni, L., Hoff, H., Jarvis, P. G., Kayane, I., Krenke, A. N., Liu, C., Meybeck, M., Nobre, C. A., Oyebande, L., Pitman, A., Pielke Sr. R. A., Raupach, M., Saugier, B., Schulze, E.D., Sellers, P. J., Tenhunen, J. D., Valentini, R., Victoria , R. L., Vorosmarty, C. J. (1998) Biospheric aspects of the hydrological cycle. *J. Hydrol.* **212–213**, 1–21.

Jian, J. (2006) Quantitative investigation of eco-water with remote sensing in the upper of Minjiang River. PhD dissertation unpublished. Chengdu University of Technology, China.

Jiang, D. J., Wang, H. X. & Li, L. J. (2003) A review on the classification and calculating methods of ecological and environmental water requirements. *Progr. Geogr.* **22**(4), 369–378.

Kang, S. Z. & Liu, X. M. (1994) *Soil-Vegetation-Atmosphere Continuum Moisture Transmission Theory and Applications*, first edn. China WaterPower Press, Beijing, China.

Lei, Z. G. & Yang, S. X. (1998) *Soil Water Dynamics*, first edn. Tsinghua University Press, Beijing, China.

Li, X. W., Wang, J. D. & Zhu, C. G. (1995) Quantitative inversion study on remote sensing models. In: *Progress in Remote Sensing Science*, first ed. Science Press, Beijing, China.

Li, Y. X. (2007) Quantitative remote sensing study of eco-water information index system at Maoergai area in the upper Minjiang River. PhD dissertation unpublished. Chengdu University of Technology, China.

Liu, C. M. & Sun, R. (1999) Ecological aspects of water cycle: advances in soil vegetation atmosphere of energy and water fuxes. *Adv. Water Sci.* **9**(3), 251–258.

Lu, J. H. (2000) On the levels of atmosphere-biosphere interactions – a review of the models. *Adv. Earth Sci.* **15**(4), 434–438.

Ma, X. N. & Ye, Z. G. (1998) *Hydrology*, third edn. China Architecture and Building Press, Beijing, China.

Mo, X. G. (1998) Modeling and validating water and energy transfer in soil vegetation atmosphere system. *Acta Meteorol. Sinica* **56**(3), 323–332.

Pei, T. F. & Yu, X. M. (2001) *Ecological Dynamics*, first edn. Science Press, Beijing, China.

Peng, X. Y., Yang, H. & Jin, R.G. (2003) The effect of environment-water absorbed in water-absorbing resin on cole crop in sand soil. *J. Sichuan Univ. (Natural Science Edition)* **40**(4), 759–762.

Sun, S. F., Niu, G. Y. & Hong, Z. X. (1998) A water and heat transport model in arid and semiarid regions. *Chinese J. Atmospheric Sci.* **21**(1), 1–10.

Tian, D. L. & Kang, W. X. (2003) *Fir Forest Ecological System*, first edn. Science Press, Beijing, China.

Wang, B. Z. (2001) Ecological water & the forest-garden unirrigated culture. *Territory & Natural Resources Study* **1**, 50–51.

Wang, G. X., Qian, J. & Cheng, G. D. (2001) Current situation and prospect of the ecological hydrology. *Adv. Earth Sci.* **16**(3), 314–323.

Wan, X. N., Yang, W. N., Wu, B F., Sun, W. D. & Huang, Q. (2004) Conception of eco-water sphere and its application. *Adv. Earth Sci.* **6**(19), 117–121.

Wang, Y., Yang, W. N. & Yuan, P. X. (2001) Application of 3S technology in Wenchuan cultivation into tree planting of the Minjiang Upper Valley. *Geoinformation Sci.* **3**(4), 37–41.

Wu, Q. & Dong, D. L. (2001) Main eco-hydrological problems and basic methodologies. *Hydrogeology and Engineering Geology* **28**(2), 69–72.

Xia, J., Feng, H. L., Tan, G., Wang, G. S. & Zhu, Y. H. (2003) Eco-hydrology: Concept, Framework and System. *J. Irrig. Drain.* **22**(1), 4–10.

Xia, J. & Zuo, Q. T. (2006) Advances in international hydrological science research. *Adv. Earth Sci.* **21**(3), 256–261.

Yang, W. N., Jian, J., *et al.* (2008) Remote sensing inversion of eco-water resource quantity. In: *International Workshop on Earth Observation and Remote Sensing Applications* (Beijing, China), 1–6.

3 Integrated Water Resources Management

Sediment simulation at Upper Sangamon River basin using the THREW model

DENGFENG LIU, FUQIANG TIAN & HEPING HU

State Key Laboratory of Hydroscience and Engineering, Department of Hydraulic Engineering, Tsinghua University, Beijing 100084, China

tianfq@tsinghua.edu.cn

Abstract The soil erosion and sediment transportation processes are incorporated into the THREW model which is based on the Representative Elementary Watershed (REW) approach, and designed to be flexible enough to incorporate more hydrological and biogeochemical processes and adopt different closure relationships as part of the multi-process coupling framework of the watershed system. The scientific purpose of this study is to explore the spatial and temporal pattern of sediment processes at the catchment scale. The current model includes hydrological processes, soil erosion and sediment transportation. The hydrological processes include ground surface depression, canopy interception, saturation and infiltration excess runoff, overland and channel routing as described in the THREW model. The sediment processes include sediment erosion from the hillslope, deposition, re-entrainment and bed degradation in the main channel. The sediment simulation model is applied to the Upper Sangamon River basin, which is 3150 km^2 in central Illinois, USA, and intensively altered by agricultural expansion and artificial drainage. The coupling of ecological processes is our future work.

Key words hydrological model; sediment; catchment scale; representative elementary watershed

INTRODUCTION

The water cycle, which consists of the interaction of myriad of processes, may be characterized as a network of feedback cycles, that is, the water cycle is a "hypercycle". The interaction between these cycles results in cooperative dynamics in the presence of a balance between the positive and negative feedbacks (Kumar, 2007). The modelling of multi-processes at the catchment scale based on the hydrological modelling has been done and is being done by many researchers, one of the representative achievements of which is SWAT (Neitsch *et al.*, 2005). But SWAT was developed to predict the impact of land management practices on water, sediment and agricultural chemical yields in large complex watersheds with varying soils, land use and management conditions over long periods of time. SWAT is a continuous time model, i.e. a long-term yield model, and is not designed to simulate detailed, single-event flood routing (Neitsch *et al.*, 2005). So a model which is flexible enough to incorporate multi-processes and simulate the details of the processes is needed to study the interactions of multi-processes in the future. In this paper, a sediment simulation model at the catchment scale based on the THREW model is proposed and applied to the Upper Sangamon River basin (USRB) in Illinois, USA, which is intensively managed, but well gauged.

MODEL DESCRIPTION

The sediment simulation model at the catchment scale is developed based on the Representative Elementary Watershed (REW) approach and the THREW model (Tian, 2006; Tian *et al.*, 2008). The model is designed to be flexible enough to incorporate more hydrological and biogeochemical processes and adopt different closure relationships. The scientific purpose is to study the spatial and temporal pattern of sediment processes at the catchment scale. In the current model, the sediment processes are coupled to hydrological processes, but the feedback to hydrological processes is excluded.

Hydrological processes

In the THREW model (Tian *et al.*, 2008) used in USRB, the REW is divided into a surface layer and a subsurface layer. Four sub-regions (or zones) are defined in the surface layer, i.e. a bare soil zone (b-zone), a vegetated zone (v-zone), a sub-stream-network (t-zone), and the main channel reach (r-zone), while two sub-regions are defined in the sub-surface layer, i.e. an unsaturated zone (u-zone) and a saturated zone (s-zone). The hydrological processes including ground surface depression, canopy interception, saturation and infiltration excess runoff, overland and channel routing are modelled as described in THREW model, please refer to Tian *et al.* (2008) for details.

Sediment processes

The sediment processes in the model include sediment erosion from the hillslope, deposition, reentrainment and bed degradation in the main channel. The sediment is generated from the hillslope and is transported to the main channel by surface runoff. In the main channel, the sediment in the water is allowed to deposit to the channel floor while the loose sediment would be removed by the streamflow. If all of the loose sediment on the channel is removed, the degradation of the channel bed will happen. In the main channel, the sediment is transported, together with water flux from the upstream, downstream by streamflow.

In the model, sediment erosion from the hillslope to the main channel is assumed to associate with surface runoff from the t-zone to the r-zone. The formula of sediment erosion rate from the t-zone to the r-zone (Q_{st}, kg/s) is a further conceptualization of the Modified Universal Soil Loss Equation (Viney *et al.*, 1999; Neitsch *et al.*, 2005), as shown in equation (1).

$$Q_{st} = C s_t Q_t^{\delta} \tag{1}$$

where C and δ are the empirical parameters, s_t is the slope of the t-zone, and Q_t is the water discharge from the t-zone to the r-zone (m^3/s).

The modelling of deposition, re-entrainment and bed degradation in the main channel followed the SWRRB model (Simulator for Water Resources in Rural Basins) (Arnold *et al.*, 1990), i.e. the original of SWAT. A new zone for the sediment storage on the main channel floor (sf-zone) is added to the THREW model to model the sediment exchange between the water and the channel floor.

The deposition in the main channel depends on the falling velocity of the sediment particles. Stokes' Law for fall velocity (v_f, m/s) is used in this study which reads:

$$v_f = \frac{1}{18} \cdot \frac{\gamma_s - \gamma}{\gamma} \cdot g \cdot \frac{d^2}{\nu} \tag{2}$$

where γ_s and γ are the density of sediment and water (kg/m^3), g is the gravity acceleration (m/s^2), d is the sediment particle diameter (m), and ν is the kinematic viscosity coefficient of water (m^2/s).

Travel time in the main channel of each REW (t, s) is:

$$t = L_c / v \tag{3}$$

where L_c is the length of channel (m), and v is the velocity (m/s).

The depth that sediment particles will fall during travel time (y_f, m) is:

$$y_f = v_f \cdot t \tag{4}$$

The instantaneous sediment delivery ratio in each main channel (DR) is:

$$DR = \begin{cases} 1 - 0.5 y_f / y_r & \text{if } y_f \le y_r \\ 0.5 y_r / y_f & \text{if } y_f > y_r \end{cases} \tag{5}$$

where y_r is water depth in main channel (m).

The deposition rate in main channel (dep, kg/s) is:

$$dep = \frac{S_{sr}}{t}(1 - DR) \tag{6}$$

where S_{sr} is the sediment storage in the water of main channel (kg).

If there is loose sediment on the channel floor, i.e. the sediment storage in sf-zone (S_{sf}, kg) is positive, the sediment re-entrainment occurs. Otherwise, the degradation of the channel floor begins. The total degradation (*deg*, kg/s) is the sum of the reentrainment and degradation, as shown in equation (7).

$$deg = k_1\ \gamma Q_r\ s_r \text{ if } S_{sf} > 0$$
$$deg = k_1\ k_2\ \gamma Q_r\ s_r \text{ if } S_{sf} \leq 0 \qquad (7)$$

where k_1 and k_2 are parameters, Q_r is the streamflow in the main channel (m^3/s), and s_r is the slope of the main channel.

STUDY AREA

The Upper Sangamon River Basin (USRB) is 3150 km^2 at the confluence of the Illinois River in central Illinois, USA. Average annual precipitation (1984–2007) was approximately 870 mm/year. Snow represents approximately 5% of the precipitation on average. Average annual potential evaporation (1984–2007) was approximately 1630 mm/year. Annual average temperature in the basin is 11°C, with monthly averages ranging from –5°C in January to 24°C in July. Average annual water yield measured at the USGS stream gauging station at Monticello (drainage area of 1425 km^2, Fig. 1) during 1971–2000 was 300 mm/year.

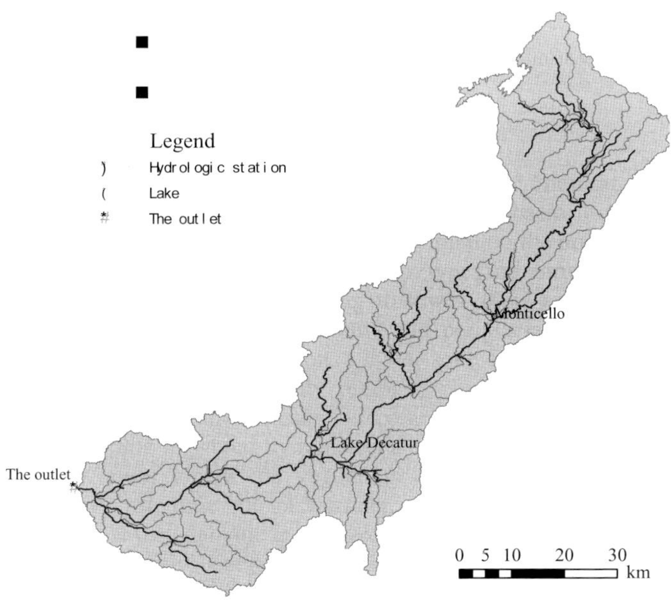

Fig. 1 The location of Monticello and Lake Decatur in USRB.

Lake Decatur is located at the middle stream of USRB with a catchment area of 2400 km^2, and is a water supply reservoir that diverts water to the City of Decatur, with a population of 86 000. The dam was modified in 1956 to increase the maximum capacity of the lake to 34.56 million m^3 (Keefer & Bauer, 2005). The operation regime of the reservoir is unknown, so the USGS stream gauging station at Monticello is selected for the calibration and validation of the model, which is at the upstream of the reservoir with the drainage area of 1425 km^2. But the whole USRB is modelled in the modelling period.

DATA

The data used in the modelling include the Digital Elevation Model (DEM), soil class, LAI, NDVI, land cover, precipitation, potential evaporation, observed streamflow, observed sediment concentration and discharge. The geographic data such as drainage area, slope, channel length, etc. are extracted from the DEM with the resolution of 1 arc second. The USRB is discretized to 51 REWs, as shown in Fig. 1. The soil class is mainly extracted from the STATSGO database. The LAI is extracted from the product of "MODIS/Terra Leaf Area Index/FPAR 8-Day L4 Global 1 km SIN Grid V004", NDVI from "MODIS/Terra Vegetation Indices Monthly L3 Global 1 km SIN Grid V004", and land cover from "MODIS/Terra Land Cover Type Yearly L3 Global 1 km SIN Grid V004". The hourly precipitation data from National Climatic Data Center (NCDC) of NOAA are used in the model. The potential evaporation is extracted from North American Regional Reanalysis (NARR) of NOAA. Observed stream flow at Monticello is downloaded from the website of USGS. The observed sediment concentration and discharge at Monticello comes from the Benchmark Sediment Monitoring Program by Illinois State Water Survey.

RESULT AND DISCUSSION

In the simulation, the water year is from October to September of next year. The period from October 1993 to September 1994 is selected for the model's warming up to eliminate the impact of the initial conditions. The period for the calibration is from October 1994 to September 1997 and the period for the validation is from October 1997 to September 2007. The simulated streamflow and sediment discharge at Monticello is used in the calibration and validation.

Calibration

The main parameters subject to calibration are listed in Table 1. Two standard indices, i.e. Nash-Sutcliffe efficiency coefficient (NSEC), and the coefficient of determination (R^2), and the regime curve are used to guide manual calibration of the hydrological model.

During the 3-year calibration period, we obtained an NSEC value falling within 0.69–0.86, and R^2 within 0.70–0.86 as shown in Table 2. For the whole calibration period, NSEC is 0.78 and R^2 is 0.80. Figure 2 presents the simulated regime curve compared with the observed. Except January and April, the simulated regime fits the observed regime curve very well.

Table 1 Main calibrated parameters.

Symbol	C	δ	k_1	k_2
Value	300.0	2.0	0.8	0.2
In equation	(1)	(1)	(7)	(7)

Table 2 Evaluation merits.

Type	Year	NSEC	R^2
Calibration	Oct. 1994–Sep. 1995	0.77	0.86
	Oct. 1995–Sep. 1996	0.86	0.86
	Oct. 1996–Sep. 1997	0.69	0.70
	Oct. 1994–Sep. 1997	0.78	0.80
Validation	Oct. 2004–Sep. 2005	0.78 (max)	0.82
	Oct. 1999–Sep. 2000	0.10 (min)	0.71
	Oct. 2004–Sep. 2005	0.78	0.82 (max)
	Oct. 2000–Sep. 2001	0.34	0.48 (min)
	Oct. 1997–Sep. 2007	0.68	0.70

The observed sediment concentration and discharge at Monticello are used to calibrate the sediment model, which is, however, not continuous; the data from October 1994 to February 1996 are unavailable. So the calibration period for the sediment is from October 1993 to September 1997. The calibrated parameters for sediment are shown in Table 1. The simulated sediment concentration and discharge at Monticello are presented in Fig. 3 and Fig. 4, respectively. The simulated values are of the same magnitude with the observed values which suggests that the simulated values are reasonable.

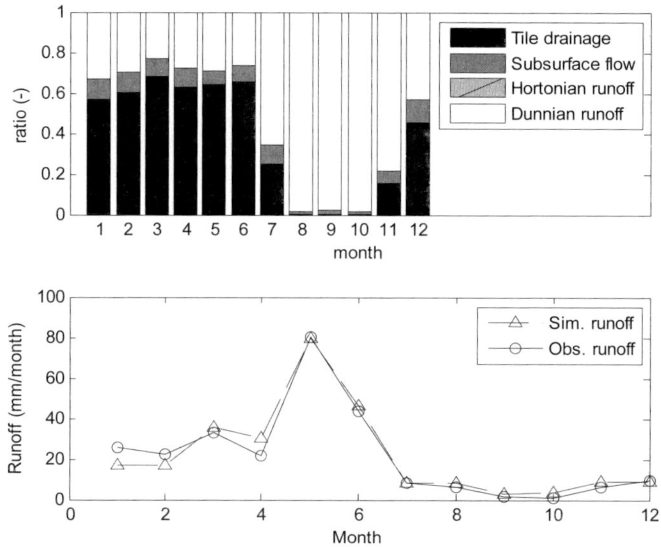

Fig. 2 The regime curve in the period for the calibration.

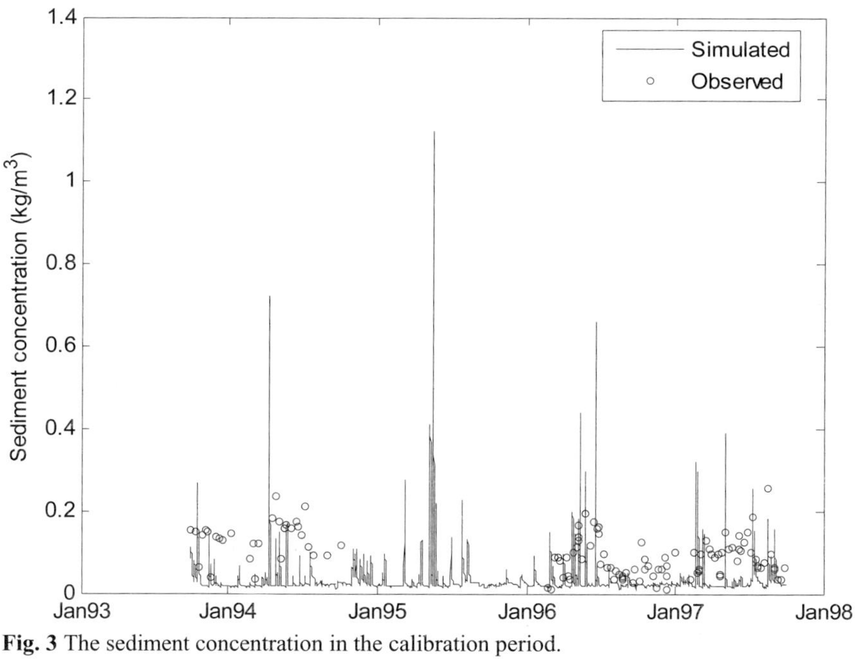

Fig. 3 The sediment concentration in the calibration period.

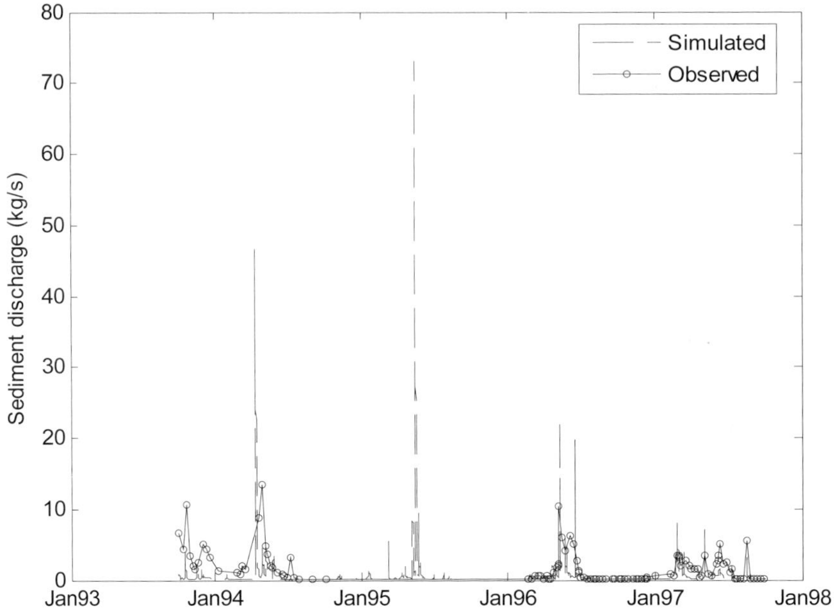

Fig. 4 The sediment discharge in the calibration period.

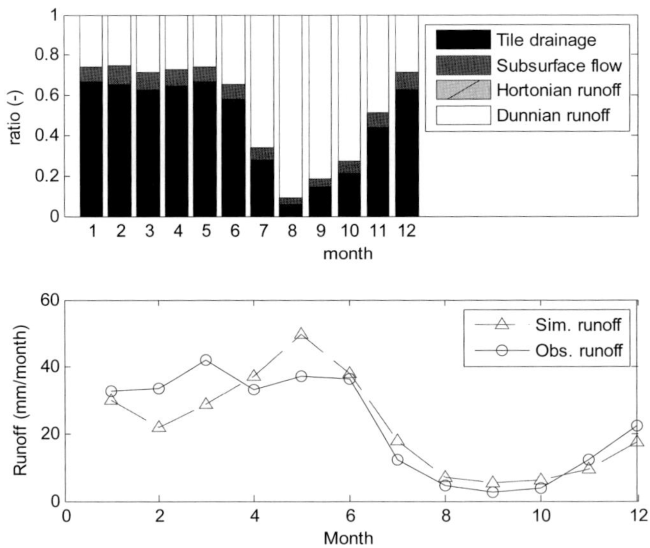

Fig. 5 The regime curve in the period for the validation.

Validation

Using the parameters obtained by calibration, the model is then validated from October 1997–September 2007. During the 10-year validation period, we obtain an NSEC value falling within 0.10–0.78, and R^2 within 0.48–0.82, as shown in Table 2. For the whole validation period, NSEC is 0.68 and R^2 is 0.70. In Fig. 5, the runoff generation is dominated by tile drainage in December and from January to June while saturation excess runoff dominates from July to November. In the regime curve, the streamflow is underestimated in February and March but overestimated in June

(Fig. 8). So the modelling of evapotranspiration should be diagnosed further. The predicted sediment concentration and discharge during the validation period are presented in Fig. 6 and Fig. 7, respectively. The performance of the sediment model in the validation period is comparable with that in the calibration period.

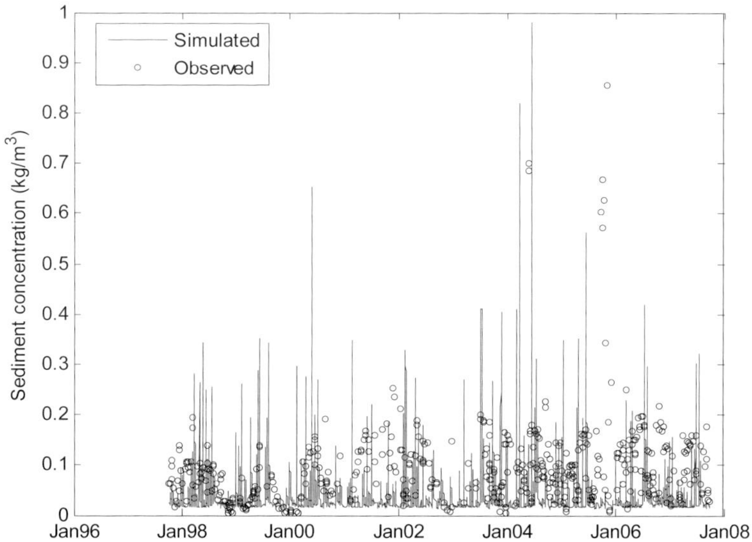

Fig. 6 The sediment concentration in the validation period.

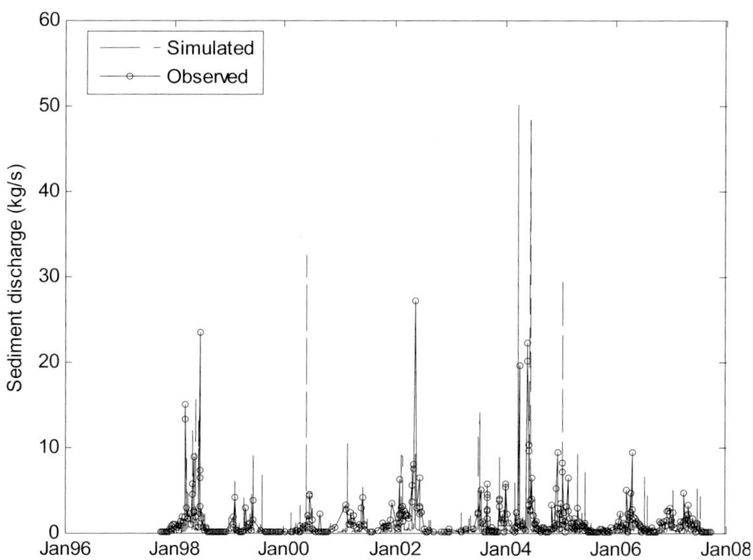

Fig. 7 The sediment discharge in the validation period.

DISCUSSION

The concept of the sediment delivery ratio (SDR) describes the complex pattern of the behaviour of sediment transfer processes at catchment scale and the spatial pattern of SDR is derived for many watersheds all over the world (Lu, 2005). The spatial pattern of SDR over all of the sub-catchments in the whole simulation period is shown in Fig. 8. The relationship of SDR *versus* drainage area is visible, excluding the two smallest sub-catchments as shown in Fig. 8. As the catchment area increases, the ratio of sediment deposited in the main channel will increase.

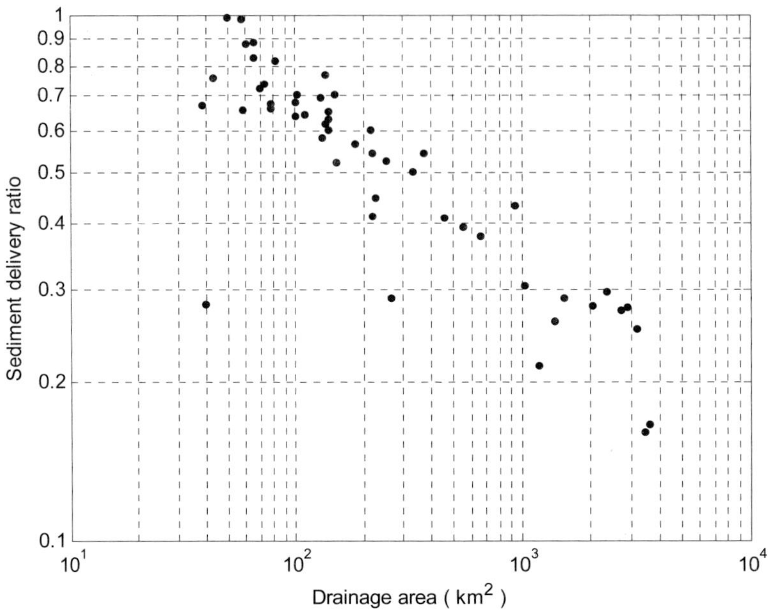

Fig. 8 Sediment delivery ratio during the whole simulation period.

CONCLUSION

A sediment simulation model at the catchment scale is developed based on the THREW model and applied to the Upper Sangamon River basin. The equations that match the spatial and temporal scale of the THREW model are developed to describe the sediment erosion, deposition and re-entrainment. After the calibration, the model works well in the validation period. The next step is to include the feedback of the sediment processes to the hydrological processes.

Acknowledgements The authors wish to thank Murugesu Sivapalan and Hongyi Li for their constructive comments. This work is funded by the projects 50779022 and 50509013 sponsored by the National Natural Science Foundation of China. With the financial support of the China Scholarship Council, Dengfeng Liu has been studied in University of Illinois at Urbana-Champaign, USA and part of the work was done during this period.

REFERENCES

Allen, R. G.., Pereira, L. S., Raes, D. & Smith, M. (1998) FAO Irrigation and Drainage Paper no. 56 Crop evapotranspiration: guidelines for computing crop water requirements, 290 pp., ISBN 92-5-104219-5.

Amenu, G. G. & Kumar, P. (2007) A model for hydraulic redistribution incorporating coupled soil-root moisture transport. *Hydrol. Earth Syst. Sci. Discuss.* **4**, 3719–3769.

Arnold, J. G., Williams, J. R., Nicks, A. D. & Sammons, N. B. (1990) *SWRRB: A Basin Scale Simulation Model for Soil and Water Resources Management.* Texas A&M University Press, College Station, Texas, USA.

Green, C. H., Tomer, M. D., Luzio, M. D. & Arnold, J. G. (2006) Hydrologic evaluation of the soil and water assessment tool for a large tile-drained watershed in Iowa. *Transactions of the ASAE* **49**(2), 413–422.

Keefer, L. & Bauer, E. (2005) Watershed Monitoring for the Lake Decatur Watershed 2000–2003. Illinois State Water Survey Contract Report 2005–09, Champaign, Illinois, USA.

Kumar, P. (2007) Variability, feedback, and cooperative process dynamics: elements of a unifying hydrologic theory. *Geography Compass* **1**(6): 1338–1360, 10.1111/j.1749-8198. 2007. 00068.x.

Lu, H., Moran, C. J., & Sivapalan, M. (2005) A theoretical exploration of catchment-scale sediment delivery. *Water Resour. Res.* **41**, W09415, doi:10.1029 /2005WR004018.

Nash, J. E, & Sutcliffe, J. V. (1970) River flow forecasting through conceptual models: Part 1-A discussion of principles. *J. Hydrol.* **10**, 282–290.

Neitsch, S. L., Arnold, J. G., Kiniry, J. R., & Williams, J. R. (2005) Soil and Water Assessment Tool theoretical Documentation Version 2005. Texas Water Resources Institute, College Station, Texas, USA.

Rupp, D. E., Owens, J. M., Warren, K. L. & Selker, J. S. (2004) Analytical methods for estimating saturated hydraulic conductivity in a tile-drained field. *J. Hydrol.* **289**, 111–127.

Tian, F. (2006) Study on thermodynamic watershed hydrological modeling (dissertation), Tsinghua University, Beijing. China.

Tian F., Hu, H. & Lei, Z. (2008) Thermodynamic watershed hydrological model (THModel): constitutive relationship. *Science in China, Series E* **51**(9), 1353–1369.

Viney N. R., & Sivapalan, M. (1999) A conceptual model of sediment transport: application to the Avon River Basin in Western Australia. *Hydrol. Processes* **13**(5), 727–743.

Evolving characteristics of groundwater resources in the Northeast Vast Plain of China

CHANG-LEI DAI[1,2,3], BAO-MING CHI[4], YONG YE[5] & ZHI-GANG WANG[6]

1 *School of Water Conservancy and Electric Power, Heilongjiang University, Heilongjiang Harbin, 150080, China*
daichanglei@126.com

2 *Institute of Frigid Zone Groundwater, Heilongjiang University, Heilongjiang Harbin 150080, China*

3 *Postdoctoral Scientific Research Station of Civil Engineering, Harbin Institute of Technology 150080, China*

4 *Institute of Disaster Prevention Science and Technology, Beijing 101601, China*

5 *Institute of Water Resources, IWHR, Beijing 100044, China*

6 *Songliao Water Resources Commission, Ministry of Water Resources, Jilin Changchun 130026, China*

Abstract With the influence of monsoon climate, precipitation in northeast China occasionally varies significantly. Therefore, a huge amount of groundwater is exploited every year, and aquifers have to supply 45% of the total water demands in northeast China. One of the most famous large plains in China, the Northeast Vast Plain, is located in an area that crosses the Liaoning Province, Jilin Province, Heilongjiang Province, and the Inner Mongolian Autonomous Region, and consists of the Liaohe Plain, the Songnen Plain and the Sanjiang Plain. With a total area of 33.70×10^4 km^2, the Northeast Vast Plain has become a famous area of industrialization, energy supply and grain production. Based on research from China and worldwide on the large-scale phreatic cycle and the result of the Second National Water Resource Integrated Evaluation by the Ministry of Water Resources, this paper gathers the statistics of groundwater recharge of every subarea in the Northeast Vast Plain during 1956–2000 (a total of 45 years). Then, groundwater resources and environment problems, especially the effect of humans against nature, is construed. The evolving characteristics of groundwater in those 45 years are pointed out. By choosing different elements to represent the natural evolution and the effect of human activities, and discussing the groundwater resources evolving process and its related degree, this paper finds out the coupling mechanism of groundwater resources in the Northeast Vast Plain of China, and illustrates the principles that the groundwater system responds to exterior factors such as natural changes and especially human activities. The paper also forecasts the evolving direction of groundwater resources in the Northeast Vast Plain, and brings forward some countermeasures and tactics to redress and control groundwater resources by humans. This paper provides a certain reference value to sustainable utilization of groundwater resources and sustainable social economic development in northeast China.

Key words Northeast Vast Plain; groundwater resources; evolution; natural evolvement; human activity

INTRODUCTION

In a monsoon climate area, as a result of the tempestuous change with time of atmospheric precipitation, the groundwater aquifer not only supplies a great amount of fresh water for manufacturing and living, it also recharges the surface water to maintain the river base flow in low flow periods. The groundwater resources have always played a significant role in the water resources system. The Northeast Vast Plain is one of the top three noted large plains in China, and is located in the Pacific Ocean monsoon climate area. The water exploitation and usage rate is very high in the Northeast Vast Plain, and 45% of the total discharge of water supply comes from groundwater exploitation.

Taking the Northeast Vast Plain as an example, according to the results of the Second National Water Resources Integrated Evaluation finished by the Ministry of Water Resources, the evolving characteristics of groundwater were analysed and both the relationship between environmental change and the influence of human activities and related factors are discussed in this paper.

HYROGEOLOGIC CHARACTERISTICS ANALYSIS

As its name suggests, the Northeast Vast Plain is located in the northeast and crosses four provinces, including Liaoning Province, Jilin Province, Heilongjiang Province, and the Inner

Mongolian Autonomous Region. Based on the rapid development, the Northeast Vast Plain has become a famous area of industrialization, energy supply and grain production in China.

The Northeast Vast Plain consists of the Liaohe Plain, the Songnen Plain and the Sanjiang Plain. The Songliao Plain is made up of the Liaohe Plain and its adjacent plain, the Songnen Plain. The Sanjiang Plain also connects to the Songliao Plain by a section of narrow plain between mountains. The Daxing'an Mountain range, of which the northwest part consists of the Xiaoxing'an Mountains and the Heilongjiang River and Wusulijiang River – the boundary rivers between China and Russia, lies northwest of the Northeast Vast Plain. To the southeast of the Northeast Vast Plain are the Changbai Mountain Chain and the Bohai Sea borders on the south side. The location and the spatial distribution are shown in Fig. 1.

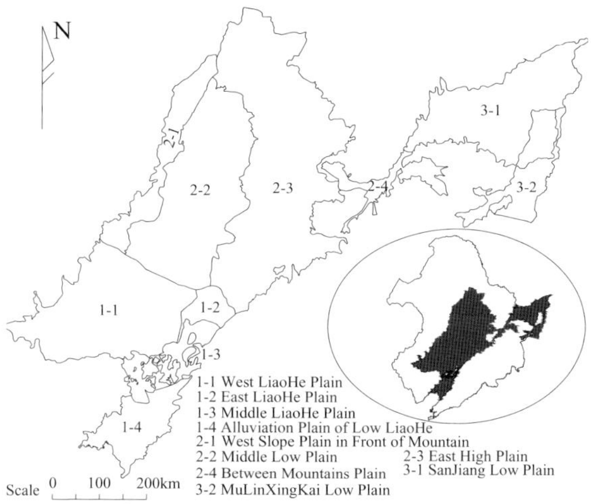

Fig. 1 location and spatial distribution of Northeast Vast Plain.

The hydrogeological characteristics of the Northeast Vast Plain can be divisionally introduced as follows:

The Liaohe Plain

The Liaohe Plain is divided into four parts: East Liaohe Plain, West Liaohe Plain, Middle Liaohe Plain and Low Liaohe Plain, the total area of which is 9.21 × 104 km^2.

The East Liaohe Plain is mainly constituted of hills, alluvial flats and first and second terraces. Groundwater occurs mostly in the pore space of sand and sandy gravel in incompact rock aquifer and the fissures of the cretaceous clastic rock aquifer.

The West Liaohe Plain is a downward basin, with the low open region lies to the southeast side, which borders on the Songnen Plain on the north and east sides. The northeast is the mountainous region of the Daxing'an Mountain range, which is a barrier for the plain, and the south side is the Nunuerhu Mountain. The basin has descended since the Mesozoic and Cenozoic Era at all times, and has occasionally ascended. The Tertiary sediment was discontiguous, but in the Quaternary period incompact terraces accumulated to a large-scale in the plain.

The Low Liaohe Plain was circled by a massif on the west, east and north side. The central section of the Low Liaohe Plain was an amplitude depressed basin, and descended all through the YanShan constructing period. Consequently, huge thick clastic fragments of the Cretaceous, Tertiary and Quaternary systems were deposited in this basin. The pore phreatic water and confined water distributed in the central section, and at the east and west sides distributed pore phreatic water that was in girdle and zonal distribution. Owing to the raising of Kangfa breck, the north also distributed pore phreatic water. The hydrogeological characteristics of the entire Low Liaohe Plain can be summarized as follows: the east, west and north sides are the zone of recharge. The central section is a pressurization runoff zone. The basin is a synclinal basin with the Tertiary system basement inclining to the southwest. The aquifer is very thick vertically. The groundwater type can be divided into pore phreatic water in the incompact rock of the Quaternary system and pore fissure water in the Tertiary system.

The Songnen Plain

The total area of the Songnen Plain is 19.91 × 104 km^2. It is an extensive reservoir basin with multilayer cascade structural confined aquifers, which include the lower Pleistocene Series and Eocene Series (Yi'an formation of Oligocene Series, Da'an formation of Miocene Series, Taikang formation of Pliocene Series) of the Tertiary System. The Songnen Plain can be divided into four hydrogeology regions: east high plain, middle low plain, west slope plain in front of the Daxing'an Mountain range, and the river valley plain. The east high plain and the west slope plain in front of DaXingAnLing Mountains are the zone of discharge for mountainous bedrock fissure water and the main area of recharge for confined water in the central basin.

The Sanjiang Plain

With a total area of 4.58 × 10^4 km^2, the Sanjiang Plain was formed in the Cenozoic and Mesozoic era. About 300 m thickness of loose sediment has been deposited since the Quaternary Period. The west, south and east sides of the Sanjiang Plain are surrounded by hills. The Sanjiang Plain is an alluviating and lacustrine plain formed by three rivers, which are the Heilongjiang River in the north, the Songhua River in the centre and the Wusuli River in the east. As a result of the alluviating and deposit effects, the grain size of the aquifer in the Sanjiang Plain is coarse and the hydraulic permeability is very high. There is no good aquiclude vertically in aquifers.

Table 1 Hydrogeological Division in the Northeast Vast Plain.

Grade 1	Grade 2	Area (10^4 km^2)	Proportion
Liao He Plain	East Liaohe Plain, West Liaohe Plain, Middle Liaohe Plain, Low Liaohe Plain	9.21	27.33%
Song Nen Plain	West slope plain in front of mountain, Middle low Plain, East high Plain, Between mountains Plain	19.91	59.08%
San Jiang Plain	Sanjiang low Plain, MuLingXinKai low Plain	4.58	13.59%
Total		33.70	100%

EVOLVING CHARACTERISTICS OF THE QUANTITY OF GROUNDWATER RESOURCES AND ITS EXPLOITATION COUNTERMEASURES

To analyse the evolving characteristics of the quantity of groundwater resources on a large-scale, the quantity of groundwater can be defined as the difference between the total quantity of groundwater recharge and quantity of groundwater irrigation. Therefore, the quantity of groundwater recharge can be gained by adding these eight items, including recharge of rainfall penetration, recharge of river channel leakage, recharge of reservoir and pond leakage, recharge of trench leakage, recharge of field ditch leakage, artificial groundwater recharge, lateral recharge in front of mountains, and recharge of groundwater irrigation regression. Then, the equation can be described as follows:

Total recharge of groundwater:

$$Q_t = Q_p + Q_r + Q_p + Q_{tl} + Q_f + Q_a + Q_l + Q_i \quad (1)$$

Quantity of groundwater resources:

$$Q_{gr} = Q_t - Q_i \quad (2)$$

Table 2 Quantity of groundwater resource sequence of the Northeast Vast Plain in 1956–2000.

Year	Groundwater Resource ($10^9 m^3$/year)			
	Liaohe Plain	Songnen Plain	Sanjiang Plain	Northeast Plain
1956	7.05	16.93	3.79	27.77
1957	6.35	16.69	4.40	27.44
1958	4.52	10.46	3.16	18.14
1959	7.44	13.65	4.77	25.86
1960	6.31	15.89	4.06	26.27
1961	4.99	12.42	2.96	20.36
1962	6.33	13.08	3.63	23.04
1963	5.77	15.51	3.77	25.04
1964	6.14	13.02	3.60	22.76
1965	4.30	13.37	4.14	21.80
1966	5.58	14.58	2.77	22.94
1967	4.04	10.93	2.52	17.49
1968	4.68	11.00	3.36	19.05
1969	7.61	14.76	3.28	25.65
1970	5.30	10.90	2.66	18.86
1971	5.85	13.21	4.39	23.45
1972	4.34	12.29	4.22	20.84
1973	5.33	12.60	3.70	21.63
1974	5.98	13.20	3.90	23.08
1975	5.22	11.13	2.28	18.64
1976	5.50	10.92	2.27	18.68
1977	6.73	13.33	2.12	22.18
1978	5.14	10.84	2.83	18.80
1979	6.47	10.89	2.40	19.75
1980	3.75	13.01	3.19	19.95
1981	5.33	15.24	4.87	25.45
1982	4.67	9.45	2.65	16.77
1983	6.05	14.35	3.26	23.66
1984	6.43	14.82	4.03	25.27
1985	7.53	16.28	3.50	27.30
1986	8.12	13.40	2.55	24.08
1987	6.21	16.00	4.14	26.34
1988	4.96	15.50	2.99	23.45
1989	3.85	10.35	3.05	17.26
1990	6.79	14.81	3.54	25.14
1991	7.54	15.33	4.02	26.89
1992	4.99	11.54	3.08	19.61
1993	5.39	14.74	3.22	23.35
1994	7.40	16.32	4.97	28.69
1995	5.44	10.17	2.91	18.53
1996	4.65	12.59	3.12	20.36
1997	4.54	13.30	3.32	21.15
1998	8.41	18.29	3.47	30.17
1999	3.82	10.80	2.10	16.72
2000	3.42	10.25	2.76	16.44
Average	5.69	13.29	3.37	22.36

Modulus of groundwater resources:

$$M_r = Q_{ri}/A_i \quad 3$$

where, in equation (1), the Q_t is the total quantity of groundwater recharge; Q_p is recharge of rainfall penetration; Q_r is recharge of river channel leakage; Q_p is recharge of reservoir and pond leakage; Q_{tl} is recharge of trench leakage; Q_f is recharge of field ditch leakage; Q_a is artificial groundwater recharge; Q_l is lateral recharge in front of the mountain; Q_i is recharge of groundwater irrigation regression. In equation (2), Q_{gr} is the quantity of groundwater resources and the other symbols have the same meaning as in equation (1). In equation (3), M_r is the modulus of groundwater resources; Q_{ri} is the groundwater resources in a particular region and A_i is the corresponding area of that region.

Based on the Second National Water Resources Integrated Evaluation result completed by the Ministry of Water Conservation, choosing corresponding data about groundwater resources in the northeast by the Song-Liao Water Resources Commission Ministry of Water Resources, this paper gathers statistics on the quantity of groundwater resources of every subarea in the Northeast Vast Plain during 1956–2000 (totalling 45 years); see Table 2. According to the data, this paper analyses the evolving characteristics about groundwater resources in the Northeast Vast Plain in the recent half century.

From Table 2, the average quantity of groundwater resources is 22.36 billion m^3. The richest groundwater resources are in Songnen Plain, with the Liaohe Plain and the Sanjiang Plain being the second and third richest, respectively. As we know, the area of Songnen Plain is the biggest, therefore these numbers just represent the total number of groundwater resources, not the groundwater yield property.

The relationship and characteristics of the quantity of groundwater resources in the Northeast Vast Plain can be seen in Fig. 2. The groundwater resources fluctuate around the average number. The general trend can be defined as a horizontal line. The time series of Liaohe Plain resembles that of Sanjiang Plain.

To determine the groundwater yield property, the modulus of groundwater resources should be given. Table 3 lists the modulus sequence of groundwater resources between 1956 and 2000. The average modulus of groundwater resources is 66.34 mm/year, not the same as the quantity of groundwater resources. The top modulus in the Northeast Plain is in Sanjiang Plain, which is 73.61 mm/year, the next is Songnen Plain, which is 66.76 mm/year, and the last is Liaohe Plain, the modulus of which is 61.83 mm/year.

In the modulus sequence graph of groundwater resources (Fig. 3), the change is a little complex. The general changing characteristics of the modulus of groundwater resources in three plains is consistent, especially since 1984. The trend is of slow reduction. The biggest amplitude of modulus is in the Sanjiang Plain.

The sum-up of quantity and modulus of groundwater resources in Northeast Vast Plain is listed in Table 4. Songnen Plain has the biggest number of average groundwater resources and its proportion, but Sanjiang Plain has the biggest number of modulus of groundwater resources. At the same time, Sanjiang Plain has the biggest square deviation of modulus. So, the change of groundwater resources is more variable in Sanjiang Plain.

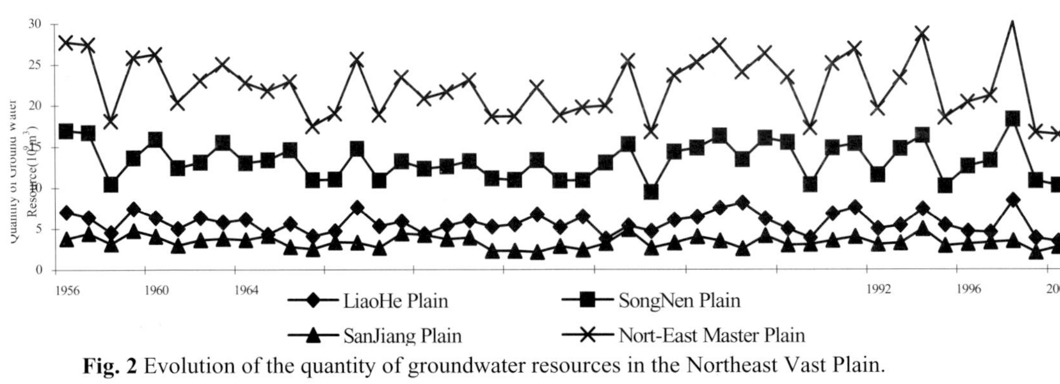

Fig. 2 Evolution of the quantity of groundwater resources in the Northeast Vast Plain.

Table 3 Modulus sequence of groundwater resources of the Northeast Vast Plain in 1956–2000.

Year	Modulus of groundwater resources (mm/year)			
	Liaohe Plain	Songnen Plain	Sanjiang Plain	Northeast Plain
1956	76.54	85.05	82.75	73.67
1957	68.97	83.84	96.06	72.80
1958	49.08	52.54	69.00	48.12
1959	80.78	68.54	104.20	68.59
1960	68.57	79.81	88.67	69.67
1961	54.16	62.38	64.53	54.02
1962	68.70	65.72	79.16	61.11
1963	62.61	77.89	82.21	66.42
1964	66.71	65.37	78.60	60.37
1965	46.67	67.13	90.29	57.83
1966	60.62	73.25	60.44	60.84
1967	43.88	54.89	55.05	46.39
1968	50.85	55.25	73.39	50.52
1969	82.58	74.15	71.70	68.04
1970	57.51	54.74	58.14	50.02
1971	63.47	66.34	95.92	62.19
1972	47.10	61.72	92.04	55.28
1973	57.92	63.29	80.68	57.37
1974	64.95	66.30	85.09	61.22
1975	56.70	55.91	49.88	49.44
1976	59.70	54.83	49.57	49.56
1977	73.09	66.96	46.30	58.84
1978	55.76	54.45	61.74	49.88
1979	70.20	54.69	52.33	52.39
1980	40.70	65.32	69.74	52.91
1981	57.90	76.55	106.38	67.50
1982	50.69	47.48	57.89	44.49
1983	65.74	72.06	71.17	62.76
1984	69.77	74.42	87.99	67.04
1985	81.72	81.75	76.38	72.42
1986	88.20	67.32	55.68	63.86
1987	67.39	80.36	90.34	69.88
1988	53.84	77.85	65.23	62.19
1989	41.83	51.98	66.69	45.77
1990	73.72	74.37	77.33	66.68
1991	81.82	77.01	87.84	71.33
1992	54.13	57.95	67.34	52.01
1993	58.52	74.05	70.22	61.93
1994	80.37	81.98	108.45	76.10
1995	59.12	51.09	63.57	49.15
1996	50.46	63.25	68.10	54.00
1997	49.25	66.78	72.49	56.11
1998	91.29	91.84	75.82	80.02
1999	41.42	54.27	45.88	44.35
2000	37.18	51.50	60.34	43.61
Average	61.83	66.76	73.61	66.34

Except the quantity of groundwater resources, the natural characteristics in the Northeast Vast Plain can be described as follows: the Northeast Vast Plain is located in the continental monsoon climate zone, where the annual average precipitation is between 300 mm and 1200 mm and decreases from southeast to northwest. The precipitation mainly occurs in summer, and the

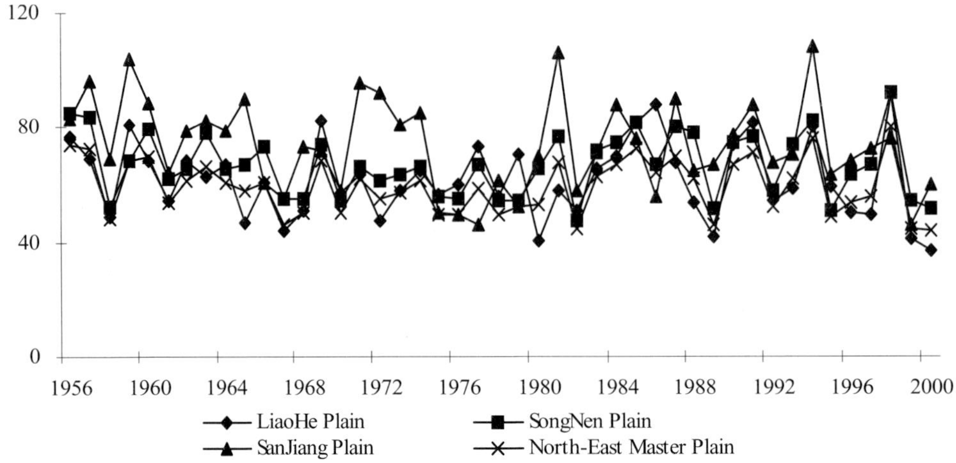

Fig. 3 the Modulus Sequence graph of groundwater resources.

Table 4 Quantity and modulus of groundwater resources in the Northeast Vast Plain.

Subarea Name	Average (10^9m^3/year)	Proportion	Modulus (mm/year)	Square deviation of modulus
LiaoHe Plain	5.69	25.4%	61.83	177.54
SongNen Plain	13.29	59.4%	66.76	119.26
SanJiang Plain	3.37	15.2%	73.61	252.49
Total	22.36	100%	66.34	139.26

precipitation from June to September is about 70–80% of the total precipitation in one year. From year to year precipitation varies significantly, and the C_v is between 0.1 and 0.3. Average evaporation is mainly between 900 mm and 2000 mm (using the 20 cm evaporating dish observed), and increases by degrees from northeast to southwest.

The water resources in the Northeast Vast Plain are indigenous. The total average water resources per capita makes up only about 1/6 of the world average and 2/3 of our nation. The water resources of one unit of infield are 588 m^3 which makes up only 1/3 of our national resources. Along with the advancement in society and economic development, the demand for water resources continually increases. Water resource deficiency and deterioration of water quality is an urgent problem in this area.

From the discussion above, several conclusions can be drawn:

(1) The groundwater resources change alternately in Northeast Vast Plain while the varying trend is consistent.

(2) The changing extent of groundwater resources in Sanjiang Plain is the biggest, Songnen Plain takes second place, and Liaohe Plain is the smallest.

(3) Songnen Plain has the richest groundwater resources, which mean 1.329 billion m^3 every year. Liaohe Plain is the second, with 0.569 billion m^3 every year. Sanjiang Plain is the smallest, which is only 0.337 billion m^3 every year. Over the entire vast plain, the groundwater resources are 2.236 billion m^3 every year.

(4) Considering to the modulus, Sanjiang Plain is 73.61 mm/year; Songnen Plain is 66.76 mm/year; Liaohe Plain is 61.83 mm/year. The modulus of the whole plain is 66.34 mm/year.

(5) Based on the data and the environmental condition in the Northeast Vast Plain, the change of groundwater resources is mainly controlled by atmospheric precipitation; the quantity is mainly influenced by the structure and boundary conditions of the hydrogeology unit; the changing extent of groundwater resources is mainly effected by human exploitation; the modulus is controlled by both precipitation and hydrogeology unit.

(6) In general, the groundwater resources vary from year to year, which makes the development of groundwater in low water years increase.

(7) In space, groundwater resources are rich in the north and lacking in the south, which means the distribution of groundwater resources does not match the economic layout in the northeast area.

Based on the analysis above, given the policy of encouraging northeast old industrial base, the demand for water resources will increase more. Therefore, the general stratagem of groundwater development in the Northeast Vast Plain should be defined as: The water resources supply the sustainable development of the economy in the northeast old industrial base and must be under the conditions of ensuring sustainable exploitation of water resources.

COUNTERMEASURES

Agricultural water Traditional agriculture, which is extensively irrigated farming and cultivation agriculture controlled by rain in dry land, should be transformed into modern agriculture, which is efficiently abstemious irrigated farming and modern dry land farming.

Municipal and industrial water Traditionally, abstemious use of water, curing pollution, and development of non-traditional water resources are not very valuable. To ensure the sustainable usage of urban water resources, a throttling policy, prevention and cure pollution policy and developing a new channel to gain water resources should be effective.

Anti dirt and reduce disaster The curing water pollution method should be changed from curing water pollution in the last point of the polluting process into controlling the pollution source and taking action in both to protect water resources from pollution.

Protection of eco-environment In the process of water resources exploitation, water use for the ecologic environment must be put apart to prevent eco-environmental destruction.

Equilibrium of supply and demand of water resources The equilibrium of how much water to supply according to the demand of water resources will change into equilibrium of supply and demand on water resources on the foundation of strengthening the water requirement management.

Water resources problem in the Liaohe Plain Based on the abstemious use of water and curing pollution and fitness using local water resources, proper measures of water transfer should be adopted to ensure the continuous increase of the economy in Liaohe Plain.

Innovation measures about water resources management Engineering and contamination of water in most of the cities cause a shortage of water supply in the Northeast Vast Plain. In most water deficient cities, groundwater has been abundantly used, an important reason for which is a shortage of funds to build waterworks and a corresponding pipe network to transfer surface water from other basins. Another problem is the lack of money to build wastewater treatment facilities. A great number of places which are short of water have a large quantity of sewage that cannot be used. More importantly, in some places the untreated sewage is not used, and this brings severe after effects.

CONCLUSION

According to the discussion above, the evolution of groundwater resources is totally affected by both natural factors and human factors. But due to the rapid development, disturbance of the groundwater system caused by humans is more and more important. Thus, man-made influences become more and more significant.

REFERENCES

Chi Bao-ming, Wang Zhi-gang, Lin Lan & Cui Xin-ying (2005) Water Resources State and Groundwater Exploitation in Songliao Basin. *Hydrogeology and Engineering Geology* **32**(3), 1–73.

Dang Lian-wen (2002) Sustainable utilization of water resources in the Songhua-Liao River Basin and corresponding modern management. *China Water Resources* **10**, 2–84.

Jiao Hong-jun & Duan Yong-hou (2005) Sustainable utilization of groundwater resource in the Yubei plain. *Hydrogeology and Engineering Geology* **1**, 1–66.

Song Xin-shan *et al.* (2001) Water resources and its environment problems and the countermeasures to sustainable development. *Territory and Natural Resources Study* **1**, 1–4.

Wang Hao *et al.* (2005) Evolutionary laws of the Yellow River Basin's water resources under the impact of human activities. *J. Nat. Resour.* **20**(2), 158–162.

Yang Yan-wu & Zhang Yan-qiu (2002) Water resources vulnerabilities and regional sustainable development. *Suzhou Institute of Urban Construction and Environment Protection* **15**(4), 85–88.

Zhang Guang-hui *et al.* (2004) Evolution of groundwater circulation in the Heihe River drainage area. *Geology in China* **31**(3), 289–293.

Zhang Zong-hu (2005) Groundwater in the vast North China Plain. *Chinese Journal of Nature* **27**(6), 311–315.

A type of the retrospective difference scheme for Richards equation

XIANGDONG CHEN[1], JUN XIA[1] & YONGGANG GE[2]

1 *Key Laboratory of Water Cycle and Related Land Surface Processes, Institute of Geographic Sciences and Natural Resources Research, Chinese Academy of Sciences, Beijing 100101, China*
chenxd.08b@igsnrr.ac.cn

2 *Yellow River Engineering Consulting Company (YREC), Zhengzhou 450000, China*

Abstract Numerical simulation is the effective method for solving Richards equation, which has a highly-nonlinear character. Most numerical solutions methods improved the spatial discretization, or were designed on the basis of physical conservation. Those methods are solved when an initial value is given. Combining the self-memory principle, the retrospective difference scheme for Richards equation is proposed. It is a new kind of time integration scheme which could include multiple time levels and absorb the character of uncertainty method, which forecasts by making use of several historical data before the initial time. This scheme is applied to simulate the fixed head vertical infiltration. The results of a single order retrospective scheme are calculated with given memory coefficients. The stable region of the memory coefficients is also given. The results show that h-based formulation using the retrospective scheme can give higher accuracy than that using the implicit scheme.

Key words soil water movement; numerical simulation; self-memory principle; retrospective difference scheme

INTRODUCTION

Soil water is the connection between surface water and groundwater and an essential element in the water cycle process. Study on the rule of soil moisture movement therefore has very important significance. The Richards equation for soil moisture movement is a partial differential equation of second order and is highly nonlinear. Analytical solutions can only be found with simple boundary conditions or simplified soil parameters (Lei *et al.*, 1988). So to deal with natural conditions, many numerical methods have been developed to solve the Richards equation, e.g. the finite difference method (Celia *et al.*, 1990; Ross *et al.*, 1990), the finite element method (Pan *et al.*, 1996), the mixed finite element method (Arnold, 1990) and the finite volume method (Farthing *et al.*, 2003; Gianmarco Manzini *et al.*, 2004).

Gu pointed out the difference between numerical prediction with an initial value and the empirical prediction with several past observations before the initial time. He proposed improving the numerical prediction by making use of the past data (Gu, 1958). After this, multiple time mode of numerical prediction became the focus of study for many scholars. Cao, based on the view that atmospheric movement is an irreversible process, introducing a memory function which can recall data in the past, proposed a concept of self-memory in atmospheric motion (Cao, 1993, 2002). The atmospheric differential equation is generalized to a self-memory equation which includes multiple time levels and initial values. The computation based on the self-memory equation is called the self-memory principle. The existing various modes of multiple time numerical prediction can be unified into one frame of the self-memory equation. The self-memory principle constructed a retrospective temporal integration method (Lu *et al.*, 2000; Feng *et al.*, 2002). The motivation of our study is to simulate the soil water movement with the self-memory principle.

THE RICHARDS EQUATION

The Richards equation is the equation most frequently used to describe soil water movement. The pressure head form of Richards equation is selected. This form is valid for both saturated and unsaturated conditions. The one-dimensional equation is expressed as follows:

$$C(h)\frac{\partial h}{\partial t} = \frac{\partial}{\partial z}\left[K(h)\frac{\partial h}{\partial z}\right] - \frac{\partial K(h)}{\partial z} \tag{1}$$

where h is the pressure head (cm), $C(h)$ is the specific water capacity (cm^{-1}), $K(h)$ is the hydraulic conductivity (cm/min), z is the depth, positively oriented downward (cm), and t is the time (min).

THE SELF-MEMORY PRINCIPLE

Introducing a memory function into a differential equation will produce a difference-integral equation, that is, the self-memory equation. This methodology of prediction based on the self-memory equation is different from that to solve differential equations governing motion of a system as an initial value problem (Cao, 2002). The self-memory equation was deduced by Cao as follows (Feng, 2002):

$$\frac{\partial x}{\partial t} = F(x, \lambda, t) \tag{2}$$

where x is the variable of the system state, λ is the parameter, t is time.

Considering a set of time T = [t_{-p}, $t_{-p+1} \ldots t_0$, t], where t_0 is an initial time, applying an inner product operation to equation (2) by introducing a memorial function $\beta(\tau)$, we can obtain:

$$\int_{t_{-p}}^{t} \beta(\tau)\frac{\partial x}{\partial \tau} d\tau = \int_{t_{-p}}^{t} \beta(\tau) F(x, \lambda, \tau) d\tau \tag{3}$$

Supposing the variables x and β are continuous, differentiable and integrable, following calculus, making an integration by parts, applying the median theorem, and performing algebra operation, the left-side of (3) can be obtained as follows:

$$\int_{t_{-p}}^{t} \beta(\tau)\frac{\partial x}{\partial \tau} d\tau = \int_{t_{-p}}^{t_{-p+1}} \beta(\tau)\frac{\partial x}{\partial \tau} d\tau + \int_{t_{-p+1}}^{t_{-p+2}} \beta(\tau)\frac{\partial x}{\partial \tau} d\tau + \ldots\ldots + \int_{t_0}^{t} \beta(\tau)\frac{\partial x}{\partial \tau} d\tau$$

$$= \beta(t)x(t) - \beta(t_{-p})x(t_{-p}) - \sum_{i=-p}^{0} \int_{t_i}^{t_{i+1}} x(\tau)\beta'(\tau) d\tau$$

$$= \beta(t)x(t) - \beta(t_{-p})x(t_{-p}) - \sum_{i=-p}^{0} x^m(t_i)[\beta(t_{i+1}) - \beta(t_i)] \tag{4}$$

where $\beta'(\tau) = \partial\beta(\tau)/\partial t$, $t_1 = t$, $x^m(t_i) = x(t_m)$, $t_i < t_m < t_{i+1}$.

Then a difference-integral equation can be obtained which is called a self-memory equation with the retrospective order p:

$$\beta(t)x(t) - \beta(t_{-p})x(t_{-p}) - \sum_{i=-p}^{0} x^m(t_i)[\beta(t_{i+1}) - \beta(t_i)] = \int_{t_{-p}}^{t} \beta(\tau) F(x, \lambda, \tau) d\tau \tag{5}$$

THE RICHARDS EQUATION COMBINED WITH THE SELF-MEMORY PRINCIPLE

Equation (1) can be written as:

$$\frac{\partial h}{\partial t} = \frac{1}{C(h)}\left[\frac{\partial}{\partial z}\left[K(h)\frac{\partial h}{\partial z}\right] - \frac{\partial K(h)}{\partial z}\right] = F(h, z, t) \tag{6}$$

Defining the memorial function $\beta(i, t)$ (i as a spatial point, t as time), and combining equation (6) with the self-memory principle, the self-memorial Richards equation can be obtained:

$$\beta(i,t)h(i,t) = \left\{\beta(i,t_{-p})h(i,t_{-p}) + \sum_{i=-p}^{0}[\beta(i,t_{i+1}) - \beta(i,t_i)]h^m(i,t_i)\right\}$$

$$+ \int_{t_{-p}}^{t} \beta(i,\tau)\frac{1}{C(h(i,\tau))}\left[\frac{\partial}{\partial z}\left[K(h(i,\tau))\frac{\partial h(i,\tau)}{\partial z}\right] - \frac{\partial K(h(i,\tau))}{\partial z}\right] d\tau \tag{7}$$

where $h^m(i,t_i) = h(i,t_m)$, $t_i < t_m < t_{i+1}$.

THE RETROSPECTIVE DIFFERENCE SCHEME OF RICHARDS EQUATION

The difference scheme of the self-memory equation is called the retrospective scheme. Some notations are used in the difference scheme as follows:

$$\beta_i^{k+1} = \beta(i,t), \beta_i^k = \beta(i,t_0), \beta_i^{k-1} = \beta(i,t_{-1}), \ldots\ldots$$

$$h_i^{k+1} = h(i,t), h_i^k = h(i,t_0), h_i^{k-1} = h(i,t_{-1}) \ldots\ldots$$

$$\frac{1}{C(h(i,t))}\left[\frac{\partial}{\partial z}\left[K(h(i,t))\frac{\partial h(i,t)}{\partial z}\right] - \frac{\partial K(h(i,t))}{\partial z}\right]$$ can be approximated by:

$$\frac{1}{C_i^{k+1}}\left[K_{i+1/2}^{k+1}\frac{h_{i+1}^{k+1} - h_i^{k+1}}{\Delta z^2} - K_{i-1/2}^{k+1}\frac{h_i^{k+1} - h_{i-1}^{k+1}}{\Delta z^2} - \frac{K_{i+1/2}^{k+1} - K_{i-1/2}^{k+1}}{\Delta z}\right]$$

Taking the arithmetic mean values of conductivities at the element interface:

$$K_{i\pm1/2} = \frac{1}{2}(K_i + K_{i\pm1})$$

Then the discretization scheme of equation (7) can be expressed as follows:

$$\beta_i^{k+1}h_i^{k+1} = \beta_i^{k-p}h_i^{k-p} + \sum_{j=-p}^{0}(\beta_i^{k+j+1} - \beta_i^{k+j})\cdot(h_i^{k+j})^m + \sum_{j=-p}^{0}\frac{\beta_i^{k+j}}{C_i^{k+j+1}}[K_{i+1/2}^{k+j+1}\frac{h_{i+1}^{k+j+1} - h_i^{k+j+1}}{\Delta z^2}$$
$$-K_{i-1/2}^{k+j+1}\frac{h_i^{k+j+1} - h_{i-1}^{k+j+1}}{\Delta z^2} - \frac{K_{i+1/2}^{k+j+1} - K_{i-1/2}^{k+j+1}}{\Delta z}]\Delta t \qquad (8)$$

assuming that memorial function $\beta(i, t)$ of each space point is equal at the same time, that is, $\beta(i, t) = \beta(i-1, t) = \beta(i+1, t)$.

Then denoting:

$$\alpha_{-p-1} = \frac{\beta_i^{k-p}}{\beta_i^{k+1}}, \alpha_j = (\beta_i^{k+j+1} - \beta_i^{k+j})/\beta_i^{k+1}$$

$$\tilde{\theta}_j = \beta_i^{k+j}/\beta_i^{k+1} \qquad (j = -p-1, \ldots, 0)$$

$(h_i^{k+j})^m$ is approximated by $1/2(h_i^{k+j+1}+h_i^{k+j})$, then retrospective difference scheme of the Richards equation can be obtained:

$$h_i^{k+1} = \alpha_{-p-1}h_i^{k-p} + \sum_{j=-p}^{0}\alpha_j\cdot\frac{h_i^{k+j+1} + h_i^{k+j}}{2} + \sum_{j=-p}^{0}\frac{\theta_j}{C_i^{k+j+1}}[K_{i+1/2}^{k+j+1}\frac{h_{i+1}^{k+j+1} - h_i^{k+j+1}}{\Delta z^2}$$
$$-K_{i-1/2}^{k+j+1}\frac{h_i^{k+j+1} - h_{i-1}^{k+j+1}}{\Delta z^2} - \frac{K_{i+1/2}^{k+j+1} - K_{i-1/2}^{k+j+1}}{\Delta z}]\Delta t \qquad (9)$$

where α_j, $\tilde{\theta}_j$ are memory coefficients.

In equation (9), setting $p = 0, \beta_i^{k+1} = \beta_i^k = 1$, equation (9) becomes:

$$h_i^{k+1} = h_i^k + \frac{1}{C_i^{k+1}}[K_{i+1/2}^{k+1}\frac{h_{i+1}^{k+1} - h_i^{k+1}}{\Delta z^2} - K_{i-1/2}^{k+1}\frac{h_i^{k+1} - h_{i-1}^{k+1}}{\Delta z^2} - \frac{K_{i+1/2}^{k+1} - K_{i-1/2}^{k+1}}{\Delta z}]\Delta t \qquad (10)$$

Equation (10) is the implicit difference scheme of the Richards equation.

Perform algebra to equation (9):

$$E_i h_{i-1}^{k+1} + F_i h_i^{k+1} + G_i h_{i+1}^{k+1} = H_i \tag{11}$$

where $\mu = \dfrac{\Delta t}{\Delta z^2}$, $E_i = -\tilde{\theta}_0 \dfrac{K_{i-1/2}^{k+1}}{C_i^{k+1}} \mu$, $F_i = 1 - \dfrac{\alpha_0}{2} + \mu \cdot \tilde{\theta}_0 \cdot \dfrac{K_{i-1/2}^{k+1} + K_{i+1/2}^{k+1}}{C_i^{k+1}}$,

$$G_i = -\tilde{\theta}_0 \frac{K_{i+1/2}^{k+1}}{C_i^{k+1}} \mu$$

$$H_i = \alpha_{-p-1} h_i^{k-p} + \sum_{j=-p}^{-1} \alpha_j \cdot \frac{h_i^{k+j+1} + h_i^{k+j}}{2} + \frac{\alpha_0}{2} h_i^k + \sum_{j=-p}^{-1} \frac{\tilde{\theta}_j}{C_i^{k+j+1}} [K_{i+1/2}^{k+j+1} \frac{h_{i+1}^{k+j+1} - h_i^{k+j+1}}{\Delta z^2}$$

$$-K_{i-1/2}^{k+j+1} \frac{h_i^{k+j+1} - h_{i-1}^{k+j+1}}{\Delta z^2} - \frac{K_{i+1/2}^{j+1} - K_{i-1/2}^{j+1}}{\Delta z}]\Delta t - \frac{\tilde{\theta}_0}{C_i^{k+1}} \Delta t \frac{K_{i+1/2}^{k+1} - K_{i-1/2}^{k+1}}{\Delta z} \quad (i = 1, \ldots, n-1)$$

Then the simultaneous tri-diagonal system can be solved by chase method. If there is no observed data before initial time, then we can calculate p steps with implicit scheme. The calculated results can be used as the initial value of the retrospective scheme.

CALCULATION EXAMPLE

The retrospective difference scheme of the Richards equation is applied to simulate the fixed head vertical infiltration. The experiment data sets were obtained from Sun (personal communication, 2008). The length of the soil cylinder is 2 m and the inner diameter is 14.5 cm. The water level in the cylinder can be controlled constantly using the connected vessel. After air dry, rolling, and sieving, the soil sample is layered into the cylinder homogeneously according to the designed soil bulk density (Zhou *et al.*, 2006). The soil sample in this experiment is obtained from the epipedon of the disturbed loess. The soil dry bulk density r_d of this sample is 1.38 (g/cm^2), saturated water content θs is 0.4688 (cm^3/cm^3).

The soil water characteristic curve and soil water diffusivity are fitted by exponential relationship as follows:

$$S = 10399e^{-12.676\theta}, \quad D = 0.00002e^{29.581\theta}$$

where S is the soil suction (cm), D is unsaturated soil water diffusivity (cm^2/min), θ is the volumetric soil water content (cm^3/cm^3).

Then the unsaturated hydraulic conductivity is obtained:

$$K = -D\frac{d\theta}{ds} = 1.52 \times 10^{-10} e^{42.257\theta}$$

where K is the unsaturated hydraulic conductivity (cm/min).

The upper boundary condition is $h(z = 0,t) = 10399e^{-12.676\theta s} + H$. H is the constant water head (5 cm). The lower boundary condition is $h(z = 2m,t) = 10399e^{-12.676\theta 0}$. The initial volumetric soil water content θ_0 is 0.08 (cm^3/cm^3). The space interval Δz is 2 cm, the time interval Δt is 0.2 min, and the given convergence precision is 0.001.

The selection of memory coefficients is important because it influences the stability of the retrospective scheme directly. The coefficients α_i ($i = -p - 1\ldots0$) can be obtained when $\tilde{\theta}_i$ ($i = -p\ldots0$) were given. So we can generate $\tilde{\theta}_i$ randomly in a given range and use Monte-Carlo method to simulate the stable region of $\tilde{\theta}_i$. Setting retrospective order $p = 1$, memory coefficients $\tilde{\theta}_{-1}$ and $\tilde{\theta}_0$ are generated randomly between 0 and 1. The coefficient combinations which result calculation convergence are recorded. The cumulative infiltration's mean square root error (ε) of each coefficient combination is also calculated. The cumulative infiltration and ε are calculated by:

$$I = A\int_0^z (\theta - \theta_0)dz\ ,\ \varepsilon = \frac{1}{n}\sqrt{\sum_{i=1}^{n}(I_i' - I_i)^2}$$

where I is the cumulative infiltration, A is the cross sectional area of test cylinder, I_i' is the calculated cumulative infiltration, I_i is the observed cumulative infiltration.

The stable region of retrospective scheme with 1 order is shown in Fig. 1. The RMSE of each point in the stable region is shown in Fig. 2. The $\varepsilon/\tilde{\theta}_0$, $\varepsilon/\tilde{\theta}_{-1}$ scatter diagrams are shown in Figs 3 and 4.

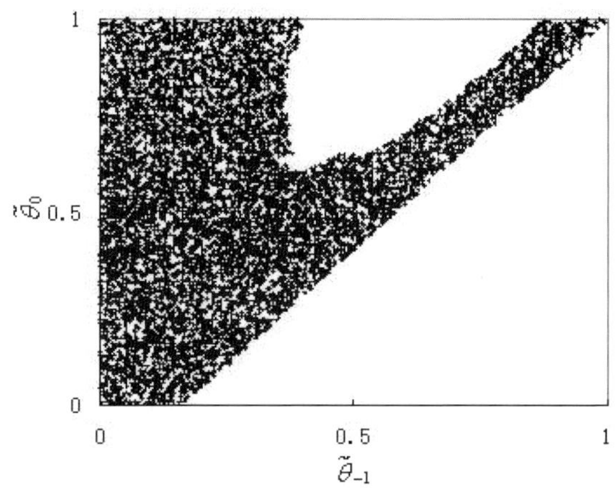

Fig. 1 The stable region.

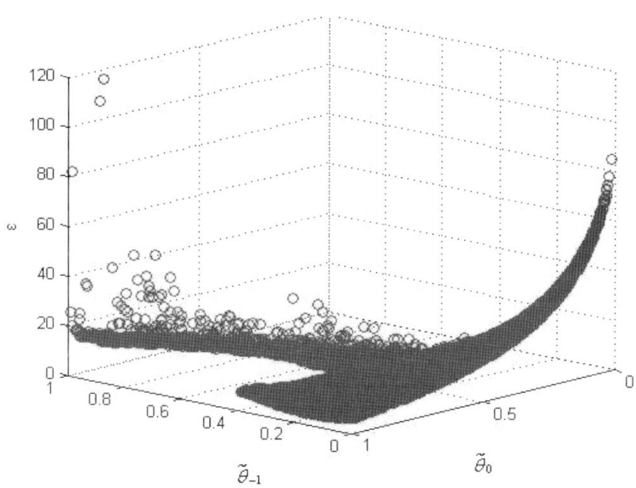

Fig. 2 RMSE of cumulative infiltration distribution.

Setting $\tilde{\theta}_{-1} = 0.13$, $\tilde{\theta}_0 = 0.99$, each cumulative infiltration curve is shown in Fig. 5.

Milly (1985) and Celia *et al.* (1990) reported a solution using the h-based formulation and a backward Euler (implicit) time discretization that produce unacceptable large mass balance error for many example calculations. It is also true for any iteration method like Picard and both finite difference and finite element approximations in space.

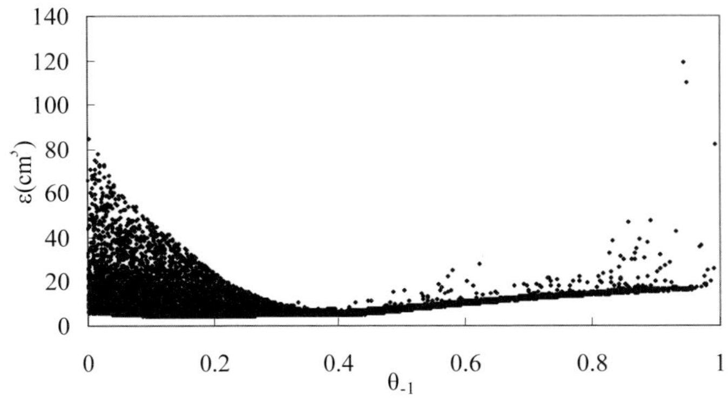

Fig. 3 $\varepsilon/\widetilde{\theta}_{-1}$ scatter diagram.

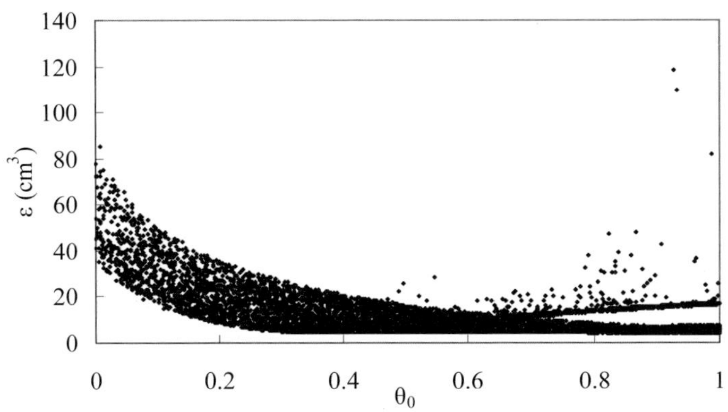

Fig. 4 $\varepsilon/\widetilde{\theta}_0$ scatter diagram.

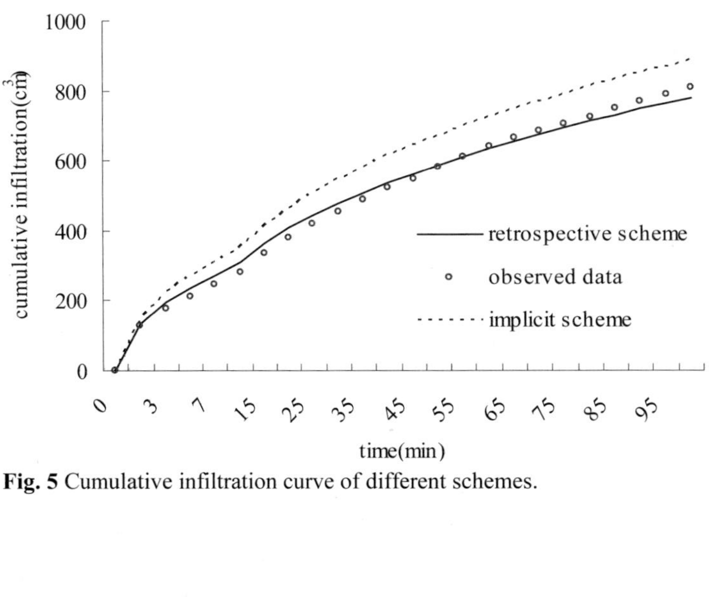

Fig. 5 Cumulative infiltration curve of different schemes.

From Fig. 5, it can be seen that the h-based equation using the retrospective approximation in time can give better results than that using simple backward Euler approximation in time. The RMSE of retrospective scheme is 4.2 (cm^3), and that of the implicit scheme is 16.8 (cm^3). Sorting each point's RMSE of cumulative infiltration in the stable region, it can compose a curve which is shown in Fig. 6.

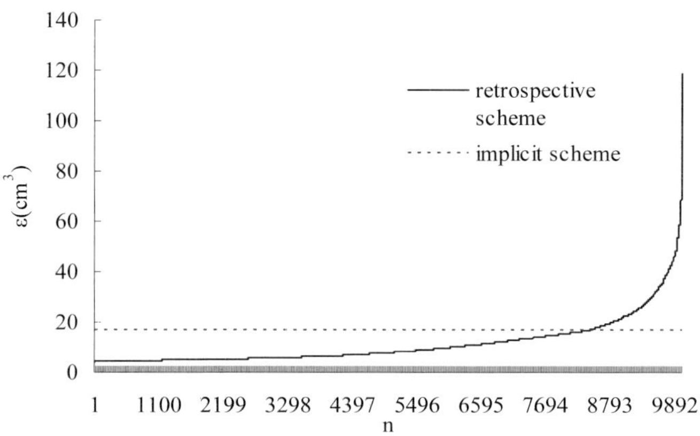

Fig. 6 The distribution curve of RMSE in the stable region.

CONCLUSIONS

The time derivative is crucial to numerical performance. A higher order correct, multi-time level approximation to time derivative $\partial h/\partial t$ can be used to produce significant improvements in the h-based methods (Bouloutas,1989; Celia *et al.*, 1990).

The retrospective difference scheme is a numerical method of multi-time level for time derivative and a higher order time approximation for h-based formulation. Its characteristic is that this scheme has different forms when given different memory coefficients, and given some special coefficients can make this scheme have a better performance.

Another characteristic of the retrospective scheme is also the important development of future in this paper. Kinds of numerical methods try to let numerical solutions approximate the analytic solutions. But the analytic solutions do not mean the true solutions because inaccuracy of parameters and inadequate consideration of the dynamics mechanism when the equation is applied to the field or the larger scale. If we can calibrate the memory coefficients with historical observed data, the retrospective scheme is able to adapt the problem itself and find the optimal solution for the differential equations. Then the scheme will turn into a dynamic-statistic mode which can set up a bridge between the deterministic method and the uncertainty method and exert their merits together.

In this paper, the finite difference method was used as the discretization approach of the self-memory equation. We also can combine other numerical method such as the finite element method with the self-memory equation in future studies.

Acknowledgements This work was supported by the National Natural Science Foundation of China (grant no. 40671035).

REFERENCES

Arnold, D. N. (1990) Mixed finite element methods for elliptic problems. *Comput Method in Appl. Mech. Engng* **82**, 281–300.

Bouloutas, E. T. (1989) Improved numerical approximations for flow and transport in the unsaturated zone. PhD Thesis, Massachusetts Institute of Technology, Cambridge, Massachusetts, USA.

Cao, H. X. (1993) Self-memorization equation in atmospheric motion. *Sci. China Ser. B*, **23**(1), 104–112. (in Chinese)

Cao, H. X. (2002) *Memorial Dynamics of Systems – Forecasting and Computation Application*. Geologic Publishing House, Beijing, China.

Celia, M. A., Bouloutas, E. T. & Zarba, R. L. (1990) A general mass-conservative numerical solution for the unsaturated flow equation, *Water Resour. Res.* **26**(7), 1483–1496.

Farthing, M. W. & Miller, C. (2003) A comparison of high-resolution, finite volume, adaptive-stencil schemes for simulating advective-dispersive transport, *Adv. Water Resour.* **25**, 29–48.

Feng, G. L., Cao, H. X., Gu, X. Q. & Chou, J. F. (2002) A new scheme for improving the accuracy of numerical prediction-retrospective time integration scheme. *J. Appl. Meteo. Sci.* **13**(2), 207–216. (in Chinese)

Gu, Z. C. (1958) On the utilization of past data in numerical weather forecasting(in Chinese), *Acta Meteorologica Sinica* **29**(3), 176–186.

Lei, Z. D., Yang, S. X. & Xie, C. S. (1988) *Soil Water Dynamics*. Tsinghua University Press, Beijing, China.

Lu, J. A., Lv, J. H. & Xia, J. (2000) Self-memory integration scheme of convection-diffusion equation. *J Comp. Phys.* **17**(6), 664–670. (in Chinese)

Manzini, G. & Ferraris, S. (2004) Mass–conservative finite volume methods on 2-D unstructured grids for the Richards equation, *Adv Water Resour.* **27**, 1199–1215.

Milly, P. C. D. (1985) A mass-conservative procedure for time-stepping in models of unsaturated flow. *Adv Water Resour.* **8**, 32–36.

Pan, L., Warrick, A. W., Wierenga, P. J. (1996) Finite elements methods for modeling water flow in variably saturated porous media: numerical oscillation and mass distributed schemes, *Water Resour. Res.* **32**, 1883–1889.

Ross, P. J. (1990) Efficient numerical methods for infiltration using Richards equation, *Water Resour. Res.* **26**(2), 279–290.

Zhou, Q. Y., Sun, X. H. & Kang, S Z. (2006) Numerical simulation for soil water movement of water storage pit irrigation. *J. Hydraul. Engng* **37**(3), 342–347. (in Chinese)

Flood resources utilization based on flood control forecast operation

GAIHONG ZHANG, KUI ZHANG & FAN ZHAO

Department of Environmental and Municipal Engineering, Henan University of Urban Construction, Henan 467044, China

zgh197768@163.com

Abstract As a key parameter of coordinating flood control and impounding, the limited level of reservoirs during the flood season is a crucial factor of flood resources utilization. With an example of the limited level analysis of Yuqiao reservoir during the flood season, a method of forecast operation judging the flood frequency by accumulated runoff is proposed to elevate the limited level of reservoirs during the flood season. Meanwhile, to utilize flood resources in security, a comprehensive evaluation system is presented to obtain an optimum limited level of reservoirs during the flood season including risk rate, damaged degree, renew degree and impounding benefit. It is shown that different flood frequency judgment indexes variously influence the impounding capacity of reservoirs, and the forecast operation mode and comprehensive evaluation system are important for the safe utilization of flood resources during the flood period.

Key words accumulated runoff; comprehensive evaluation; flood control forecast operation; flood resources; limited level of reservoirs during the flood season

INTRODUCTION

Yuqiao reservoir is located east of Jixian town, Tianjin city. The basin area above the dam is 2060 km^2, and the maximum reservoir capacity and the common capacity of flood control and impounding benefit are 15.59×10^8 m^3 and 1.08×10^8 m^3, respectively. Its benefit water level is 21.16 m, and the flood limited water level is 19.87 m in the flood season. Yuqiao reservoir has become an important supply source since the water-carriage project from Luanhe River was built in 1983.

With the shortage of water resources, more and more problems have emerged in reservoir operation. Not considering the information of meteorology forecasts and flood forecasts, the flood limited water level is on the low side and remains constant during the whole flood season. These are the main problems unfavourably impacting flood resource utilization. These problems exist in flood operation of not only Yuqiao reservoir, but also many other reservoirs in China. There are three means to solve the problems (Dalian University of technology, 1996; Wan *et al.*, 2000; Qiu *et al.*, 2004; Wang *et al.*, 2004): (1) Divide the whole period of flood season into several periods according to the season characteristic and orderliness of rainstorm and flood. (2) Elevate flood limited water level based on forecast information. (3) Control dynamically the flood limited water level in real-time operation.

Flood limited water level is a key parameter for guaranteeing safe utilization of flood resources in the flood season. The rational flood limited water level should meet the optimal combination of flood control risk and impounding benefit. The flood control system is generally complex as it involves a large number of uncertain factors, including hydrological phenomenon, engineering structure and hydraulic calculation, etc. The overall evaluation system, including operation risk, resulted from a variety factors and economic benefit should be adopted to weigh the advantages or disadvantages of raising the flood limited water level. Therefore, hydrological risk should be taken into account in conventional flood risk evaluation.

Taking Yuqiao reservoir as an example, this paper adopted different frequency design floods in relation to the dam and downstream, and studied the forecast operation mode and rule of classifying flood by accumulated runoff. Finally, the preferred flood limited water level was obtained by comparing the flood control risk and impounding benefit of five alternatives of flood limited water level.

DESCRIPTION OF YUQIAO RESERVOIR

The main role of Yuqiao reservoir is to guarantee the safety of the dam and downstream in the flood season. At present, flood with frequency equal to 1% is taken as a design standard, and probable maximum flood (PMF) as check criterion. The flood control assignment of the downstream is to protect the downstream channel encountering flood with frequency less than 2% and the downstream railway encountering flood with frequency between 1% and 2%. Presently, the actual water level is taken as classifying flood index in practical operation. The operation rule of Yuqiao reservoir is as follows: (1) encountering flood with frequency less than 5%, all gates are closed until 7 days, and then the discharge flow is dominated at 150 m^3/s; (2) encountering flood with frequency between 2% and 5%, the discharge flow is controlled at 150 m^3/s; (3) encountering flood with frequency between 1% and 2%, the discharge flow is kept at 500 m^3/s; and (4) encountering flood with frequency more than 1%, all outlets discharge freely for the safety of the dam.

FORECAST OPERATION RULE

Forecast operation and conventional operation are distinguished foremost for classifying the flood index. Forecast operation is implemented according to forecast data, but conventional operation according to the actual flood process or level. The forecast period of the former is longer than that of the latter, achieving pre-discharge and expanding flood control capacity. Nevertheless the operation risks resulting from forecast error have to be considered in forecast operation. All shows that the forecast operation scheme should be enacted warily. Meanwhile, the classifying flood index should be provided with reliability, efficiency and sustainability.

Classifying flood index

The hydrological automatically measuring system of Yuqiao reservoir was built and applied in 1983, consisting of a main control station, a relay station and 12 remote stations. The operating records show that the availability degree is 99.6% and the expediter degree is 99.8%. The system can collect the information of precipitation and water inflow correctly and in a timely fashion, and supply science and credible data for the Yuqiao reservoir operation.

With the development and improvement of flood forecast methodology, the flood forecast scheme of Yuqiao reservoir has been established, and the Dahuofang runoff model and Instant Unit-Hydrograph model are employed to predict net rainfall and flood processes, respectively. The outcome of assessing total net rainfall and flow process demonstrates that the flood forecast model reaches second rank level and may be applied in practice according to the "standard for hydrological information and hydrological forecasting". Furthermore, heavy rainstorms have a great effect on reservoir operation and the safety of defended objects. It is concluded that the runoff coefficient will increase and forecast error will decrease with the saturation of the antecedent soil moisture by analysing the precipitation–net rainfall relationship of Yuqiao reservoir. All of the above demonstrate that it is feasible to implement the flood control forecast operation for Yuqiao reservoir due to the available flood forecast scheme, but also it is a correct and timely information source. In terms of the mass balance in practical operating, accumulated net rainfall corresponds to flood inflow, flood inflow with flood storage, and flood storage with flood level. In the light of the sequence of obtaining information, actual rainfall happens before forecast net rainfall, forecast net rainfall before peak inflow, and peak inflow before peak level. Therefore, the classifying flood index of accumulated net rainfall can meet the principle dominated water level and prolong the prediction period. Moreover, most rainstorms commonly last less than 24 hours. The accumulated net rainfall within 24 hours of the whole basin is taken as the first classifying flood index in forecast operation. Owing to forecast error, the flood index is 20% more than the accumulated net rainfall. At the same time, for the sake of operation safety, the current water level is taken as the second classifying flood index in forecast operation.

Accumulated runoff process of design flood should be known in forecast operation planning, which is an important parameter judging how to discharge. They are commonly computed according to design flood process by simulating runoff-flow.

Approach of planning operation rule

First assuming an operation rule, different frequency design floods are calculated according to the chosen limited water level. If the results satisfy the outlet capacity and could guarantee the safety of the dam and the downstream, the operation rule is feasible. Otherwise, the operation rule is unfeasible. Through recalculating and reanalysing different alternatives, a satisfactory forecast operation rule and an appropriate risen value of flood limited water level can be obtained.

COMPREHENSIVE EVALUATION

Comprehensively considering the effect on flood control operation of raising limited water level, a multi-factors risk system is established, aiming at the complex operation objects of the Three Gorges reservoir (Wang & Guo, 2004). Based on the flood control criteria of Yuqiao reservoir, the probability, ponderance and renew capacity of the defended objects as the assessing factors are proposed.

According to the operation criteria of Yuqiao reservoir, the dam would be dangerous when the flood control water level is higher than 23.89 m and the downstream would be submerged when the discharge flow is more than 500 m^3/s.

Description of evaluation index

(a) Risk rate. It shows the probability of the dam or the downstream being destroyed. The flood duration is denoted by *TT*; *T*(*Z*) and *T(Q)* denote the lasting time of the dam and the downstream being destroyed, respectively. Therefore, the risk rate of the dam is expressed as:

$$r_z = T(Z)\,/TT \tag{1}$$

The risk rate of the downstream is shown as follows:

$$r_z = T(Q)/\,TT \tag{2}$$

(b) Destroyed rate. It presents destroyed degree of the dam and the downstream. The higher the water level is, the more seriously the dam is destroyed. To defended objects in the downstream valley, destroyed degree is also in proportion to the discharge flow. Therefore, the average water level and discharge flow are adopted to describe the damaged status of the dam and the downstream, respectively.

(c) Renew rate. It expresses the capacity and rate of recovering from damaged state to normal state, with the inverse of the damaged duration of the dam or the downstream denoted.

(d) Economic benefit. Water supply is the main assignment of Yuqiao reservoir, thus water–price method is employed to estimate the water supply benefit, which is regarded as the index of the increased impounding benefit of the elevated limited water level higher than 19.87 m. (Unit-price of the citizen life water in Tianjin city is 2.9 Yuan / m^3).

Standardization of evaluating index

The evaluating index needs be standardized to assess uniformly whether an index is superior or inferior. It is divided into two states:

(a) The bigger the objectives value, the larger the optimal membership degree, such as renew rate and benefit. The formula is as follows:

$$r_{ij} = \left(y_{ij} - y_j^{\min}\right) / \left(y_j^{\max} - y_j^{\min}\right) \tag{3}$$

(b) The bigger the objectives value, the less the optimal membership degree, such as risk rate and destroyed rate. The formula is expressed as:

$$r_{ij} = \left(y_j^{\max} - y_{ij}\right) / \left(y_j^{\max} - y_j^{\min}\right) \tag{4}$$

In equations (3) and (4), r_{ij} represents the *j*th characteristic value of the *i*th alternative after standardizing. y_{ij} represents the *j*th characteristic value of the *i*th alternative before standardizing. $y_j^{\max}$ and $y_j^{\min}$ represent the maximum and minimum characteristic value of the *j*th index among all alternatives before standardizing, respectively.

Index weight

The weight represents relative importance of all indexes at certain criteria. According to the analysis methodology based on non-structural fuzzy sets (Chen, 1998), the assessing indexes may be sorted by priority and recognized importance by comparing two indexes. Then the index weight vector is ascertained as $(w_1, w_2, \ldots w_n)$.

Objective

The operation risk of flood with frequency less than 1% is taken into account according to analysing the flood control criteria of Yuqiao reservoir. The formula is defined by:

$$H_i = \sum_{j=1}^{n} w_j r_{ij} \tag{5}$$

In equation (5) H_i is the synthetic evaluation value of the *i*th alternative.

OUTCOME ANALYSIS

Forecast operation of Yuqiao reservoir

Adopting fourth-order Runge-kutta numerical solution in flood routing, the forecast operation rule is simulated and optimized, which takes 24 h accumulated runoff as classifying flood index of Yuqiqo reservoir. The contrasting outcome of the preferred forecast operation rule and the conventional operation rule controlled by flood level is obtained as in Table 1.

Table 1 The contrasting outcome of forecast operation rule and conventional operation rule of Yuqiao reservoir.

Project	Frequency	Mass runoff (mm)	Water level (m)	Outlet manner
Forecast operation rule	>1%	>415	>23.95	discharge freely
	2–1%	332–415	23.92–23.95	discharge flow limited 500 m^3/s
	2–5%	228–332	23.40–23.92	discharge flow limited 150m^3/s
	<5%	<228	<23.40	gate closed 7days, then discharge flow limited 150 m^3/s
Conventional operation rule	>1%		>24.25	discharge freely
	2–1%		23.86–24.25	discharge flow limited 500 m^3/s
	2–5%		23.12–23.86	discharge flow limited 150 m^3/s
	<5%		<23.12	gate closed 7days, then discharge flow limited 150 m^3/s

Keeping the flood limited water level at 19.87 m, which is taken as beginning water level, the collecting outcomes of flood routing adopting forecast operation rule and conventional operation rule, respectively, is enumerated in Table 2. From Table 2, the peak level routing by forecast operation rule is higher than that by conventional operation rule encountering flood with frequency more than 1%, so is the maximum discharge flow. Therefore, it is more reliable and favourable to adopt forecast operation than conventional operation for Yuqiao reservoir.

Table 2 The results of forecast routing and conventional routing (limited water level at 19.87 m).

Project	Frequency	Inflow (m^3/s)	Maximum level (m)	Corresponding volume (10^8m^3)	Maximum discharge (m^3/s)
Forecast operation	PMF	17960	27.74	16.39	4386
	0.01%	11505	26.49	13.37	3621
	0.1%	8327	24.72	9.69	2385
	1%	5216	23.66	7.84	500
	2%	4300	23.67	7.86	150
	5%	3130	23.12	6.96	150
Conventional operation	PMF	17960	28.00	17.06	5204
	0.01%	11505	26.96	14.49	3680
	0.1%	8327	25.26	10.75	3137
	1%	5216	24.24	8.83	500
	2%	4300	23.85	8.16	150
	5%	3130	23.12	6.96	150

Comprehensive evaluation

Encountering flood with frequency more than 5%, the sluice gate must be kept closed for 7 days to avoid peak flow together with the downstream flood. The forecast operation rule is not prior to the conventional operation rule because the two discharge manner is uniform. Therefore, the values are taken as a constraint condition, which is required so that the peak level of forecast operation rule is under that of normal operation rule encountering flood with frequency less than 5%. The limited water level may be elevated to 20.27 m by flood routing. Selecting five alternatives as the limited water level as 19.87 m, 19.97 m, 20.07 m, 20.17 m, 20.27 m, the preferred flood limited water level is determined through evaluating comprehensively, which is rationally combined flood control with impounding benefit. The index characteristic values and assessing values corresponding to five alternatives are shown as Table 3 and Table 4; the indexes are standardized according to formula (3) and (4).The integrative weight is expressed as Table 5.

Table 3 The evaluation index characteristic values.

NO	Limited water level (m)	Risk rate		Damaged degree		Renew degree		Benefit (10^6 Yuan)
		dam (%)	downstream (%)	dam (m)	downstream (m^3/s)	dam	downstream	
1	19.87	13.70	34.72	25.22	2209	0.0101	0.0040	0
2	19.97	13.80	34.17	25.21	2225	0.0101	0.0041	0.23
3	20.07	13.94	33.66	25.34	1998	0.0100	0.0041	0.46
4	20.17	13.98	33.15	25.35	2045	0.0099	0.0042	0.70
5	20.27	14.03	32.69	25.34	2064	0.0099	0.0042	0.93

Table 4 The relative weights of evaluation indexes.

Objects	Risk rate		Damaged degree		Renew degree		Benefit
	dam	downstream	dam	downstream	dam	downstream	
a*	0.330	0.086	0.370	0.061	0.043	0.010	0.10
b*	0.235	0.121	0.323	0.098	0.030	0.023	0.17

Note: a* denotes considering strongly flood control safety; b* denotes considering strongly impounding benefit.

Aiming at the contradiction of flood control and water supply, integrating diverse decision-maker preference, the weights of flood control risk and impounding benefit are different. As a

Table 5 The evaluation values of flood limited water level.

Alternatives		1	2	3	4	5
Evaluation value	a*	0.721	0.666	0.309	0.284	0.306
	b*	0.572	0.601	0.357	0.355	0.399

result, the optimum alternative is also varied. For example, if decision-makers favour flood control safety, the index weights vector will be {0.330, 0.086, 0.370, 0.061, 0.043, 0.010, 0.100}, and then the assessing values vector will be {0.721, 0.666, 0.309, 0.284, 0.306}. Knowing all the above, the relative membership value of the first scheme is maximum, so it is optimum, i.e. it is appropriate for the flood limited water level to remain at 19.87 m. However, if decision-makers are in favour of impounding benefit, the index weights vector will be {0.235, 0.120, 0.323, 0.098, 0.030, 0.023, 0.170}, and then the assessing results vector will be {0.572, 0.601, 0.357, 0.355, 0.399}. Knowing all the above, the relative membership value of the second scheme is maximum, so it is optimum, i.e. it is suitable for the flood limited water level to remain at 19.97 m.

CONCLUSION

By studying the flood control forecast rule and comprehensively evaluating the alternatives of flood limited water level, it is concluded that: (1) Forecast operation is a crucial measure for alleviating the balance of flood control and impounding, of which the different distinguish index impact straight on pre-discharge capacity of the reservoir. The classifying flood index with longer prediction period may gain greater pre-discharge capacity and increase flood prevention storage, but the operators have to consider the veracity of adopting the recognition index and the remedy measures for forecast errors. (2) Implementing forecast operation can bring double benefit. If keeping the limited water level immobile, the capacity protecting dam and downstream is strengthened for the reservoir, decreasing the flood loss. However, the flood control standard remained unchangeable, the flood limited water level may be elevated, enlarging the utilization rate of flood resource. (3) Index weights are varied due to the choice of the decision-maker, thus the optimum alternative is changeable.

This paper is devoted to the study of forecast operation programming for reservoirs. It is merely different frequency period designed flood process to be simulated. While there are several medium or small floods are encountered commonly, a variety of historical flood processes should be analysed to ascertain the effective forecast operation mode. Moreover, the forecast operation may be combined with meteorological forecast if the meteorology data is reliable. The tail-water in flood fall should be impounded via controlling dynamically the limited water level to utilize the flood resource.

Acknowledgements I would like to thank Ke Jin for his assistance.

REFERENCES

Chen, S. Y. (1998) *Engineering Fuzzy Set Theory and Application*. National Defense and Industry Press, Peking, China (in Chinese).

Dalian University of technology and Flood Prevention and Drought Resistant Command Bureaux (1996) *Flood Control Forecast Operation Methodology and Application*. Chinese Hydrology Press, Peking, China (in Chinese).

Qiu, R. T., Wang, B. D. & Zhou, H. C. (2004) An approach on the theory and idea of controlling the limited reservoir elevation in flood season. *Adv. Water Sci. J.* **1**, 68–72.

Wan, J. & Gao, G. M. (2000) Operation manner researches of water storage of Baibenzhu reservoir during flood season concerned flood forecast and flood pre-discharge. *J. Wuhan University of Hydraulic and Electric Engineer* **33**, 10–13.

Wang, C. J. & Guo, S. L. (2004) Risk criteria and comprehensive evaluation model for the operation of three gorges reservoir under dynamic flood limit water level, *Adv. Water Sci.* **15**, 376–381.

Wang, X. Y. & Zhang, X. (2004) Over standard integrated risk analysis of flood control system. *J. Hydraul. Engng* **2**, 83–87.

The proportion of water usable distribution for sustainable development in Haihe River basins

YONGHUA ZHU[1], LILIANG REN[1], JUN XIA[3], HAISHEN LÜ[1,2], ZHONGBO YU[1,4] & HUALI FENG[5]

1 *State Key Laboratory of Hydrology-Water Resources and Hydraulic Engineering, Hohai University, Nanjing 210098, China*
yonghua321@yahoo.com.cn

2 *Department of Applied Mathematics, Hohai University, Naijing 210098, China*

3 *Institute of Geographical Sciences and Natural Resources Research, CAS, Beijing 100101, China.*

4 *Department of Geoscience, University of Nevada, Las Vegas, Nevada 89154, USA*

5 *Nanjing Hydraulic Research Institute, Naijing 210029, China*

Abstract Rational distribution of usable water resources is the key to resolving the contradiction between water use of ecology and that of agriculture, industry and life, especially to reach the aim of eco-environments restoration around 2040. Multi-objectives optimization and time series analysis was used to calculate the optimized proportion of water usable distribution (PWUD) in Haihe River basins under the conditions of sustainable development, the plan of society and economy normal development, the water supply of "South-to-North Water Transfer Project" and eco-environments gradually being meliorated in the river basins. We concluded that if the river basins sustainably develop from 1998 to 2040 under the appointed scheme, the water usage in life will not greatly change, that in ecology will gradually increase, that in agriculture will show a trend of first decreasing and after 2030 maintaining a steady value, while that in industry shows a trend of first increasing and then decreasing.

Key words Haihe River basins; multi-objectives optimization; optimized PWUD; sustainable development; time series

INTRODUCTION

Sustainable development aims to make mankind and the socio-economy develop sustainably, stably and harmoniously under the pre-requisite of protecting and maintaining a high quality environment (Chen, 2002). Since the 1970s the socio-economy has developed quickly in the Haihe River basins, and the population has increased to a large extent, making the proportion of water usage in industrial and agricultural production and for life increase greatly, correspondingly making water usage in ecology reduce continuously. At the same time, pollution increases continuously during the development of socio-economy, therefore, serious eco-environmental problems (EEP) related to water such as water and soil loss, river channel shortening or river water drying, etc. are caused, and have hindered the sustainable development of river basins. In a region, its sustainable development concerns two aspects: eco-environment and socio-economy. In a region where its eco-environment has been destroyed, its sustainable development means socio-economy develop sustainably under the limit of the regional carrying capacity, with the eco-environment being gradually restored to a good state. In a region where water is a restricting factor, the optimized distribution of limited water resources is the basis for it to realize its sustainable development. In this paper, the sustainable development is where water usage is distributed in a maximum optimized way, i.e. the Haihe River basins should develop to a maximum bearable state every year.

The eco-environmental state is expressed by eco-environmental quality, and the socio-economic state is expressed by socio-economic development level. Then, the sustainable development is that the socio-economic level is gradually raised sustainably, with eco-environmental quality improving steadily until it is a good state, and then maintaining the good state.

Today, there are many papers about sustainable development (e.g. Du, 2003; Martha & Yang, 2003; Zhu, 2003), but there are only a few of papers dealing with it by quantitative analysis; the quantitative analysis approach usually used integrated evaluation and, in the quantitative model time series was not used to describe the change process of a regions sustainable development (Wang *et al.*, 2003). In this paper, multi-objectives optimization and time series analysis is used to study the change process of the optimized PWUD from 1998 to 2040 in Haihe River basins under conditions of sustainable development, the plan of society and economy normal development, the water supply of "South-to-North Water Transfer Project" and eco-environments gradually being meliorated in the river basins. This will supply a scientific basis to the eco-environmental restoration of Haihe River basins.

MATERIALS AND METHOD

General situation of Haihe River basins

The Haihe River basins, as shown in Fig. 1, include the Luan River, Haihe River and Tuhaimajia River systems, and are located in northern China, 35°–43°N and 112°–120°E, bordering the Bo Sea in the east, connected with Liaoning Province in the northeast, with the Inner Mongolia Autonomous Region in the northwest, with Shandong Province in the southeast, and Shanxi Province in the west. Included are Peking, Tianjin Municipality, most of Hebei Province, eastern Shanxi Province, northern parts of Henan and Shandong Provinces, a small part of the Inner Mongolia Autonomous Region and Liaoning Province. The total area is 318 000 km^2, among which hills and plateau comprise 189 000 km^2 or 60% of the whole area, and plains comprise 129 000 km^2 or 40% of the area. Haihe River basins lie in the temperate zone of the continental monsoon climate. Mean annual precipitation is 539 mm, and mean annual free water surface evaporation is 1100 mm. In 1998, the population was 122 million, near 10% of China's total. Urban population was 33 650 000 and the rate of urbanization was 28% according to the census. The mean population density was 384 people per square kilometre, 608 in the plain regions.

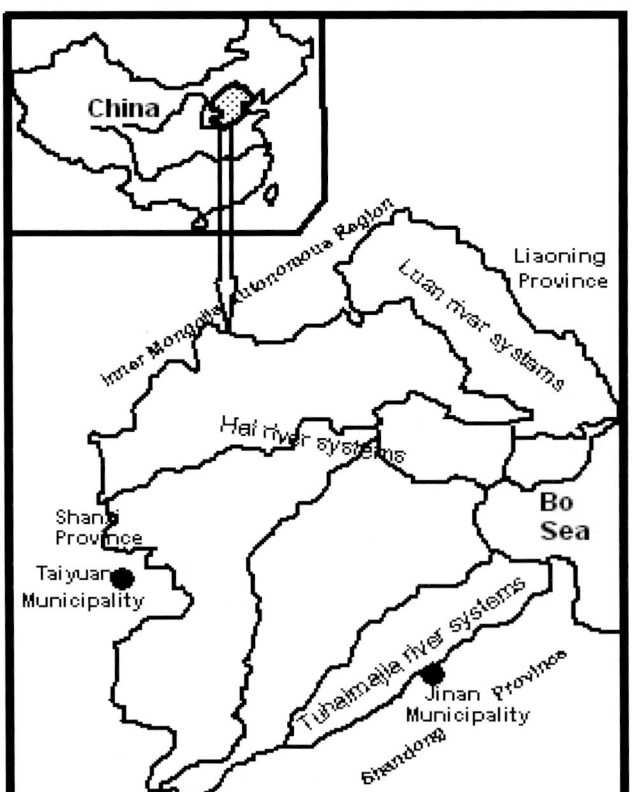

Fig. 1 Sketch of Haihe River basins in China.

CALCULATION OF THE OPTIMIZED PWUD

Quantization model

The model includes an objective function and restraining conditions, among the restraining conditions, water-soil resources–eco-environments–social-economy compound system interaction relationship restraining is the key.

The objective function describes sustainable development, it is expressed by sustainable development index (*SDI*), calculated in the following:

$$\begin{cases} SDI = Max\prod_{T=1}^{N}(ES(T))^{\frac{1}{N}} \\ ES(T) \geq ES(T-1) \end{cases} \tag{1}$$

where $ES(T) = SE(T)^{\beta_1} \cdot EE(T)^{\beta_2}$, $SE(T) = \prod_{i=1}^{m} U_i(T)^{a_i}$, $EE(T) = \prod_{j=1}^{n} H_j(T)^{b_j}$.

ES(*T*) is the measuring value of eco-environmental quality and social-economy level synthetic evaluation during the period *T*; *SE*(*T*), *EE*(*T*) are the measuring value of social-economy level and that of eco-environmental quality during period *T*, respectively. They are all scalar assignments in [0, 1] by grade of membership. β_1 expresses the weight of social economy level in *ES*(*T*), β_2 expresses the weight of eco-environmental quality in *ES*(*T*). $U_i(T)$ expresses the grade of membership of social-economy level index *i* during period *T*, $H_j(T)$ expresses the grade of membership of eco-environmental quality index *j* during period *T*; *m* and *n* express the number of social-economy level indexes and eco-environmental quality indexes, respectively; a_i expresses the weight of social-economy level index *i* in *SE*(*T*); b_j expresses the weight of eco-environmental quality index *j* in *EE*(*T*). All the weights are determined by stratification analysis.

In the Haihe River basins, the indexes of socio-economic level are per capita GDP (Gross Domestic Product), percentage of tertiary industry to GDP, urbanization ratio and per capita grain yield, chosen by their importance in the socio-economic system; the indexes of eco-environmental quality are the area ratio of water and soil loss, the discharge amount of COD (chemical oxygen demand), extraction coefficient of groundwater, length ratio of river channel shortening, area ratio of wetland, area ratio of city river or lake, and volume of water reaching sea outfall, chosen according to the main EEP in Haihe River basins.

The restraining conditions include the restraining of water-soil resources–eco-environments–social-economy compound system interaction relationship and other restraining conditions. The first restraining model includes a water budget sub-model, socio-economy–water relationship sub-model and environment–water relationship sub-model. All the sub-models are connected by water resources and the hydrological cycle. The other restraining conditions involve the restraint of water resources, environments related to water and socio-economy specially described in the papers by Zhu *et al.* (2005).

Judging whether or not eco-environments is in a good state

In Haihe River basins, whether or not the eco-environment is in a good state is decided by the eco-environmental quality estimating index *EE*. If *EE* = 0.8, eco-environment would be in a bearable state and *EE* = 1, eco-environment would be in an optimization state. It should be noted: *EE* is determined by various eco-environmental quality indexes, and *EE* = 0.8 is the minimum value at which men can endure.

Calculation scheme and material source

The calculation scheme is that of South-to-North Water Transfer Project, the local society and economy developing in the usual style and the eco-environments gradually ameliorate, the scheme is determined according to the report on the programme of water resources in Haihe River basins (Haihe Water Conservancy Committee, State Water Conservancy Ministry, 2000). The year to begin calculations was chosen as 1998.

The materials needed are those on water and soil resources, eco-environments and social economy, except for the programmed values of sewage treatment ratio and percentage of tertiary industry to GDP; other materials are from the report of water resources program in Haihe River basins (Haihe Water Conservancy Committee, State Water Conservancy Ministry, 2000) and the statistical yearbooks of Peking Municipality, Tianjin Municipality, Hebei, Shanxi, Shandong, Henan Province and the Inner Mongolia Autonomous Region (1998–2003). The programmed value of sewage treatment ratio is determined by the proportion of capital investment for environmental protection in GDP, and that of percentage of tertiary industry to GDP is determined by the corresponding industrial structure relationship theory on various economy development phases developed by H. Qiannule, the famous American economist, and revised by the development centre of State Council in 2001, according to the conditions of China.

RESULTS

Model results are shown in Table 1 and Fig. 2.

Table 1 The change of optimized PWUD for sustainable development in Haihe River basins.

Year	IP (%)	AP (%)	EP (%)	LP (%)	EE	SDI
1998	16.18	71.0909	1.8200	10.9100	0.0772	0.26
2000	17.26	69.8800	1.9100	10.9500	0.0804	0.27
2010	24.15	57.9900	7.6700	10.1900	0.2469	0.48
2020	26.95	50.3600	12.5000	10.1800	0.4276	0.64
2030	29.51	43.7800	16.4000	10.3100	0.7193	0.83
2033	27.76	44.0800	17.6200	10.5400	0.7988	0.88
2035	26.58	44.3000	18.4200	10.6900	0.8518	0.91
2040	23.55	44.9100	20.4100	11.1200	0.9739	0.98

IP, AP, EP and LP are the percentage of water usage in industry, agriculture, ecology and life, respectively.

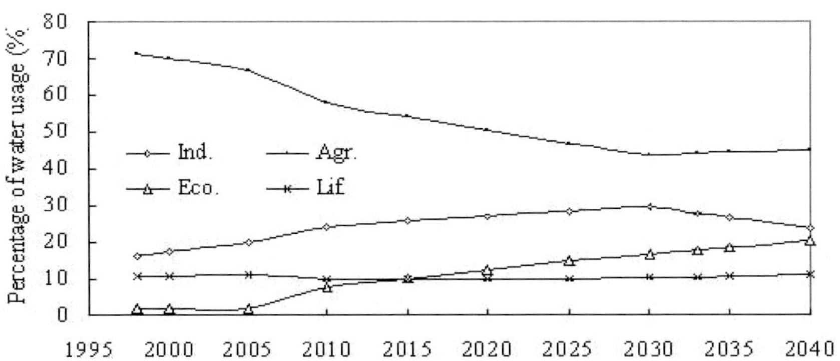

Fig. 2 The change of optimised PWUD for sustainable development in Haihe River basins.

Under the water supply of the South-to-North Water Transfer Project, assuming that the eco-environments gradually ameliorate, and society and economic development are normal. The Haihe River basins develop sustainably:

(1) At present, the percentage of water usage in ecology to the total water usable is least, only 1.82%.

(2) The change of percentage of water usage in life to the total water usable is not obvious, about 10%; percentage of water usage in industry to the total water usable will increase gradually

before 2030, and after 2030 will not increase continually, but reduce gradually; percentage of water usage in ecology to the total water usable will gradually increase; percentage of water usage in agriculture to the total water usable will reduce gradually before 2030, and after 2030 it will basically maintain a steady value, about 10.5%, and shows a weak increasing trend.

(3) During the change course of optimised PWUD for sustainable development in Haihe River basins from 1998 to 2040, the percentage of water usage in agriculture to the total water usable remains the biggest; that in industry is the second biggest, the *status quo* of that in ecology is lower than that in life, but after 2015 that in ecology will surpass that in life and then continuously be bigger.

CONCLUSION

In this paper, a scientific and rational computation model is set up, and the change of course of optimized distribution of water usage is calculated under the appointed scheme. While the theory and methods for studying sustainable development are well worked out, there are many new problems for future study:

- How to consider all the factors influencing sustainable development of a region. In this paper only water resources are considered, how to combine other resources with the quantification model on sustainable development is needed to be further studied.
- How to accurately express the quota of water usage in agricultural irrigation and that in industry as variables in the future, because they all change with the improvement of socio-economy and technical level.
- How to popularize the model. How to apply the model to study the sustainable development for other river basins should be done in the future.

Acknowledgements This research is supported by the National Basic Research Program of China (2006CB400502), supported by the 111 Project under grant no. B08048, Ministry of Education and State Administration of Foreign Experts Affairs, China, supported by the Program for Changjiang Scholars and Innovative Research Team in University under grant no. IRT0717, Ministry of Education, China, supported by the Grand Sci-Tech Research Project of Ministry of Education under grant no. 308012, and also supported by the Ministry of Water Resources Special Public Sector Research (200701031). The authors are grateful to the reviewer and editor, the comments and suggestions of the reviewer have contributed significantly to the improvement of the manuscript.

REFERENCES

Chen, Y. W. (2002) *Natural Resources Science*. Eastern China Normal University Press, Shanghai, China.

Du, L. Q. (2003) Resources, environment and sustainable development. *Journal of Peking University (Philosophy and Social Sciences)* **40**(3), 117–123.

Haihe Water Conservancy Committee, State Water Conservancy Ministry (2000) Report of Water Resources Layout in Haihe River Basins.

Martha, G. R. & Yang, G. A. (2003) The international progress of sustainable development research: a comparison of vulnerability analysis and the sustainable livelihoods approach. *Progr. Geogr.* **22**(1), 11–21.

Wang, H. F., Dong, Z. C. & Zuo, Z. G. (2003) Index system and evaluating method for sustainable development of regional complex system. *Journal of Hohai University (Natural Sciences)* **31**(2), 212–215.

Zhu, G. H. (2003) Sustainable development of China from a globalization perspective. *Population Research* **27**(3), 12–20.

Zhu, Y. H., Ren, L. L. Xia, J., Feng, H. L. & Huang, H. (2005) The study of carrying capacity of eco-environments related to water in Haihe river Basins. *Journal of Lanzhou University (Natural Sciences)* **41**(4), 11–15.

Watershed hydro-ecological zoning in the Upper Reaches of Yangtze River

CHUAN LIANG & YU-BANG LIU
College of Water Resources and Hydropower, Sichuan University, Chengdu, Sichuan 610065, China
lchester@sohu.com

Abstract Defining functional water resource zones and hydro-environmental zones play a critical role in water resources utilization, water environment improvement, and hydro-ecological preservation. Further studies are necessary when making a reference to advanced international ideology of contemporary water environment management, e.g. zoning based on the integrity evaluation, full consideration of regional differentiations, and suitability for local circumstances. This paper, employing relevant theories, takes into full account the natural and ecological features of the water system in the Upper Reaches of Yangtze River, develops the concept of the watershed hydro-ecosystem, and elaborates the preliminary idea of watershed hydro-ecosystem zoning. The definition and basic principles of watershed hydro-ecosystem zoning are proposed as the integrity evaluation and the comprehensive assessment index system are established, and a practical zoning scheme of the hydro-ecosystem in the Upper Reaches of Yangtze River is presented, which to a great extent embodies international advances and domestic management trends, with holistic considera-tion of watersheds, coordination conception, suitability for local circumstances, and dynamic property. The study provides fundamental ground work for making scientific hydro-environment management strategies and constructing a basic platform for applied studies in the upper Yangtze River basin.

Key words the Upper Reaches of Yangtze River; watershed hydro-ecosystem zoning; zoning principles

INTRODUCTION

The watershed hydro-ecosystem, a synthesis made of hydrogeologic aspects, topographic features, ecological types, and environment, has become a research field describing and studying various hydrological and ecological processes of a river drainage basin. As an integrated hydrological and meteorological, ecological and environmental, social and economic system, the system has the characteristic of its temporal and spatial hierarchy, evolutional, dynamic and flexible composition, limited power to resist ecological pressure and keep its balance, and the non-total predictability of its evolution of configuration and processes (Fu Bo-jie, 2001; Fitzhugh, 2002; USEPA, 2003; Xia Jun *et al.*, 2003).

Why do we develop the concept of watershed-based hydro-ecological zoning? With the defining demand of the basic features of the hydro-ecological system of a river drainage basin, an understandable and effective expression that can be shown on a map is needed to reflect the configuration and construction of the hydro-ecological system. The most severe environmental problems and ecological disasters are characterized on the scale of a watershed, and at the same time are in direct relationships with the preservation of water supply, water and soil, and water environment (Andrew, 1999; Miao Hong, 2001; Zhao Wen-zhi, 2001; Cheng Gen-wei, 2006). Watersheds of a river are just functional in these respects, and the watershed-based zoning consequently serves as a coupling platform in describing the environmental changes. The concept of watershed hydro-ecosystem primarily refers to:

(a) Definition of the natural boundaries between watersheds, and between climatologically, hydrological, topographical and ecological zones in terms of hierarchical structure and scale features.
(b) Typological description of river drainage system, hydrological climate and ecosystem in terms of ecological features and environmental units.
(c) Identification of the impact of human activities upon ecological environment, elimination of regional differentiations of the coordination of watershed basins, and the assessment of the impact of human activities.

Watershed-based hydro-ecological zoning is supposed to be in accordance with the functional zones of natural conditions in terms of hydrological, meteorological, geological, ecological features or environmental units. In the light of values, a holistic view is applied with emphasis on the resource attributes of such essentials like water, climate, vegetation, and even more on ecological attributes of these essentials so that the composition, structure, and functions of the ecosystem are better observed (CAEP, 2001; NBEP, 2002; Yan Nai-ling, 2004). Hydrologically-speaking, a river drainage basin, collecting all the surface and groundwater in the catchment, is an integral unit of water resources management as well water and soil conservation. Water is the most important ecologically in that freshwater resources in the drainage basin support the ecosystem. Ecologically-speaking, the drainage basin serves as the connection between land and oceans. Wetlands, rivers, and lakes in the basin are fish habitats; forests and grasslands not only function as habitats for terrestrial plants and animals, but also play a decisive role in terms of water conservation, water quality prevention, flood detention, soil nutrients restoration and circulation; Besides, regional ecological problems, because of the material and energy flow in the basin, become problems, and the zoning and management of ecosystem based on a watershed unit have been widely accepted. The watershed based zoning, therefore, is in accordance with the following conditions:

(a) The watershed is a unique ecological unit with abundant water system, forests, mountains, swampy land, agricultural and other artificial ecosystems, which are interconnected through material and energy flows, biological movement, and human activities so that a complex system is established, synthesizing the research objects of hydrology, meteorology, geology, ecology, and environmental sciences, involving ecosystem life and physical environment and their relationship.
(b) The watershed is an independent physical unit with widely accepted and well-defined boundaries, having relatively long-term and wide-space stability.
(c) The watershed has a legible hierarchy assuming flexibility, agreeing with the hierarchical structure of its ecosystem property, which is adequate to achieve realistic and effective research of the hydrological ecosystem of the watershed.

BASIC CONCEPTS AND PRINCIPLES OF WATERSHED HYDRO-ECOSYSTEM ZONING

Watershed hydro-ecosystem zoning refers to the division of a natural region or environmental unit in terms of the conformability between natural conditions (including hydrological, meteorological and geological conditions) and natural ecological features in the light of the comparability of the productive potentials of the ecosystem types. The essential purpose is to divide regional hydrological, meteorological and ecological environments into varied integral functional regions for the convenience of the research of the environmental bearing capacity (environmental capacity), the present situation and the developing trend of the environment, in addition to the revelation of coupling mechanisms of the environmental changes in the drainage basin (Zhou Feng, 2007).

Watershed hydro-ecosystem zoning is based on the analysis of the essential features of watershed hydrological environment and ecosystem, with its concepts, connotations, characteristics and technical research methods established on the water system and eco-types on the watershed scale. The denotation of watershed hydro-ecosystem and eco-hydrology are not distinguished in terms of which has more emphasis on which aspect, and the two terms both refer to cross-disciplinary fields of ecology and hydrology, namely the influence of water processes upon the structure, distribution, configuration, and growth of ecosystem, and the influence of ecosystem – its vegetation types, configuration, and distribution – upon water circulation, and vice versa. In the research process, certain research aspects of watershed hydro-ecosystems are emphasized according to research needs.

Watershed hydro-ecosystem denotes that the hydro-ecosystem zoning is based on hierarchical position and on different scales, and that the zoning principle of hydro-ecosystems in different river basins of varied hierarchical structures must be consistent, in addition to differences on certain hierarchy. The general zoning principles are as follows:

- Regional relativity – dominating factors and zoning assumption are defined in accordance with the relationships between hydrological, meteorological, geological, ecological structure, distribution, configuration and regional differentiation regularities.
- Regional integrality – each region is defined in the light of its integrally hydro-geological space with acknowledgement of and respect for integrity and uniqueness of the natural ecosystem boundary.
- Spatial continuity – uninterrupted hydro-ecological zoning on certain hierarchy in space, neither repeated nor overlapped, form a continuum with the higher hierarchy.
- Comparability principle – although zoning is in accordance with regional differentiations, the internal zoning features on different hierarchies are comparable, and ecosystems in terms of time and space are justifiably analysed on different scales.
- Furthermore, the collocation of partial river basin with ecological region and conformity to administrative boundary also apply to certain hierarchical zoning.

In the case of an emphasis on hydrology, watershed hydro-ecosystem zoning:

(a) Mainly refers to the study of changed laws of the water cycle when vegetation types, configuration, allocation and other factors in the ecosystem keep changing, and non-animate material remains relatively stable. Among the changing factors, the variation in vegetation type is the leading factor which is to be studied from the angle of production, assembling, and circulation of its classification, distribution, and allocation in conditions of the same physical factors of climate, slope and soil.
(b) Or refers to the study of water cycle response mechanism – mainly in the case of the same vegetation belt and different climate zones – when vegetation types, distribution and allocation remain stable, and such non-animate factors like precipitation, temperature, wind speed and soil keep changing.
(c) Or refers to the study of the changed laws of water cycle when animate and non-animate material both keeps changing. For instance, GIS-based hydro-ecological method, when applied to the research into a river basin made of ecotones and basic belts, is advantageous in obtaining microscopic hydro-ecological information, which is further integrated on a macro scale so as to better explain the water configuration in the river basin.

INTEGRALITY EVALUATION OF WATERSHEHED HYDRO-ECOSYSTEM AND THE COMPREHENSIVE ASSESSMENT INDEX FOR ZONING

Integrity evaluation of watershed hydro-ecosystem

Integrality evaluation of watershed hydro-ecosystem refers to a holistic evaluation of the natural condition and ecological environment in the watershed. As a concept to measure the self-renewability of the river-water and ecological system, watershed hydro-ecosystem integrality denotes the maintenance and function capacity of the whole hydrological ecosystem – river-water system, hydrological climate, topography, ecological environment, and the natural processes, which guarantee sustainable and extended adaptability to environmental changes and human utilization demand.

Integrity in space

Watershed, the physical catchments of water system and the continuum with hydrological functions, determines the manners of water runoff and convergence. From the perspective of landscape ecology, watershed is a slope-plain-lake patch-unity with river-water system as corridors. Correspondingly, point source and non-point source pollutants from natural processes and human activities in different patches converge through the straight flow corridors into the main river course, exerting great influence upon the water environment and ecological system. From the perspective of river basin management, the watershed, as a relatively integral management unit and the concentration space for social and economical activities, is not only a

Table 1 The comprehensive assessment indexes of watershed hydro-ecosystem in the Upper Reaches of the Yangtze River.

The first rank assessment indexes	The second rank assessment indexes	The third rank assessment indexes
(a) differences in topography; (b) climate type(divided by climatic zones); (c) natural water system(divided by the distribution of the trunk river and its tributaries)	(a) indicators of hydrological characteristics; (b) hydrothermal climate indexes; (c) habitat indicators; (d) geomorphic type; (e) the second rank division of water resources	(a) geomorphic type; (b) climate type(divided by climate zones); (c) vegetation types(subdivided by vegetation in their subjects); (d) indicators of human activity

Based on these three frank indexes, watershed-based hydro-ecological zoning in the Upper Reaches of Yangtze River can be done. See the zoning results and the hydro-ecological characteristics of every district through Table 2 and Fig. 1.

contradiction carrier of water resources and eco-environment issues, but also the very place where contradictions in ecological and environmental preservation are most intensely centralized. The study of river basin and the research method, therefore, are helpful in internalizing regional stakeholder's conflicts and maximizing the water system's multiple functions and the potentials of the scale economy.

Dynamic property in time

Watershed hydro-ecosystem zoning, as a spatial unit of ecological management, reflects the mutual impacts of human activities and ecosystem, and in entirety plans and measures human needs function and ecological needs function on watershed and temporal scale so that inter-coordination and basin management in the light of local circumstances are achieved.

Watershed hydro-ecosystem zoning, a watershed as a unit, aims at the coordination between the watershed and the ecological environment. In a watershed basin, not only human impact upon water volume changes in a river, but also human impact upon water quality changes can be well-observed. Watershed hydro-ecosystem zoning can reveal the tolerance and bearing capacity of certain featured area, and reflect structural and organizational disparities in different ecosystems. The integration of watershed and hydro-ecosystem zoning is favourable for defining the preservation rankings of hydrological and the ecological environment.

Comprehensive assessment indexes for watershed hydro-ecosystem zoning

Watershed hydro-ecosystem, a grand system, is a synthesis made of hydrological features, topographic forms, eco-types and eco-environment, and other factors. Its zoning indexes also constitute a scientific system. According to the primary conception and principles of watershed hydro-ecosystem zoning and the terrain features and hydrological characteristics and ecological features functional in the hydro-ecosystem and the impact of human activities, the comprehensive assessment indexes of watershed hydro-ecosystem evaluation should be built in three levels:

(1) The first rank involves three indexes with main reference to the natural distribution of water system, the comparability of climate features, including:
(a) Differences in topography, such as pattern of large terrain and altitude;
(b) Climate type (divided by climatic zones), such as northern subtropical zones;
(c) The distribution of natural water system (divided by the distribution of the trunk river and its tributaries, such as Jiangyuan Water System, Chuanjiang Water System.

(2) The second rank refers to sub-watershed units with main reference to the integrity of geomorphic unit and the conformity of hydrological characteristics and eco-system, including:
(a) Indicators of hydrological characteristics, such as annual rainfall, annual runoff, and annual land surface evaporation;
(b) Hydrothermal climate indicators, such as dryness, average annual temperature;
(c) Habitat indicators, such as eco-type, regional vegetation (differentiation of regional vege-

tation in longitude direction and latitude direction), regional soil;
(d) Geomorphic type, such as plateau, plain, hills, basin, etc.

(3) The third rank refers to further division of the second rank with main reference to the impact of human activities on the hydrological cycle and ecological environment, including:
(a) Geomorphic type, that is, homology landform segmentation;
(b) Climate type (divided by climate zones), such as Qingnan Zone, Chuanxi Zone, etc.
(c) Vegetation types (subdivided by vegetation in their subjects);
(d) Indicators of human activity, such as land-use types, soil erosion, population density, etc. See the system of specific indicators through Table 1.

Table 2 Watershed for hydro-ecosystem in the Upper Reaches of Yangtze River.

Zoning scheme			The hydrological and ecological features according to each hydro-ecological zone
The first rank	The second rank	The third rank	
I River-head alpine plateau-grassland and prairie zone	I-1 Tuotuo River basin of the head of Yangtze River	I-1-1 Geladandong glaciers and Hoh Xil main river course section	The district of I-1 hides inland area, with such characters as resounding terrain, bare mountain, strong frost crack, cold and dry climate and regional distribution of lakes, swamps and glacial. It also lies in the arid region of plateau climate zones, with its annual average temperature being –4.2°C and its absolute minimum temperature being –33.8°C, and its freeze period being 7 months. The vegetation in this area is mainly woody plants, while the eco-type in this district includes alpine tundra desert, alpine meadow, sparse semi-arid grassland, and sub-humid alpine meadow. The type of soil in this area consists of alpine desert soil and alpine meadow soil, which has less organic matter content. At the same time, the district of I-1 also lies in a low runoff area, with its average annual rainfall being less than 400 mm, its surface evaporation being more than 900 mm, its land surface evaporation being 200 mm, and its relative humidity being very small.
	I-2 Watershed of Tongtian River's main course	I-2-1 Dangqu River main course section I-2-2 Qumalai River and Chumaer River main course section I-2-3 Qumalai-Zhimenda main river course section	This district of I-2 lies in the transition zone from the Jiangyuan hilly plains to high mountain valleys. Its peak elevation gradually increases from 100 m to 500 m which is a.m.s.l for 400–600 m. River terraces develop at the two sides of rivers. Its climate type lies in sub-humid region of highland climate zones which is cold, with its annual average temperature being from –2.6°C to 2.9°C and its extreme minimum temperature being –34.8°C. The economy in this district is mainly based on agriculture and animal husbandry. Its vegetation mainly consists of shrubs, grass and forest. Trees arise gradually in the Lower Reaches of the Yangtze River. This district also lies in low runoff area, with its annual average precipitation raging from 400 mm to 500 mm, its annual evaporation being less than Jiangyuan area, and its relative humidity being very small.
II Alpine and sub alpine valley and forest zone	II-1 Xerothermic basin of Jinsha River	II-1-1 Zhimenda-Shigu main river course section II-1-2 Reaches below Shigu main river course section II-1-3 Yalong River main course section II-1-4 Dadu River main course section	The district of II-1 is located in the southern Qinghai-Tibet Plateau and in the northern Yunnan Plateau. Due to strong seismicity, there is the development of collapse, landslide, debris everywhere. The main features of the landscape patterns of this district is high mountains-deep valley, while rivers form shows the characteristics of "high, deep, steep, narrow, curved". Because of the large gap between the region lying, the district is rich in water resources, and most of the rivers contain a high amount of sediment. The weathering material of this district is mainly alpine weathering critic and the impact of the floods, etc. The district also lies in the humid subtropical region, with its climate type changing alternatively from cold to warm, with its annual average temperature raging from 20°C to 22°C, with its annual average precipitation ranging from 400 mm to 800 mm and with its land surface evaporation raging from 200 mm to 700 mm. Runoff of this district shows regional distribution. The species of plant is tree-based, and the forest cover is beyond 20%.
	II-2 Basin-and-hilly area of Tuojiang River	II-2-1 Minjiang River main course section II-2-2 Tuojiang River main course section	Most of the district of II-2 is located in the Sichuan Basin in the west of which distributes low mountains. The Upper Reaches of Min River and Tuo River lie in Dianbei-Longnan mountains, while the lower reaches lie in plains and hilly. The mountain area in the Upper Reaches lies in humid plateau region, while the plains and hilly in the downstream area lies in the humid subtropical region. This district is also rich in abundant rainfall, with its average annual precipitation being almost over 800 mm, and its annual largest precipitation measured at 2434.8 mm. In the Dadu River basin, there is the largest centre of Emei in rainfall. Due to its plenty of water resources, this district is the main hydropower-development area and important farming area.
III Basin-plain farming and forest zone	III-1 Mountainous area of Jialing River	III-1-1 Fujiang River main course section III-1-2 Jialing River	Most of the district of III-1 is located in the Sichuan Basin in the west of which distributes parallel range-gorge, with the terrain tilting from N to S. This district mostly lies in north sub-tropical humid region, with its annual average temperature being 16–18°C, and with its maximum temperature reaching 42°C. The mountain area of this district lies in arid plateau region, with its annual average temperature being 11–16°C, and the minimum temperature being below –10°C. The precipitation

	main course section III-1-3 Qujiang River main course section	consists of rainfall and snowfall, ranging from 600 mm to 1000 mm. The Upper Reaches of the Jialing River and its tributaries Bailong basin are the main sediment regions. This district is also the main farming area, which has fertile soil and suitable climate.
III-2 Watershed of Chuanjiang River's main course	III-2-1 Three Gorges Reservoir area III-2-2 Chishui River main course section	Most of this district is located in Sichuan Basin. Terrain in the region increased gradually from the west to the east. The trunk stream of the Yangtze River is the golden path for water transportation. This district also lies in the humid subtropical climate zone in which cold wave is not easy to invasive, with its annual average temperature reaching above 18°C, with its average temperature of the coldest month being 5–8°C, with its average temperature of the hottest month being 25–29°C and with its maximum temperature being above 40°C. Along the main stream of the Yangtze River, rainfall is abundant. The volume of runoff and sediment of the main stream of the Yangtze River is relatively large, since the main stream undertakes the runoff and sediment transport capacity of the tributary. The world-famous hydropower project Yangtze River Three Gorges Hydropower Station and Gezhouba Hydropower Station just lie in this district. This district is also the area of comprehensive utilization of water function and the erosion area. Water pollution of the city along the Yangtze River is more serious.
III-3 Plateau area of Wujiang River	III-3-1 Wujiang River main course section	This district of III-3 is mostly located in the transition zone from the northeastern part of the Yunnan-Guizhou Plateau to the Xiangxi Hills in which mountain undulates and valley deeply down cuts and karsts and large underground river grows everywhere. Topography of this district marks the layered development characteristics. At the same time, the district lies in the humid subtropical region in which seasonal alternation is fuzzifying. Temperature gradually increases from upstream areas to downstream areas, with its minimum monthly average temperature being 3–8°C, and its highest monthly average temperature being 18–24°C. Multi-year average precipitation of this district ranges from 900 mm to 1400 mm, while its multi-year average runoff is 50 billion m^3. In this district, mountain area covers more than 85% of the total, while the arable land is very limited. This district is the region which has a certain potential for hydropower development.

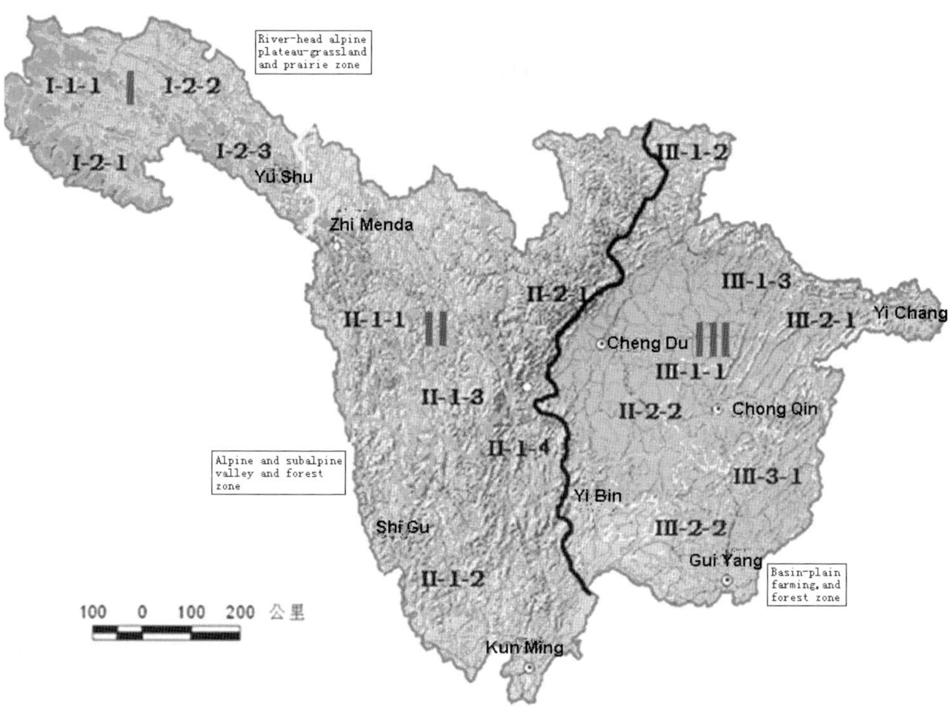

Fig. 1 Three ranks zoning of the hydro-ecosystem in the upper Yangtze River basin.

CONCLUSIONS

(a) The hydro-ecosystem in the Upper Reaches of Yangtze River is a large complex system, and the comprehensive assessment indexes of integrity evaluation and zoning is a measure of river water system and eco-environmental sustainability. Watershed hydro-ecosystem zoning, therefore, is a fundamental research program with great value in terms of the exploration of environmental changes and water-sediment productivity mechanism.

(b) By employing advanced concepts and relevant theories and adequate studies of regional natural condition and eco-environment, this paper delves into the hydro-ecosystem zoning in the Upper Reaches of Yangtze River by taking full consideration of human interference with the ecological environment and making full use of theoretical methods of contemporary bio-hydrology, environmental sciences, and eco-economics. The conceptual watershed hydro-ecosystem is proposed, its denotation and critical issues defined with special emphasis on the basic zoning idea of watershed, coordination ideology, and suitability for the local circumstances. The research has revealed the comparability and differentiation laws between river water systems and natural ecosystems, practiced a zoning scheme in the area and recognized hydro-ecological units. Watershed hydro-ecosystem zoning refers to the ecosystem zoning in the natural watershed, and the zoning practice, recognizing neither pure nature zones nor simple eco-function zones, puts stress on the composite influence exerted by water, climate, and eco-environment. The zoning principles are established not only on water-climate factors — regional differences of natural system, but also on the hierarchical structure of river water ecosystem, and most important of all, on the watershed ecosystem.

(c) Hydro-ecosystem zoning in the Upper Reaches of Yangtze River is intended to further delineate the typological compositions and regional differentiations of its watershed hydro-ecosystem for the sake of necessary hydro-ecological reconstruction. The zoning analysis, fundamental in the research of water and sediment production regularities, is of great value in terms of scientific monitoring and management of the watershed hydro-ecosystem to finally achieve sustainable use of natural resources in the Upper Reaches and water resources lustration and supply in the Mid-and-Lower reaches of Yangtze River.

REFERENCES

Baird, A. J. & Wilby, R. L. (1999) *Eco-Hydrology: Plants and Water in Terrestrial and Aquatic Environments*. Routledge, London, UK.

Cheng Gen-wei & Wang Jin-xi (2006) *Theories and Models of the Construction of Ecological Functions in the Three-Parallel Rivers Region*. Sichuan Publishing House of Science and Technology, China.

Chinese Academy for Environmental Planning. (2001) *Technical Guidelines for Zoning Water Environment Functions* [R/OL]. Beijing: National Bureau of Environmental Protection.

Fu Bo-jie (2001) *Principles and Applications of Landscape Ecology* [M]. Beijing Science & Technology Press, 43–72.

Miao Hong, Wang Xiao-Ke & Ouyang Zhi-Yun (2001) Study on regionalization of eco-environmental stress process in China. *Acta Ecological Sinica* **21**(1), 7–13.

Ouyang Zhi-yun, Wang Xiao-ke & Miao Hong (2000) China's eco-environmental sensitivity and its spatial heterogeneity. *Acta Ecological Sinica* **20**(1), 9–12.

Fitzhugh, T. (2002) Watershed characteristics and aquatic ecological integrity: a literature review. http://www.freshwaters.org/docs/litrev.pdf.

The editorial board of *Archives of Yangtze River*. (2003) *Water Systems* [M], 16–192]. China Encyclopedia Publishing House, Beijing, China.

The Office of West China Development Leading Group of State Council, National Bureau of Environmental Protection. (2002) *Tentative Technical Specifications of Ecological Function Regionalization*[R]. http://www.sepa.gov.cn/download/1055352716377.doc.

USEPA (2003) Watershed Analysis and Management (WAM) Guide for States and Communities[R]. Washington DC, US EPA.

Xia Jun *et al.* (2003) Eco-hydrology – concept, framework and system. *J. Irrign Drainage* **22**(1).

Yan Nai-ling & Yu Xiao-gan. (2004) The research methodology of China's Ecologically functional zoning based on river basins. *J. Lake Sciences* **16**(Suppl.), 143–152.

Zhao Wen-zhi & Cheng Guo-dong (2001) Ecohydrology – A science for studying the hydrologic mechanisms of ecological patterns and processes. *J. Glaciol. Geocryology* **23**(4).

Zhou Feng, Liu Yong, Huang Kai, Guo Huai-cheng & Yang Ping-jian (2007) Water environmental function zoning at watershed scale and its key problems. *Adv. Water Sci.* **18**(2), 216–222.

Numerical experiment on water and soil losses for the application in ungauged basins

XIN-HUA ZHANG, YU-CHENG XIAO, RUI-PING YAN, NENG-CHUAN LI & XIAO-RONG HUANG
State Key Laboratory of Hydraulics & Mountain River Engineering, Sichuan University, Chengdu 610065, China
xhzhang@scu.edu.cn; hxiaorong@tom.com

Abstract Water and soil losses are dominated by the conditions of regional geomorphologics, land-uses, and landcovers, in addition to strong influences of rainfall characteristics in the region. An ideal approach to deal with the problem of water and soil losses in the study should consider all conditions together. Therefore, a coupled 2-D numerical model of water, sediments, and morphological change based on the governing equations of 2-D shallow water, 2-D unbalanced sediment transportation, and ground surface variation in elevations was proposed in this study. Considering the general complexities of regional land covers and land-uses, the coupled 2-D numerical model was developed by an unstructured FVM method based on irregular polygonal mesh systems. Throughout its application the experimental station of water and soil losses in Suining Prefecture, Sichuan Province, the coupled 2-D numerical model was validated. Results indicate that the model can perform stably and reliably with satisfactory accuracy for the evaluation of the distribution of water and soil losses, both in temporal and spatial domains. Furthermore, the results also imply that the model can be applied to ungauged basins for the evaluation of the sources of water and soil losses coming from within a basin so that adequate and practical methods for treatment can be adopted.

Key words water and soil conservation; shallow water equations; sediment; erosion

INTRODUCTION

Soil erosion is generally controlled by spatial distributions of flow variables such as velocities and water-depth, as well as by the topographical characteristics in a watershed when rainfall takes place. The rules of water and soil losses obtained from the study of a single variable can provide some guidelines for the practices of water and soil controls. Besides the strong influences of rainfall properties, however, water and soil erosion will be controlled by topography, land use conditions and land covers. Ideal approaches for the study of soil erosion should integrate all of the impact variables into a system.

Although the runoff variation on a slope is unstable and the flow belongs to a 3-D movement, shallow water characteristics with a small scale of water depths and wide range of flow can be found. Therefore, the flow in a watershed can be formulated by 2-D shallow water equations which are deduced by integrating the 3-D Navier Stokes equations along the vertical direction of water depths. Some terms in the shallow water equations can reflect the influences of variations in ground elevations and land use conditions. In other words, 2-D shallow water equations can be used to consider the complicated regional topography, spatial distributions of flow states, water depths, velocities, land covers and resistances, and rainfall intensities. Adding the non-equilibrium sediment transport equation and ground variation equation into the 2-D shallow water equations, runoff and erosion on the slopes of a river-basin can be simulated.

From a brief review of the literature, we find four categories of numerical discretization for the 2-D shallow water equations: (1) finite different methods (FDM) such as Garcia & Kahawita (1986) for the numerical solution of St Venant equations with MacCormack finite difference scheme; and Zhang *et al.* (2003) for the simulation of flood inundation using 2-D shallow water equations in river basins with complex land-covers; (2) finite element methods used by Zhang & Takeuchi (2004) for modelling the groundwater flow in a large area; and Wood (1996) for the study on how to avoid spurious oscillation in the finite element solution of unsaturated flow equation; (3) characteristic method used by Katopodes & Strelkoff (1978); and (4) finite volume method (FVM) such as Zhao *et al.* (1996), Tan (1998), and Zhang *et al.* (2007).

Generally the land-covers in a watershed are quite complicated. The land-use-conditions and roughness for water flows are not uniform. The finite difference method may not be suitable for the need to reflect these influences in the computations because it uses structured meshes. To avoid the limitation of FDM, we can consider using FEM because all of the schemes based on this method were developed in the unstructured mesh system. However, for the stability of simulation, the scales of grid size and time step Δt must be properly selected so that the numerical oscillation can be avoided (Wood, 1996; Zhang & Takeuchi, 2004).

In order to investigate water and soil loss mechanisms systemically, an integrated numerical model that consists of 2-D shallow water equations, a 2-D non-equilibrium sediment transport equation and ground evolution equation is proposed in this paper. Considering the complexity of land-cover, land-use conditions and topographies, a FVM model based on an unstructured grid system was adopted to discretize the governing equations for the coupled model. The numerical model was validated through its application at An-ju Experimental Station located in Suining Prefecture, Sichuan Province. Results suggest that the model was stable, accurate and it can be used to investigate the mechanism of water and soil losses in a watershed.

GOVERNING EQUATIONS

2-D Shallow water equations

Supposing x–y represents the horizontal plan, z represents the vertical direction, and ignoring the viscosity in the x–y plan, we can deduce the 2-D shallow water equations when the gravity is dominant and water pressures in slope flows can be approximated by hydrostatic pressures. The resulting 2-D shallow water equations used in this study can be expressed as:

$$\frac{\partial h}{\partial t} + \frac{\partial M}{\partial x} + \frac{\partial N}{\partial y} = 0 \tag{1}$$

$$\frac{\partial M}{\partial t} + \frac{\partial}{\partial x}(u_0 M) + \frac{\partial}{\partial y}(v_0 M) = -gh\frac{\partial \eta}{\partial x} - \frac{1}{\rho}\tau_{bx} \tag{2}$$

$$\frac{\partial N}{\partial t} + \frac{\partial}{\partial x}(u_0 N) + \frac{\partial}{\partial y}(v_0 N) = -gh\frac{\partial \eta}{\partial y} - \frac{1}{\rho}\tau_{by} \tag{3}$$

where h is the flow depth; u and v are flow velocities in the x- and y- directions, respectively; M(= uh) and N(vh) are the fluxes in the x- and y-directions, respectively; g is gravitational acceleration; H is surface water level which equals to $Z_b + h$, here, Z_b is the ground elevation. ρ is mass density of water flow; τ_{bx}, τ_{by} are ground shear stresses in the x- and y- directions, respectively. They can be estimated by the equations:

$$\tau_{bx}/\rho = \frac{gn^2 u_0\sqrt{u_0^2+v_0^2}}{h^{1/3}}, \quad \tau_{by}/\rho = \frac{gn^2 v_0\sqrt{u_0^2+v_0^2}}{h^{1/3}} \tag{4}$$

where n is Manning roughness coefficient.

Sediment transport and land surface evolution

Because suspended sediment concentration is not generally equal to the carrying capacity of inundation flow, we use the non-equilibrium 2-D sediment transport equation, which is given by:

$$\frac{\partial (hS)}{\partial t} + \frac{\partial (huS)}{\partial x} + \frac{\partial (hvS)}{\partial y} = \varepsilon_s\left[\frac{\partial^2 (hS)}{\partial x^2} + \frac{\partial^2 (hS)}{\partial y^2}\right] + \alpha\omega(S_* - S) \tag{5}$$

where S means suspended sediment concentrations; S_* is sediment carrying capacity; ε_s is sediment diffusion coefficient; α is a coefficient; ω is the settling velocity; and other variables are the same as those in the 2-D SWE. In this study, sediment diffusion coefficient ε_s is considered to be the same as eddy viscosity, which can be written in the depth-averaged form:

$$\varepsilon_s = \kappa u_* h/6 \tag{6}$$

where κ is the Karman constant and u_* is the bed shear velocity. The shear velocity u_* can be estimated by the equation:

$$u_* = \frac{\sqrt{g}}{C}\sqrt{u^2 + v^2} \tag{7}$$

where C is the Chezy coefficient.

The carry capacity of uniform sediments can be expressed as:

$$S_* = K_c\left(\frac{(\sqrt{u^2 + v^2})^3}{h\omega}\right)^m \tag{8}$$

where K_c and m are coefficients. Equation (8) is not dimensionally homogeneous because the units of S_* (or K_c), u (or v, or ω), and h are kg/m^3, m/s and m, respectively.

In conditions dominated by suspended sediment, the changes of land surface elevation due to sediment settling and soil erosion can be estimated by:

$$\gamma'\frac{\partial Z}{\partial t} = \alpha\omega(S - S_*) \tag{9}$$

where γ' means the dry density of sediments; Z is the ground surface elevation; and t is the time. Therefore, on the basis of equation (9), the variation of ground surface elevation, ΔZ, resulting from sediment movement, in which some erosion or deposition will occur, during a single time step, Δt, can be calculated by:

$$\Delta Z = \frac{1}{\gamma'}\alpha\omega(S - S_*)\Delta t \tag{10}$$

NUMERICAL SCHEMES AND STABILITY CONDITIONS

Numerical schemes of the governing equations

In order to get a simulation scheme, which is accurate and without numerical oscillation, it is very important to design the variables of governing equations. If scalar and vector variables in the governing equations are designed irrationally, numerical oscillation or diffusion will occur during the simulation process. It is well known that a numerical scheme with variables designed in a staggered style in a spatial domain is stable and has a second order accuracy. Therefore, the variables in the numerical schemes of the governing equations have been defined spatially in a staggered style to guarantee the stability and accuracy of the scheme.

To ensure the numerical scheme worked out has a second-order accuracy in temporal domain, a leap-frog scheme for advancing in time has been adopted. On the basis of these methodologies, a 2-D unstructured FVM scheme based on an arbitrarily polygonal mesh system has been developed (Zhang *et al.*, 2007).

Stability conditions

In order to guarantee the stability of the scheme during the simulation process, time step Δt must be selected based on its relationship with grid sizes of a study region and meet the requirements of Courant-Friedrichs-Lewy (CFL) conditions. For the conciseness of this paper, the deduction process and the final expressions for the 2-D unstructured FVM scheme will not be given here (see Zhang *et al.* (2007) for more details).

CASE STUDY AND DISCUSSION

Study area description

In order to validate the coupled numerical model, which consists of 2-D shallow water equations, non-equilibrium 2-D sediment transport equation and evolution equation of land surface elevation, a case study was conducted at An-ju Experimental Station located in Suining Prefecture, Sichuan Province. The station is located at 30°21′51″N and 105°28′51″E and within the middle undulating hills of Sichuan Basin, belonging to Qiongjiang watershed of the first river branch of Fujiang River. The station is within the wet climate zone with an average annual rainfall of 902.00 mm based on the observed data from 1985 to 2001. Before the establishment of the station, vegetation coverage was very low and the coverage-rate was only about 6.1%. After a 3-fold expansion, the experimental station has finally reached an area of 38 mu, approximately 2.53 hectares. The highest location of Youfangpo has an elevation of 330.58 m while the elevation at the lowest point is 288.00 m. The elevation difference in this station is about 42.58 m.

The area of the experimental station has been divided into 87 arbitrary polygonal meshes with 112 nodal points based on the variation of topography and land-use conditions. After meshing, we can get the contour map of the region based on the elevation of each mesh. Figure 1 displays the topography of the watershed and the location of the gauge stations.

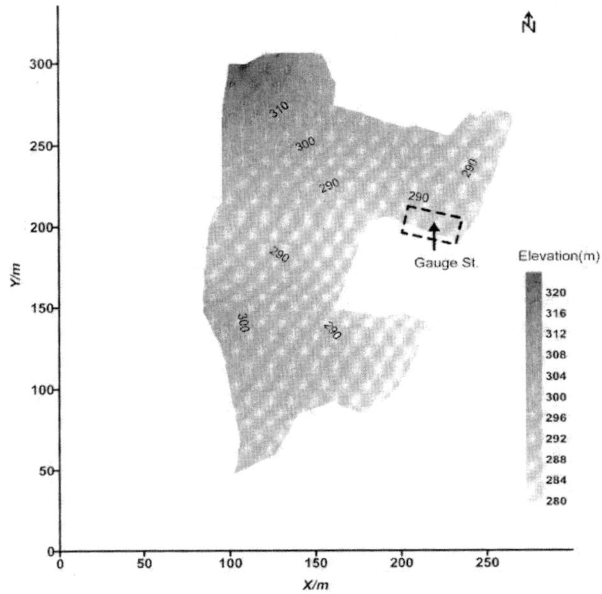

Fig. 1 Elevation map of the study region being meshed.

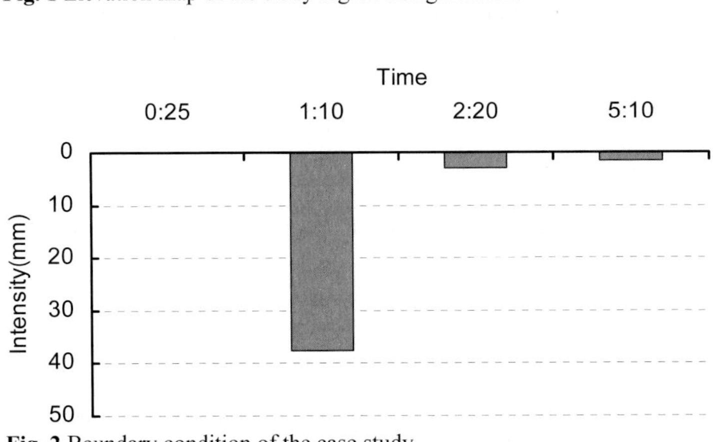

Fig. 2 Boundary condition of the case study.

Boundary conditions of the simulation

Based on the observed data series of rainfall and runoff, the rainfall event that occurred from 27 to 28 June 2004 was selected for this study. Its rainfall intensity is displayed in Fig. 2. This rainfall process was treated as the boundary condition for the case study.

Initial conditions

For the simplicity of simulation, it was supposed that the ground was dry when a rainfall scale occurred on 27–28 June 2004 at the experimental station, i.e. the initial water depth in the study region was regarded as zero.

Results of the simulation

The Spatial Distribution of the Water Depth Numerical simulation was carried out based on the given boundary condition and initial conditions. Results obtained are displayed in Fig. 3.

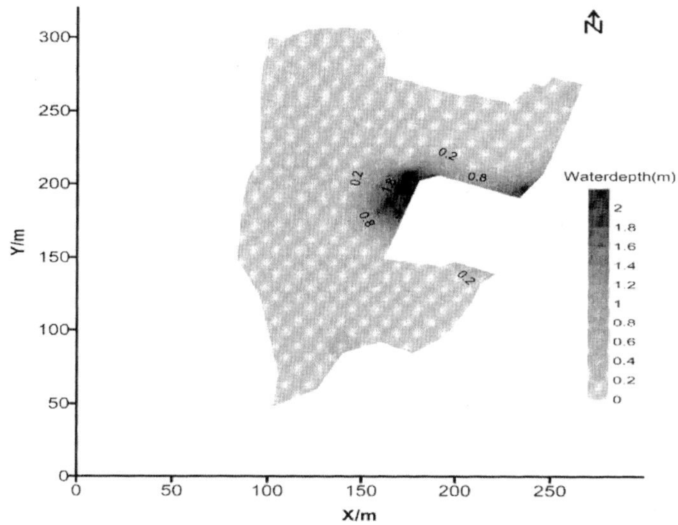

Fig. 3 Spatial distribution of water depth simulated.

The Spatial Distribution of Erosion Depth Erosion depths simulated are displayed in Fig 4(a). When compared with Fig. 1, erosion is seen to occur at a range of elevations from 290 m to 304 m, which are located in the transition zone of slopes from the step to the flat of the watershed. From a runoff formation point of view, the erosion simulated is reasonably located at the places where the runoff accumulates and a certain thickness of water-depths and flow velocities have been generated.

Spatial Distribution of Sediment Concentrations In this case study, water depth, erosion depth and deposition have been simulated and updated simultaneously at each time step. Sediment concentrations at the time when runoff reaches its maximum value are shown in Fig. 4(b). As shown in Fig. 4(b), the distribution of sediment concentration is closely related to the distribution of water depth, i.e. sediment transports are accompanied by runoff movement.

DISCUSSION

Water depth greater than a certain thickness on the slope of catchment is very important for the transport of sediments eroded by a rainfall event. The variations of water depth are a function of

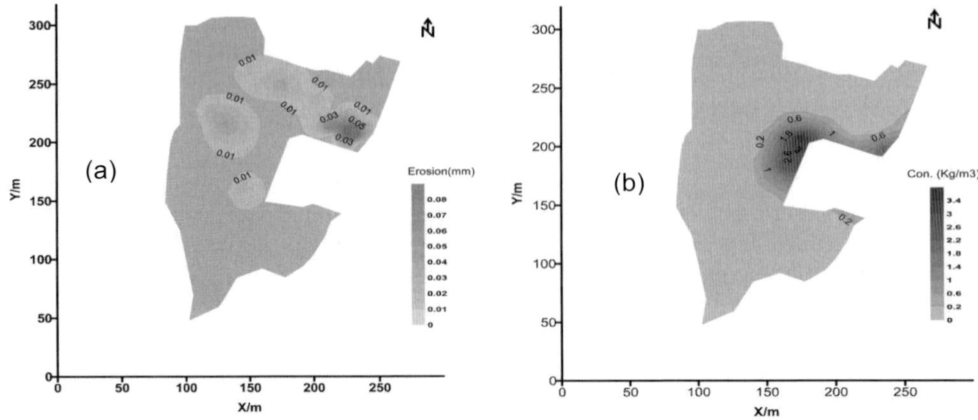

Fig. 4 Spatial distributions of erosion depth and sediment concentrations: (a) spatial distribution of erosion depths, (b) spatial distribution of sediment concentrations.

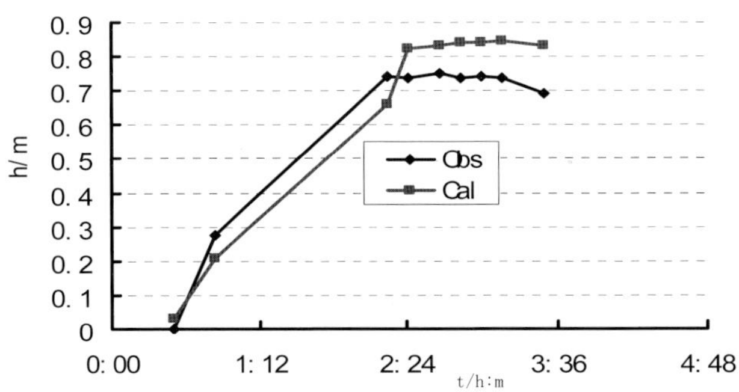

Fig. 5 Water depth variation processed at the gauge station.

time and the intensity of rainfall. Figure 5 illustrates the change of water depth with the time at the gauge station (simulated and observed). Comparing the simulated result with the observed as shown in Fig. 5, we can find out that simulated results agree with the observed quite well. The maximum water depth simulated is 0.84 m while the observed is 0.74 m. The difference of maximum water depth is only 0.1 m.

The reason for this difference is that the simulated result is the average value of five different observation-slopes, ranging from 5°, 10°, 15°, 20°, and 25° while the observed, as indicated in Fig. 5, is that of observation slope 15°. Therefore, some differences existing in water depth are reasonable.

CONCLUSIONS

In this study, we proposed a fully coupled simulation model, consisting of the 2-D shallow water equations, 2-D non-equilibrium sediment transport equation, and a ground elevation variation equation due to soil erosion and sediment deposition for a rainfall event on a watershed. After its application at An-ju Experimental Station in Suining prefecture, Sichuan province, the feasibility and accuracy of the model have been validated. Results indicate that:

- The coupled numerical model can accurately simulate the variation processes and the spatial distribution of water depth, soil erosion and sediment concentration in a rainfall event of a watershed under the integrated effects of land use condition, regional topography, slope variation, and rainfall characteristics.

- The model has a flexibility to adapt for the study of water and soil losses systematically at the watershed scale so that the limitations of traditional approaches, such as the typical experimental unit, in which only a single influence factor such as slope or others is considered, can be overcome. In addition, this model is physically distributed, and we can consider the rainfall intensity, infiltration and evaporation in each grid of a study region.
- The accuracy of the model can be further improved if we can accurately assess land-use conditions, landcovers, rainfall patterns, infiltration and evaporation before the simulation. It is also possible to apply the model to ungauged basins for the evaluation of the sources where the water and soil losses are coming from, because geo-related data and rainfall data can now be obtained by modern technologies such as 3S techniques.

Acknowledgements This work has been supported in part by the State Key Laboratory of Hydraulics and Mountain River Engineering through Grant no. 0805 and the National Key Fundamental Research and Development Program (namely National 973 Program of China) with Grant no. 2007CB714105. The authors are grateful for the useful comments and suggestions provided by anonymous reviewers.

REFERENCES

Garcia, R. & Kahawita, R. A. (1986) Numerical solution of the St. Venant equations with MacCormack Finite Difference Scheme. *Numerical Methods in Fluids* **6**, 507–527.

Katopodes, N. D. & Strelkoff, R. (1978) Computing two-dimensional dam-break flood waves. *J. Hydraul. Div. ASCE.* **104**(9), 1269–1288.

Tan, Weiyan (1998) *Computational Shallow Water Hydrodynamics*. Press of Tsinghua University, China.

Wood, W. L. A. (1996) Note on how to avoid spurious oscillation in the finite element solution of the unsaturated flow equation. *J. Hydrol.* **176**, 205–218.

Zhang, X. H., Oishi, S., Ishidaira, H. & Takeuchi, K. (2003) Practical aspects in the simulation of flood inundation using 2-D shallow water equations on a complex land-cover. *Japan Soc. Hydrol. Water Resour.* **16**(5), 501–517.

Zhang Xiang-wei & Takeuchi Kuniyoshi (2004) Methodology for modeling of groundwater flow in large area. *Shuili Xuebao* **6**, 7–13.

Zhang, Xin-hua, Long, Wen-fei, Lei, Xiao-zhang *et al.* (2007) Quantitative approach to influences of high sediment laden inundation flows on the morphological variation of floodplains. *J. Hydrodynamics, Ser. B.* **19**(1), 041–049.

Zhao, D. H., Shen, H. W., Lai, J. S. & Tabios, G. Q. III (1996) Approximate Riemann Solvers in FVM for 2-D hydraulic shock wave problems. *J. Hydraul. Engng ASCE* **122**, 692–702.

Rational allocation of water resources for economy and ecosystem in Ningxia

XIAO-RONG HUANG[1,2], XIN-HUA ZHANG[1], WEN-JIANG ZHANG[1,2] & TIAN-QI AO[1,2]

1 *State Key Laboratory of Hydraulics and Mountain River Engineering, Sichuan University, Chengdu 610065, China*
hxiarong@tom.com; xhzhang@scu.edu.cn

2 *College of Hydraulic & Hydroelectric Engineering, Sichuan University, Chengdu 610065, China*

Abstract Water is one of the most sensitive and restricted factors in natural social systems. In arid and semi-arid zones water resources are a major factor in economic development and ecosystem conservation. The aim of advanced water resources planning is that the virtuous circle of ecosystem should be kept in the process of supporting economic and social development by water resource development. The rational allocation of water resources between the economy and ecosystem is investigated in Ningxia Autonomous Region through systematically analysing the inter-relationship between economic-social-eco-environmental system and water resources with the help of an integrated multiple objective optimization and simulation model.

Key words Ningxia; rational allocation of water resources

INTRODUCTION

Water is an important natural resource, indispensable to the life and production activities of mankind. With economic development, population growth and the promotion of levels of material and cultural life, mankind's need for water continuously increases. In recent years, a water crisis has occurred in some areas in China. The gap between water supply and demand has not only become an important restricting factor for the sustainable social and economic development, but has also markedly affected regional ecosystem construction, conservancy and maintenance of healthy rivers. Water resources have become a major factor in economic development and ecosystem conservancy, especially in arid and semi-arid zones. The essence of advanced water resources planning is that the circle of the ecosystem should be kept in the process of supporting economic society development by water resources development.

The Ningxia Hui Autonomous Region is located in the central northwest of the country, the northern part of the province straddles the 350-km section of the Yellow River between Gansu and Inner Mongolia, Its total population is around 5.8 million, with the Hui minority making up about one third. About 37% of the population lives in urban areas, principally the provincial capital (Yinchuan) and the second largest city, Shizuishan, in the north of the province. The total area is 51 800 km^2. In terms of geographical features, the southern part is mountainous and hilly, the middle part is an arid desert area, and the northern part is the traditional irrigation area by the Yellow River. The landform declines from south to north over an elevation of 1000 m. Apart from some small inland basins, all the territory of Ningxia is within the watershed of the Yellow River. Three quarters of Ningxia is arid or semi-arid with an annual precipitation of around 289 mm. Evaporation is about 1300 mm annually. Excluding the Yellow River, the annual water resource of Ningxia is only 0.949 billion m^3, but its effective utilization is limited due to uneven distribution in terms of time and space. Ningxia is among the provinces with the smallest internal water resource. Without irrigation, there would be no agriculture in Ningxia. In accordance with the national scheme, Ningxia is allocated a net use of 4 billion m^3 water annually from the Yellow River. This is defined as the balance between the inflow measurement at Xiaheyan Hydrological Station in Zhongwei County and the outflow measurement of Shizuishan Hydrological Station. The actual use of water resources in Ningxia has reached 3.8 billion m^3, the proportion of agricultural water use was 87.8% and the proportion of industrial water use was 2.8%, which was 17% lower than the country average. The agricultural water use is the dominant water consumer among all sectors: Industry, People's Livelihood, Urban Public Utilities etc. Industrialization, city planning and ecosystem construction lack essential water resources. Thus the sustainable social and economic development of Ningxia will depend on the rational allocation of water resource consumption between the economy and ecosystem.

MODEL DEVELOPMENT

In planning for large-scale regional water resource systems, a variety of technical, social, economic and ecosystem-environment factors must be considered. Obviously, a comprehensive plan for regional water resources is usually characterized by a number of preference criteria, the usual problems encountered in modelling and optimization are complicated and tend to confuse the analysis. This is due to the high dimensionality of the model in question, the large number of variables as well as their complex interactions and semi-structured problems, therefore, the last solutions for the model belong to the non-inferior solution, without the best solution. The main characteristics of the approach in this model are the interactive utilization of qualitative and quantitative analysis, optimization and simulation techniques. The optimization model results in development scenarios of the future, which are constrained by the overall availability of water resources. The simulation model provides a detailed calculation in terms of different scenarios obtained by optimization model. The quantitative calculation is computer-based, and the qualitative semi-structured problems can be solved by interactions between human and computers. The main structure of the model, as shown by Xu & Wang (1997), is given in Fig. 1.

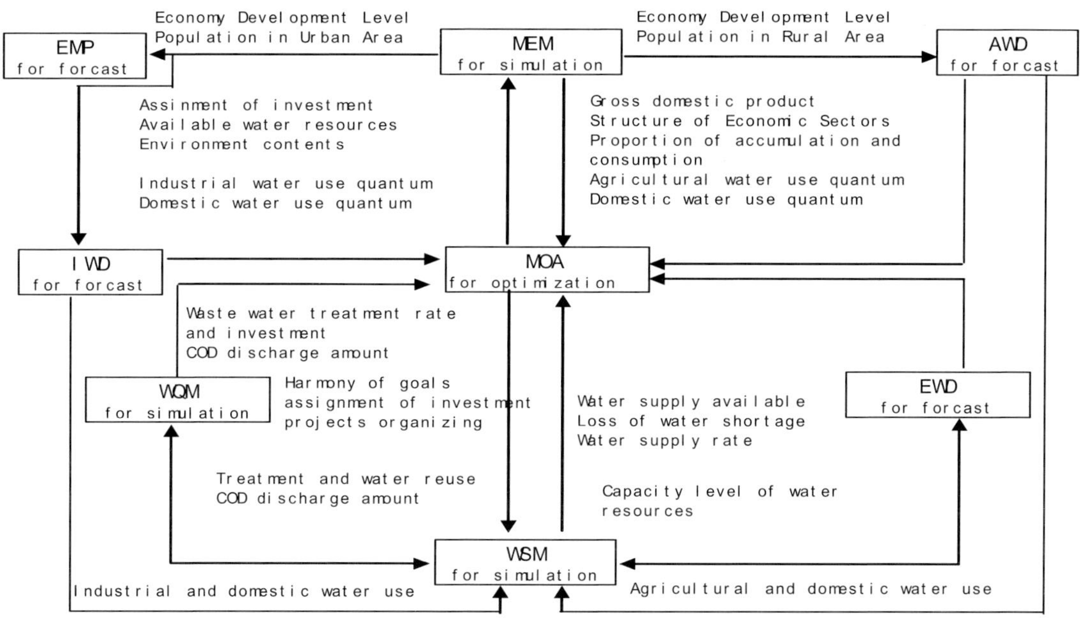

Fig. 1 Main structure of the model. Notes: MEM: macroeconomic model; EMP: employment; IWD: industrial water demand; MOA: multi-objective analysis model; WQA: water quality sub-model; EWD: environment water demand; WSM: water resources simulation model

The system has employed an integrated approach by applying macroeconomic forecasting techniques, multi-objective optimization techniques and a water resources simulation technique as stated by Weng *et al.* (1995), and Xu & Wang (1997). The Macroeconomic Model (MEM) will provide the economic development patterns to the Multiple Objective Analysis Model (MOA) for overall water resources planning, optimization and allocation and the detailed industrial and agricultural development indexes at different target years for different social, economic consideration. These three models, namely MEM, MOA and WSM, are interconnected, directly and indirectly-either utilizing a joint database, or via feedback relations. The results from MOA, such as the industrial production value, irrigation areas, crop pattern, etc. will provide input information to WSM for water resources allocation, regulation, management study and risk analysis.

Macroeconomic model (MEM)

A primary component of MEM is the input-output (I-O) model providing information for sectoral structure, economic output as well as direct and indirect industrial water demand coefficients. The prevailing method of macroeconomic water demand analysis in China is to calculate the direct water consumption coefficients of each sector, then rank these sectors by their coefficients to derive sectoral priorities. The advantage of this approach is simplicity and easy use by decision makers. Its weakness is that it can not reveal the interaction among different economic sectors and does not provide overall and comprehensive conclusions for the marginal effect of the water input of each industry. The I/0 multiplier analysis provides these aspects. Due to water resources non-profit peculiarity, the financial input-out sheet of water resources can not be directly constructed until traditional financial input-output sheet and physical input-output sheet of water resources can be combined. With the help of this approach as proposed by Huang *et al.* (2005), the characteristics and variation of water utilization in various sectors of economy and the influence on demand and supply of water resources are studied with a macroeconomic perspective, therefore, finding out the result of water shortage affecting economic development, and its countermeasures are to be proposed.

Multiple objective analysis model (MOA)

Based on the input-output model, the multiple-objective analysis model is constructed. Multiple criteria approaches go beyond more restrictive traditional techniques such as single objective optimization or goal programming methods, which embody the level of local socio-economic conditions, ecosystem, and agricultural self sufficiency. It includes five objectives for regional development, as stated in Weng *et al.* (2004): GDP per capita, BOD discharge per capita, grain production, urban employment rate and green equivalent area, which summarize the general situation of regional sustainable development. Due to competitive relations among different objectives, it may not be feasible to attain the best solution to each or all of the management objectives simultaneously. They provide a greater range of information on the tradeoff between multiple policy objectives and competing resource requirements. In addition, technical analysts and decision makers can interactively explore multiple management alternatives and their effects on the policies and water resources utilization. The main structure of the model is given in Fig. 2.

Water resources simulation model (WSM)

The WSM, as a water supply and allocation model, identifies total surface water and groundwater availability and describes the water management system. This model describes the hydrological conditions and quantifies the water supply available for industrial, agricultural, and domestic use. Based on the 1999 version, the International Water Management Institute has improved PODIUM – the policy dialogue simulated model in terms of water resources development and food security. The model assists policy makers in analysing regional water balances and increasing food requirements for the future between supply and demand under different options of policies or hypothesis with a variety of considerations of growth rate in population, changes in diets, developments in agriculture and sustainable water use. With the help of PODIUMSim model, the recommendation schemes from multiple-objective analysis model are simulated. The PODIUMSim model provides water demand computations for agricultural, domestic and industrial sectors in Ningxia for different planning years, which are shown in Fig. 3 in 2030.

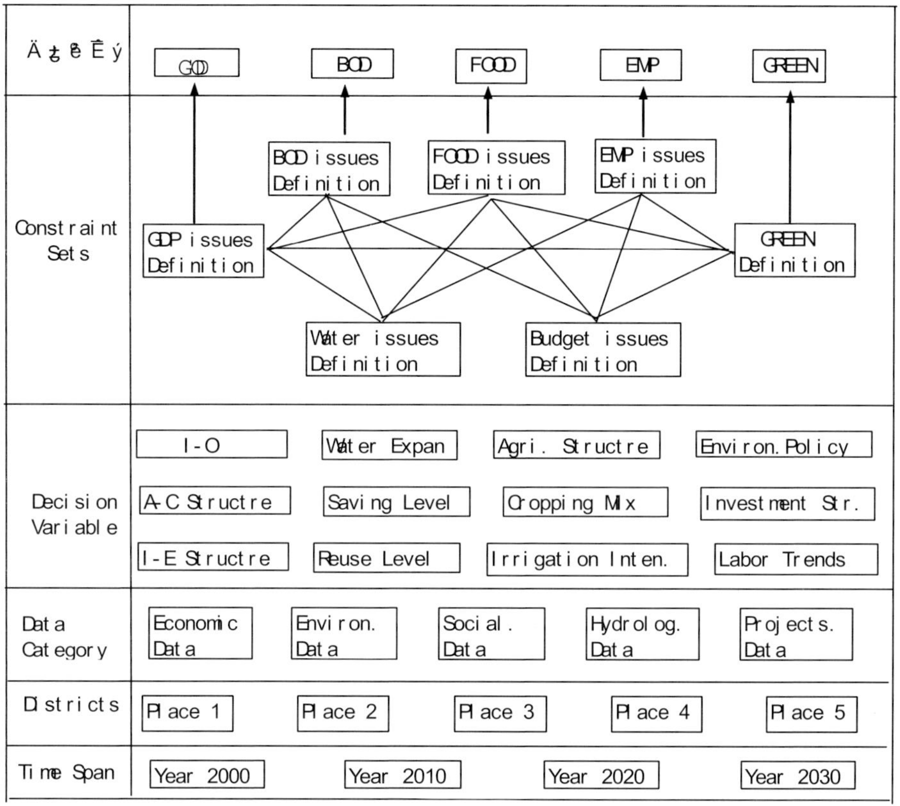

Fig. 2 MOA Model structure diagram. Notes : I-O: Input-Output A-C: Accumulation-Consuming I-E: Import-Export.

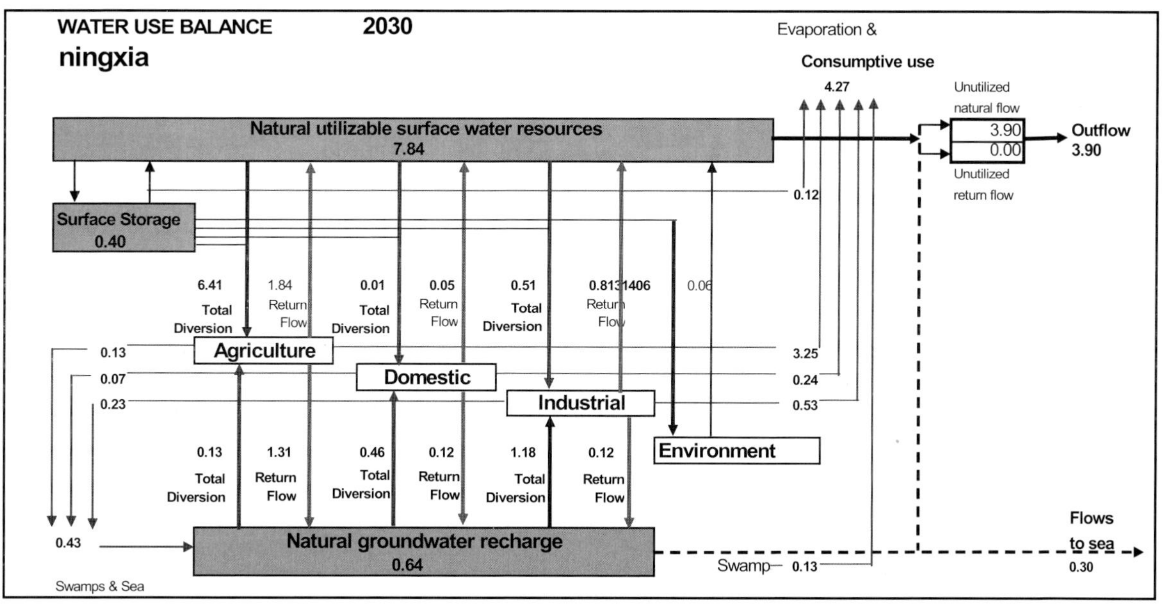

Fig. 3 Water use balance in Ningxia in 2030.

Ecosystem water resources requirement

The main natural habitat types in Ningxia are: grasslands with desert plants, meadow, forests, xeric shrublands and wetlands. Land reclamation and prolonged deforestation due to the long-horned beetle since the mid-1980s has led to an 8.7% reduction in forested areas. The natural flora and fauna around lakes have been altered due to the development of fish ponds and pollution from agricultural lands. Several lakes have become seasonally dry due to lower rainfall and groundwater pumping. "Artificial" habitats have become a very important feature in Ningxia as a major means of desert control. Artificial habitats refer to the plantings of trees, shrublands and grasslands along the desert margins and the Baolan railway (Gansu to Inner Mongolia) to protect the irrigation areas and railway from the advance of the mobile sand dunes. The trees have to be irrigated, either from the irrigation canals fed by the Yellow River, or from groundwater sources.

Around the proper objective and scale of ecological environment and with the core task of reasonable utilization of water resources using RS and GIS techniques, the project studies ecological evolution and ecological water demand in Ningxia, deriving some creative suggestions as mentioned by Liu *et al.* (2004) and Wan *et al.* (2002). After the analysis of ecological significance, existing ecological conditions and water resources in the watershed, theories relating to the ecological evolution and ecological evolution study are put forward with new ideas of ecological water demand as in Huang *et al.* (2006). The process of calculating ecological water demand in Ningxia is given in the literature, e.g. Huang *et al.* (2008).

RESULTS ANALYSIS

Emphasizing both optimization and simulation techniques, in terms of equality and impartiality, safely using water, validity and sustainable principles, it scientifically designs water resources planning of Ningxia by considering the interactive impacts between regional economic system, ecosystem and water resources system, the way of water resources exploitation and the coordinated development of regional economy. Table 1 shows the results of calculation.

Table 1 Results of calculation for water resources planning in Ningxia.

Year	Area	AWD	DWD	IWD	EWD	AWS	DWS	IWS	EWS	AWSR	DWSR	IWSR	EWSR
		0.1^*Bm^3								(%)			
2000		76.50	1.60	4.70	2.45	76.50	1.60	4.70	2.45	100	100	100	100
2010	Ningxia	68.70	2.20	9.30	2.45	68.70	2.20	9.30	2.45	100	100	100	100
	Yinchuan	18.45	0.59	3.34	0.44	18.45	0.59	3.34	0.44				
	Shizuishan	9.23	0.40	2.78	0.50	9.23	0.40	2.78	0.50				
	Wushong	38.03	0.79	2.97	1.44	38.03	0.79	2.97	1.44				
	Guyuan	2.85	0.42	0.19	0.07	2.85	0.42	0.19	0.07				
2020	Ningxia	67.00	3.30	13.70	3.54	64.82	3.30	13.70	3.54	96.74	100	100	100
	Yinchuan	17.91	0.83	4.66	0.44	17.33	0.83	4.66	0.44				
	Shizuishan	8.96	0.55	4.24	0.50	8.67	0.55	4.24	0.50				
	Wushong	36.92	1.26	4.47	1.44	35.72	1.26	4.47	1.44				
	Guyuan	3.06	0.67	0.32	0.11	2.96	0.67	0.32	0.11				
2030	Ningxia	65.40	5.70	17.90	4.10	56.14	5.70	17.90	4.10	85.84	100	100	100
	Yinchuan	17.48	1.44	6.09	0.74	15.01	1.44	6.09	0.74				
	Shizuishan	8.75	0.95	5.54	0.83	7.51	0.95	5.54	0.83				
	Wushong	36.04	2.17	5.84	2.40	30.94	2.17	5.84	2.40				
	Guyuan	2.98	1.16	0.42	0.12	2.56	1.16	0.42	0.12				

Notes: AWD: agricultural water demand; DWD: domestic water demand; IWD: industrial water demand; EWD: ecosystem water demand; AWS: agricultural water supply DWS: domestic water demand; EWS: ecosystem water supply IWS: industrial water supply; AWSR: agricultural water supply rate DWSR: domestic water demand rate; EWSR: ecosystem water supply rate IWSR: industrial water supply rate.

The water consumption of main stream of the Yellow River in Ningxia was 3.66 billion cubic metres in 2000. From 1991 to 2000, water consumption of the Yellow River increased. The quantity of water consumption of the Yellow River in Ningxia was 3.14 billion cubic metres in 1991, and 3.81 billion cubic metres in 1999. In future, the gap between water supply and demand will become wider and wider in Ningxia. Water-saving in agriculture/irrigation has great potential for increasing water availability through the completion of on-farm work and land consolidation for improving the irrigation application efficiency. Therefore agricultural water demand will reduce from 7.65 km^3 in 2000 to 6.87 km^3 in 2010, 6.70 km^3 in 2020, and 6.50 km^3 in 2030. Conversely, industrial water demand, domestic water demand and ecosystem water demand has risen since 2000, and the total water demand will still increase for each target year. At the same time, domestic water supply rate, industrial water supply rate and ecosystem water supply rate always remain 100% in different calculation year, but the agricultural water supply rate can not reach 100%, decreasing with time due to water insufficiency.

CONCLUSIONS

(1) The model raises the Rational Allocation of Water Resources based on the Macroeconomic Structure Rationalization, mainly including: (a) analysing the utilization efficiency and output value of water resources in the current economic structure of Ningxia Province; (b) calculating the direct and indirect water consumption coefficients of each sector, and (c) constructing the best regional economic structure depending on the relevant relationship among different National Economic Sectors in order to make good use of water resources.

(2) It analyses the essence of ecological water demand based on sustainable development. Targets of ecological protection and scale of ecological construction were identified based on ecological protection and restoration mode.

(3) Effective measures emphasize both the optimization and simulation techniques and integrate qualitative and quantitative analysis. Providing the decision makers with adequate information not only depended on the optimization model, but also depended on the simulation model.

REFERENCES

Huang Xiaorong, Wang Dangxian & Pei Yuansheng (2005) Input-output analysis of water consumption for economic development in Ningxia. *Resour. Sci.* **3**, 135–139.

Huang Xiaorong, Jiang Jianjun & Pei Yuansheng (2006) Research on ecological water demand of NingXia oasis based on ecological conservancy. *Adv. Water Sci.* **3**, 312–316.

Huang Xiaorong, Zhang Xinhai & Zhai Caiwang (2008) Ecological water resources requirement of the Ningxia natural oasis using the groundwater level variation amplitude method. In: *Hydrological Research in China: Process Studies, Modelling Approaches and Applications* (ed. by Dawen Yang), 171–177. IAHS Publ. 322. IAHS Press, Wallingford, UK.

Liu Changming, Wang Lixian & Xia Jun (2004) Water resource allocation & ecosystem conservancy and research of sustainable development strategy in northwest area in China. In: *Ecosystem Environment*. China Science Press, Beijing, China.

Wan Fang, Liang Rui-ju & Yang Xiao-liu (2002) A study of ecological water requirements in Northwest China. *J. Nat. Resour.* **17**(2), 129–137.

Weng Wenbin, Cai Ximing & Shi Huibin (1995) Development of the macro-economic based water resources multiple objective analysis approach and its applications. *Shui Li Xue Bao*. **2**, 3–7.

Weng Wenbin, Wan Zhongjing & Zhao Jianshi (2004) *Advanced Water Resources Planning – Theory. Method and Technology*, 292–295. Tsinghua University Press, Beijing, China.

Xu Xinyi & Wang Hao (1997) *Water Resources Planning based the Macro-economic in North China – Theory, Method.* Yellow River Water Conservancy Press, Zheng Zhou, China.

4 Eco-environmental Protection

Water chemistry and nitrate pollution in the Liangtan River basin in the Three Gorges Area

XIANGCHAO LIU[1,2], CHANGYUAN TANG[2,3], JUN XIA[2], XIANFANG SONG[2] & YIZHANG ZHANG[2]

1 *Key Laboratory for Hydraulic and Water Transport Engineering of Chongqing City, Chongqing Jiaotong University, Chongqing 400074, China*
liuxc@igsnrr.ac.cn

2 *Faculty of Horticulture, Chiba University, Chiba 271-8510, Japan*

3 *Key Laboratory of Water Cycle and Related Land Surface Processes, Institute of Geographical Sciences and Natural Resources Research, Chinese Academy of Sciences, Beijing 100101, China*

Abstract Water chemistry and nitrate pollution loadings in the surface water and groundwater bodies in the Liangtan River basin influenced by human activities in the Three Gorges Area were analysed by water chemistry analysis technology. The results show that: the surface water bodies from the middle-upper reaches of the basin were polluted by nitrate nitrogen, and an increasing tendency is shown in the lower reaches; serious nitrate nitrogen pollution exceeding the standard of 10 mg/L is present in the groundwater bodies in the area between the upper and middle reaches because of the influence of stream water, and decreases downstream towards the outlet.

Key words Liangtan River; water chemistry; nitrate

INTRODUCTION

The construction of the Three Gorges Project has been producing comprehensive benefits for flood control, power generation, navigation improvement, fisheries and water supply. At the same time, widespread and profound influences are imposed upon the ecology and environment in the Three Gorges Area. A reservoir with an area of 58 000 km^2 will be created when the Three Gorges Project is finished in 2010. With the development of industry, increase of human activities, intensive application of chemical fertilizer, and the increasing application of animal manure in agriculture and the disposal of human waste via septic systems, the non-point source pollution will be more serious, and the problem for ecology and environment will be more severe. For most countries, cities and villages, surface water and groundwater are important sources of drinking water. These are vulnerable to contamination, especially by nitrate, a widespread pollutant with ongoing levels caused by agricultural and residential activities. As such, the US Environmental Protection Agency has established a maximum contaminant level of 10 mg/L as NO_3^--N (Mohammad *et al.*, 2004). Studies concerning anthropogenic influences in mixed land use watersheds generally focus on assessing the impact of land use. Mason (1990), Owens (1991), Nearing (1993), Thomas (1993) and Edwards (1996) reported on the effects of agricultural land use on stream water nutrient concentrations. Valiela (1990), Hanson (1995) and Portnoy (1998) studied nitrogen loading to surface waters caused by urban and residential developments. Much work has been done to study the processes affecting pollutants in groundwater systems receiving fertilizers or wastewater from livestock production. Previous studies have identified nitrate as a widespread contaminant in groundwater systems affected by agricultural activities, and nitrate has attracted considerable attention because of its toxicity to infants at high concentrations (Rak-Hyeon *et al.*, 2003).

Water bodies in the Liangtan River basin in the Three Gorges Area have been polluted severely because of duck domestication and livestock production by residents, the groundwater and surface water quality especially have deteriorated in the last few decades due to an increase of nitrate.

The main purposes of this study are to examine the spatial distribution of water chemistry and nitrate-nitrogen (NO_3–N) contamination in the water bodies of the Liangtan River basin by technologies of water chemistry and water quality, and reveal the sources of nitrate in the water bodies and the circulation process, and also to provide a theoretical foundation for revealing water pollution, for revealing the relationships between water resources and water power construction and ecology protection.

DESCRIPTION OF THE STUDY AREA

The Liangtan River basin is one of the main tributaries lying on the right side of the lower reaches of Jialing River located in Chongqing (Fig. 1). The length of the main channel of Liangtan River is 88 km, and the area of the basin is 511.8 km^2. The Liangtan River originates from the Liaojiagou Reservoir located at Baishiyi in Jiulongpo District, then meanders through 15 towns in Jiulongpo, Shapingba and Beibei districts, and at last converges its flow into the Jialing River.

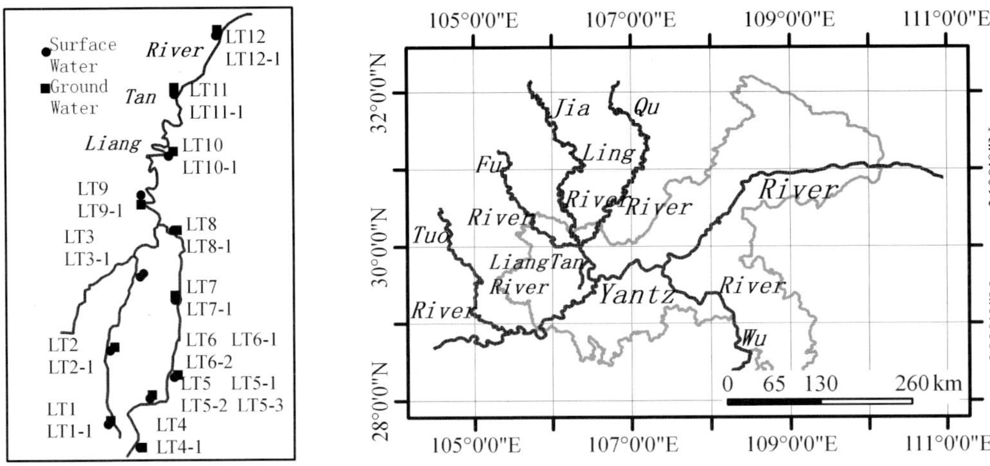

Fig. 1 Location of the Liangtan River and the distribution of water sample sites.

From the 1990s of the last century, lots of livestock production originated in Chongqing, from individual cultivation at home to semi-factory farming. The water bodies in some tributaries in the Three Gorges Area have been seriously polluted because of duck and livestock production. According to the special survey of the water quality of some sub-tributaries, including Liangtan River directed by the National Department of Ecology Protection of China, the water quality of the Liangtan River was below V class. Water pollution in the Liangtan River not only poses a severe influence on the development of industry and agriculture in the basin, it also seriously restricts the construction of ecological agriculture in the west of Chongqing and threatens the water supply for the 100 000 people inhabiting the lower reaches.

SITE SURVEY AND ANALYTICAL METHODS

For illuminating the spatial distribution of nitrate pollution in the Liangtan River affected by human activities, field surveying, surface water and groundwater sampling along the river channels were carried out in January 2008 (Fig. 1). Water temperature, pH, electrical conductivity (EC) were measured with a WM-22EP handled electrical meter at the field sites during sampling. A total of 14 surface water samples and 16 groundwater samples were obtained for this study.

Samples were filtered through 0.45-μm cellulose membranes and kept at 4°C before analysing. Water chemistry was measured for major anions, cations and nitrate concentration with a Shimadzu CTO-10ACvp ion chromatograph with the analytical precision of 1% in the

Hydrological Science Laboratory of Science Department of Chiba University. Carbonate and hydro-carbonate concentration were measured by diluted vitrol-methylic titration method in the Key Laboratory for Hydraulic and Water Transport Engineering of Chongqing Jiaotong University.

RESULTS AND DISCUSSION

Electricity conductivity (EC) depicts the anion and cation concentration in water bodies, and reflects the length of flow paths and residence times in water circulation. Along these surface water and groundwater flow paths, the EC values of surface water and groundwater bodies increase gradually (Fig. 2).

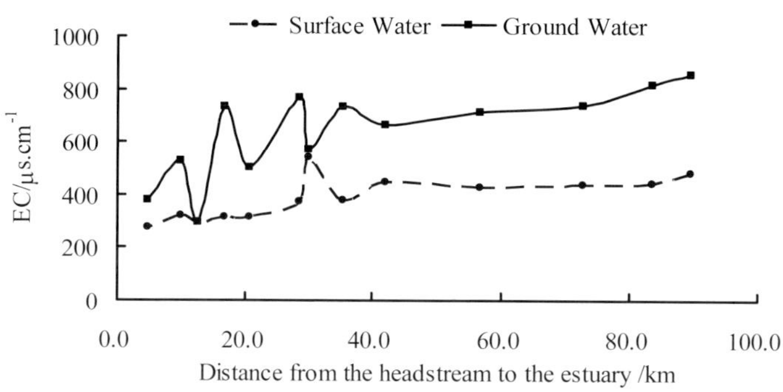

Fig. 2 Variation of EC values along the flow paths.

For describing the spatial distribution tendency of water chemistry and nitrate nitrogen concentration in the study area, the Liangtan River was delineated into three segments as follows: (1) the upper reaches of the river located in the South region of the study area (which includes the surface water sample sites of LT1, LT2, LT4, LT5 and LT6, and the groundwater sample sites of LT1-1, LT2-1, LT4-1, LT5-1 and LT6-1); (2) the middle reaches located in the middle region (which includes the surface water sample sites of LT3, LT7, LT8 and LT9, and the groundwater sample sites of LT3-1, LT7-1, LT8-1 and LT9-1); and (3) the lower reaches located in the North region (which includes the surface water sample sites of LT10, LT11 and LT12, and the groundwater sample sites of LT10-1, LT11-1 and LT12-1).

The water chemistry of each sample was plotted in a Piper diagram (Fig. 3). In the study area, four water types could easily be identified. A freshwater type of Ca-HCO_3 was located in the headstream area of the Liangtan River basin, which characterized that the water bodies in the sources region of the basin were slightly affected by human activities. The main source of recharge water of the study area is rainfall, so the water is Ca-HCO_3 type. When this water infiltrates into the shallow aquifer and water quality is affected by fertilizers or animal excrement, the groundwater character may shift to the Ca-Cl or Ca-SO_4 type.

In the upper reaches area of the basin, higher mineralization water of Ca + Na-SO_4 + Cl gradually emerges, which indicates that the surface water and groundwater bodies have evidently been polluted by human activities. From the site survey, sewage production from livestock, especially from ducks, is the main cause of the deterioration of water quality. Groundwater bodies in this area affected by livestock were mainly located in the Ca-Cl area in the Piper diagram.

Obvious differences were found between the surface water and groundwater bodies located in the middle and lower reaches of the Liangtan River basin. Higher mineralization water of Na-SO_4+Cl type dominates the surface water bodies in this region, which agrees with the evolving tendency of water chemistry in the study basin. However, relatively lower mineralization water of

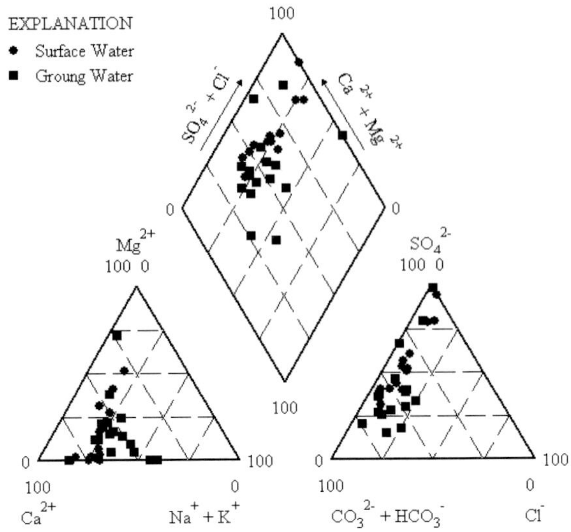

Fig. 3 Piper's diagram.

Ca+Na–CO_3+HCO_3 type dominates the groundwater bodies. Consequently, the distinct water chemistry indicates that the groundwater bodies in this region were not affected by the seriously polluted surface water bodies.

Considering the conservative behaviour of chloride in water bodies, it is usually used as an indicator of the relative amount of mixing *versus* denitrification in surface and subsurface environments. Figure 4 contrasts the concentrations of NO_3-N with those of chloride, thus four groups of water bodies were classified into types with various NO_3-N and chloride characteristics: (1) surface water bodies with lower NO_3-N and lower chloride concentrations dominate the whole basin; (2) groundwater bodies with lower NO_3-N and lower chloride concentrations mainly located in the headstream and upper reaches of the river; (3) groundwater bodies with higher NO_3-N and higher chloride concentrations mainly located in the area between the upper and middle reaches of the river; (4) groundwater bodies with lower NO_3-N and higher chloride concentrations mainly located in the middle and lower reaches of the river. Headstream areas located in the mountain regions, where surface water and groundwater bodies kept their natural state with lower NO_3-N and lower chloride concentrations. The freshwater group was characterized by low concentrations of chloride and NO_3-N (below 3 mg/L). NO_3-N and chloride concentrations in the surface and groundwater bodies located in the upper reaches of the river increase gradually, contrasting with those in the headstream areas, thus indicating that the water bodies, especially the surface water bodies, have been polluted by human activities. In the area between the upper and middle reaches of the river, shallow aquifers in the study area are the most important sources of drinking or irrigation water. Groundwater bodies were characterized by high concentrations of chloride and NO_3-N (above 10 mg/L). However, this shallow aquifer is highly susceptible to pollution due to land-use practices, owing to the shallow aquifer being within a thickness of approx. 30 m. Land-use type is often the dominant factor in the contribution of contaminants to shallow groundwater. Therefore, agricultural fertilizer or animal excrement is thought to be the major source of both nitrate and chloride in the study area. Higher NO_3-N in the groundwater bodies located in this area illuminates that intense nitrification has occurred, while surface water has infiltrated through the unsaturated zone to the groundwater table. Groundwater bodies in the middle and lower reaches of the river are characterized by higher chloride but lower NO_3-N concentrations, and NO_3-N concentration is lower than that of surface water, indicating that groundwater in this region has not been affected by surface water.

Many studies have demonstrated that groundwater NO_3-N concentrations often decrease substantially along groundwater flow paths (Simmons *et al.*, 1992; Groffman *et al.*, 1996). There

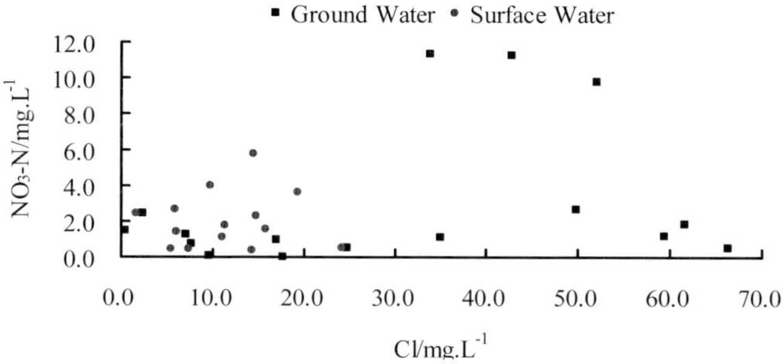

Fig. 4 Nitrate nitrogen concentration *versus* chloride concentration.

are two main biological processes involved in the removal of NO_3^- from groundwater: (1) uptake by plants and microorganisms, representing temporary retention; (2) microbiological denitrification, representing permanent nitrogen loss from soils and subsurface waters. Stream water NO_3-N concentrations increased downstream from the most upstream sampling sites. Generally, this pattern coincides with the negative relationship between NO_3-N concentration and mean catchment elevation, which could be observed in the whole basin.

However, NO_3-N concentrations in the groundwater bodies along the groundwater flow paths in the study basin of Liangtan River disobey the rule mentioned above. As described in Fig. 5, along the groundwater flow paths there is a sharp increase for the groundwater NO_3-N concentrations from the headstream area (with concentrations vary from 0.96 to 2.67 mg/L) to the middle reaches (from 9.81 to 11.34 mg/L) of the river, while a sharp decrease appears from the middle reaches area (from 9.81 to 11.34 mg/L) to the lower reaches (from 0.71 to 2.62 mg/L) of the river. While stream water NO_3-N concentrations increase gradually with a mild fluctuation from the headstream area to the outlet along the stream lines, which exhibited a narrower range of values (from 0.44 to 5.77 mg/L) contrary to those of the groundwater (from 0.71 to 11.34 mg/L). Fluctuation of surface water and groundwater NO_3-N concentrations along the stream lines may be due to a number of factors. Firstly, NO_3-N concentrations may decrease along the stream lines if the stream water is being diluted by groundwater with lower NO_3-N concentrations; being assimilated by plants during transport along the stream lines; or being denitrified by denitrifying bacteria in streams during transport. Secondly, NO_3-N concentrations may increase along the stream lines if the stream water is being recharged by higher NO_3-N concentration; wastewater discharged directly from agricultural runoff, animal wastes, septic systems and industrial processes to the river flows; or due to nitrification in the unsaturated zone after infiltration of surface water with higher organic mater from leaching of fertilizer, animal waste, or human waste to the aquifer.

Spatial distribution of contamination for surface water and groundwater NO_3-N was delineated below:

- In the region of headstream and upper reaches of the Liangtan River, both surface water and groundwater pollution from nitrate are mild (surface water NO_3-N concentrations vary from 0.44 to 1.74 mg/L, groundwater NO_3-N concentrations vary from 0.71 to 2.67 mg/L). Lower mineralization with lower chloride concentration is the distinct hydrochemistry attribute of both surface water and groundwater bodies, which account for a minor effect from human activities.
- In the area between the upper and middle reaches, nitrate pollution of stream water is aggravated slightly (from 1.85 to 3.63 mg/L), and NO_3-N concentration of water from the farmland is only 0.54 mg/L, indicating that contamination in the stream is primarily from duck excrement along the river sides, but not from farmland leaching. However, NO_3-N concentrations of groundwater bodies reach the highest level in the basin (from 9.81 to 11.34 mg/L) and exceed the maximum contaminant level of 10 mg/L for drinking water supplies. For groundwater NO_3-N concentrations in the middle reaches, a distinct difference exists

between sample site LT3-1 and the other three sample sites (including LT7-1, LT8-1 and LT9-1). The main reason is that they are located in two different tributaries, and causally testified that the distinct pollution condition from the two main tributaries is in the headstream area. The increasing proportion of agriculture in the land use patterns downstream and lots of duck excrement is certainly the dominant control on the observed increase in groundwater NO_3-N concentrations. This hypothesis is supported by the thick unsaturated soil layer of about 30 m in this region, where the organic matter from the upstream was oxidized into nitrate contamination; and the close interrelationship existing between surface water and groundwater because of higher annual precipitation varies from 700 to 1100 mm in the study area. The analysis showed that high nitrate concentrations in groundwater can be expected in areas that have both high groundwater recharge and high on-ground nitrogen loading. Also, the analysis indicated that denitrification is not a significant process in the shallow depths due to the relatively high dissolved oxygen concentrations. As such, denitrification processes may not be sufficient to naturally attenuate the nitrate mass in groundwater.

– From the middle reaches to the lower reaches of the river, NO_3-N concentrations of stream water continue increasing until the outlet, with the concentration of 5.77 mg/L, whereas those of the groundwater decrease along the flow paths (from 11.34 to 0.71 mg/L). Nitrate leaching from agricultural fields in the recharge area and subsequently denitrification in some of the geological formations yields this spatial pattern. The travel time of groundwater flows in the unsaturated and saturated zone might also cause the low nitrate concentrations in the discharge area, because some water could have infiltrated prior to the higher fertilizer applications.

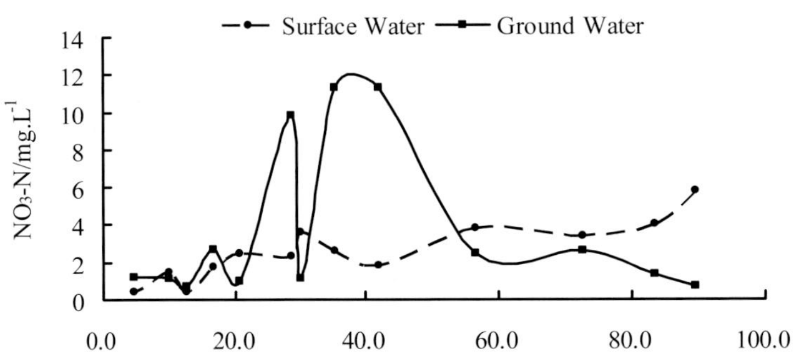

Fig. 5 Variation of nitrate nitrogen concentrations along the flow paths.

CONCLUDING REMARKS

In light of the above results and observations, the following general conclusions can be made for the Liangtan River basin with heavy livestock production and agricultural activities:

(1) The analysis of nitrate occurrences in the surface water and groundwater bodies of the study basin showed that agricultural application of excrement of livestock manure and widespread use of nitrogen fertilizers are potential sources of nitrate contamination. In particular, excrement of livestock manure to the streams is a major nitrate source for severe contamination of groundwater.

(2) The link between the water chemistry and water quality is a useful tool to delineate the nitrate pollution distribution and spatial trends. In areas where the water bodies have been polluted by nitrate with a long-term trend of nitrate concentration in groundwater of agricultural watersheds, there is also generally higher mineralized water from agriculture-related land use practices.

(3) In a basin with many human activities, the trend of groundwater nitrate may disobey the rule that groundwater nitrate concentrations decrease substantially along groundwater flow paths. As for the study basin of the Liangtan River basin, along the groundwater flow paths, there is a sharp increase in the groundwater nitrate concentrations from the headstream area to the area between the upper and middle reaches of the river, while a sharp decrease appears from the area between the upper and middle reaches to the middle and lower reaches of the river.

Acknowledgements This work was financially supported by the Natural Science Foundation Project of CQ CSTC (CSTC2007BB7420) and the Opening Foundation of the Key Laboratory for Hydraulic and Water Transport Engineering of Chongqing City, Chongqing Jiaotong University (SLK2006A02).

REFERENCES

Edwards, D. R., Daniel, T. C. & Scott, H. D. (1996) Stream quality impacts of best management practices in a northwestern Arkansas basin. *Water Resour. Bull.* **32**(3), 499–510.

Groffman, P. M., Howard, G. & Gold, A. J. (1996) Microbial nitrate processing in shallow groundwater in a riparian forest. *J. Environ. Qual.* **25**, 1309–1316.

Hanson, G. C., Groffman, P. M. & Gold, A. J. (1995) Symptoms of nitrogen saturation in a riparian wetland. *Ecol. Appl.* **4**, 750–756.

Mason, J. W., Wegner, G. D. & Quinn, G. I. (1990) Nutrient loss via groundwater discharge from small watersheds in southwestern and south central Wisconsin. *J. Soil Water Conserv.* **45**, 327–331.

Mohammad, N. A., Jagath, J. K. (2004) Assessment and management of long-term nitrate pollution of ground water in agriculture-dominated watersheds. *J. Hydrol.* **295**, 225–245.

Nearing, M. A., Risse, R. M. & Rogers, L. F. (1993) Estimating daily nutrient fluxes to a large Piedmont reservoir from limited tributary data. *J. Environ. Qual.* **22**, 666–671.

Owens, L. B., Edwards, W. M. & Keuren, R. W. (1991) Baseflow and stormflow transport of nutrients from mixed agricultural watersheds. *J. Environ. Qual.* **21**, 147–150.

Portnoy, J. W., Nowicki, B. L. & Roman, C. T. (1998) The discharge of nitrate-contaminated groundwater from developed shoreline to marsh-fringed estuary. *Water Resour. Res.* **34**(11), 3095–3104.

Rak-Hyeon, K., Jeongho, L. & Ho-Wan, C. (2003) Characteristics of organic matter as indicators of pollution from small-scale livestock and nitrate contamination of shallow groundwater in an agricultural area. *Hydrol. Processes* **17**, 2485–2496.

Simmons, R. C., Gold, A. J. & Groffman, P. M. (1992) Nitrate dynamics in riparian forests: groundwater studies. *J. Environ. Qual.* **21**, 659–665.

Thomas, G. W., Haszler, G. R. & Crutchfield, J. D. (1993) Nitrate-nitrogen and phosphate-phosphorus in seven Kentucky streams draining small agricultural watersheds: eighteen years later. *J. Environ. Qual.* **22**(4), 666–671.

Valiela, I., Costa, J. & Foreman, K. (1990) Transport of groundwater-borne nutrients from watersheds and their effects on coastal waters. *Biogeochemistry* **10**, 177–197.

Field observation of total dissolved gas supersaturation downstream of Ertan high dam

RAN LI, JIA LI, KEFENG LI, LU QU & JINGJIE FENG
The State Key Laboratory of Hydraulics and Mountain River Engineering, Sichuan University, Chengdu 610065, China
liran@china.com

Abstract The problem of total dissolved gas (TDG) supersaturation downstream of high dams and its effects on aquatic life has been recognized and has been paid much attention in China. Field observations were carried out in July 2007, when the Ertan high dam discharge flooded. It shows that TDG supersaturation occurs downstream of the Ertan high dam at the large flood rates. The TDG levels relate closely to the discharge mode. The TDG dissipation rate is slow in Yalongjiang River. The convergence with other branches can speed up the TDG dissipation rate, but deep water depth offers less-effective gas dissipation. The study is of great help for the understanding of the TDG supersaturation problem downstream of high dams, and can also provide the basis and references for further theoretical study.

Key words high dam; total dissolved gas; supersaturation; field observation

INTRODUCTION

The 20th century was a period of great development of dam building technology. Countries across the world, including China, have built a number of high dam or ultra-high dams. According to incomplete statistics China has built, or is building, about 35 high dams more than 100 m high and about 16 high dams more than 150 m high. With the development of the western regions and the strategy of the electricity diversion from the western to the eastern regions, China will build a number of high dams with heights of more than 200 m, such as Shuangjiangkou, Baihetan. These high dams play an influential role in China's hydropower industry and also in the development of the national economy. Along with the construction of high dams, the problem of the total dissolved gas (TDG) supersaturation downstream of the high dam during the flood period and its effect on aquatic life have been recognized and paid much attention in China.

A lot of research on the problem of TDG supersaturation and gas bubble disease (GBD), resulting from TDG supersaturation, has been done abroad. Study in the United States began in the 1960s. Much of the research centred on the impact of the TDG supersaturation problem on salmon and other species of fish in the Columbia and Snake River during the flood period of the water conservancy project. TDG field observations, model prediction and fish impact surveys were conducted. In recent years, the United States has built a TDG and fish impact monitoring network in the Columbia and Snake River. Besides, C. M. Angelaccio also studied the TDG problem of the special project. Owing to the rich research results on TDG supersaturation, the United States Environmental Protection Agency (EPA) stipulated the TDG upper limit of 110% in the Water Quality Criteria. On this basis the TDG upper limit is also stipulated in the quality standard of some States, such as Washington and Oregon.

It can be seen that research abroad is mainly centred on dams of less than 50 m height. The results are also limited to specific projects, and less common. As there are many high or ultra-high dams with unique flood, aeration and energy dissipation characteristics in China, the TDG supersaturation research results from abroad cannot apply directly to the dams in China. The TDG study in China should be conducted based on the actual needs and characteristics of China's high dams.

According to the literature, China noted the TDG supersaturation problem in the 1980s. For example, GBD symptoms were found in fish fry and juveniles downstream of the Gezhouba Project when it flooded in 1982, and some fish died of GBD. In 1983, the Water Resources Protection Bureau of Yangtze River conducted a preliminary investigation for dissolved oxygen (DO) supersaturation and GBD downstream of Gezhouba dam (Water Resources Protection Bureau of Yangtze River, 1983). In June, 1994, trout in a fish farm in Jiande City, 3 km downstream from Xinanjiang dam in Zhejiang Province, commonly suffered from gas bubble disease during floods. In summary, domestic research is still at the initial stage and confined to the dissolved oxygen; there is a lack of research on the total dissolved gas.

According to world research on bubble disease (Ebel, 1975), many researchers believe that the total dissolved gas leads GBD more easily than dissolved oxygen. In most research about dissolved gas supersaturation abroad, TDG is studied rather than DO. In 2005, the State Key Laboratory of Hydraulics and Mountain River Engineering, Sichuan University, applied to study TDG supersaturation. The application was granted by the National Natural Science Foundation of China and the study on total dissolved gas, rather than DO, was initiated in China.

Based on the analysis of research in China and abroad, the TDG field observations of Ertan high dam are summarised in the paper. The TDG level and its dissipation process are explored.

TDG FIELD OBSERVATION OF ERTAN HIGH DAM

Project sketch and observation program

The hyperbolic arch dam is taken in Ertan Hydropower Station, with a maximum height of 240 m. The flood buildings include seven upper spillways, six middle spillways, four bottom spillways and two spillway tunnels. The water plunge pool, second dam and apron downstream the second dam consist of the energy dissipation and erosion control structures downstream of the arch dam.

The State Key Laboratory of Hydraulics and Mountain River Engineering, Sichuan University, conducted TDG field observations on Yalongjiang River and Jinshajiang River from 24 to 29 July in 2007, when the Ertan dam flooded. The TDG is measured by an YSI 5566 TGP probe (made in the US) and a PT4 Tracker, with a TGP Smart Probe (made in Canada). The atmospheric pressure was 656–660 mmHg and the water temperature was 19.6–20.7°C during the observation period. The layout of the observation section is shown in Fig. 1.

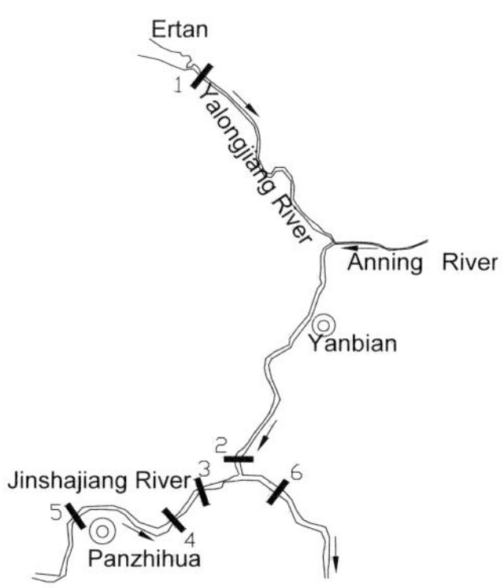

Fig. 1 Layout of the observation section.

Analysis of observations

(1) TDG level downstream of the dam:
The maximum TDG supersaturation level observed is 127.87% when the flood rate is 4000 m^3/s, discharged by no. 3, no. 4 and no. 5 upper spillway and no. 1 flood tunnel. When the discharge mode changed so that no. 2 flood tunnel worked and the discharge flow was about 1265 m^3/s, the maximum TDG supersaturation level observed is 140.04%. It shows that a TDG supersaturation phenomenon exists downstream of Ertan high dam when the dam floods. The TDG supersaturation level is closely related to the discharge mode, the flow distribution between the discharge structures. It is concluded in previous studies that in the same discharge

way, the greater the flood rate is, the higher the degree of supersaturation. Furthermore, it is indicated in the paper that the supersaturation level is not only associated with the flood rate, but also directly associated with the discharge mode. In other words, in different discharge modes, a small discharge flow may lead to high TDG supersaturation level.

(2) The TDG dissipation process in Yalong River:
Figure 2 shows the comparison of TDG supersaturation level between different observed sections, section 1# downstream of Ertan dam, section 2# in Yalongjiang River upstream of the mouth, section 6# Sanduizi Hydrological Station in Jinshajiang River downstream of the mouth. It shows that the supersaturated TDG has recovered to a certain extent, about 120%, when it flows into the observation section 2#, about 32 km downstream of the dam. But after it is analysed and the entrance effect of Anning River is excluded, it shows that the dissipation process in Yalongjiang River is very slow.

(3) Impact of branch confluence on TDG supersaturation level:
At the mouth of Yalongjing River, water with supersaturated TDG converges with the Jinshajiang River. Because of Jinshajiang's dilution effect, the TDG supersaturation level obviously decreases, as shown in observation section 6#, Sanduizi hydrological station.
According the TDG levels before and after convergence, it is summarized that the variation of TDG follows the flow-weighted well-mixed formula:

$$C=(C_1Q_1+C_2Q_2)/(Q_1+Q_2) \tag{1}$$

where C is TDG supersaturation level after mixed, (%); C_1 and C_2 illustrate the TDG supersaturation levels of branches before mixed (%); Q_1 and Q_2 illustrate the flow rates of branches before mixed (m^3/s).

(4) Experiment on supersaturated TDG dissipation process:
To explore the dissipation process of the supersaturated water under different conditions, the supersaturated water samples were taken from the river during the observation period. The experiment was done in a 2000-ml beaker, and blended by a blender under controlled rotation speed. TDG level was measured during the experiment.

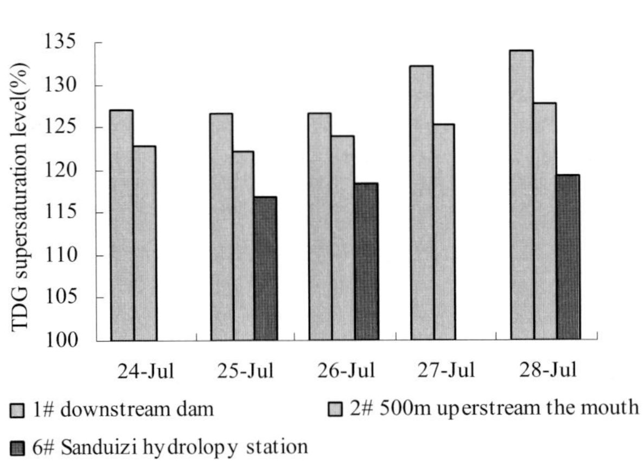

Fig. 2 Comparison of TDG on different sections.

Figure 3 illustrates the various TDG levels with time in 200 RPM and 600 RPM rotation speed. The comparison of the TDG dissipation process under two rotation speeds shows that the higher the blender's rotation speed is, the faster the supersaturated TDG dissipated. It means that the turbulence characteristic is an important factor which influences the TDG dissipation process, i.e. the higher the turbulence is, the faster the supersaturated TDG dissipated.

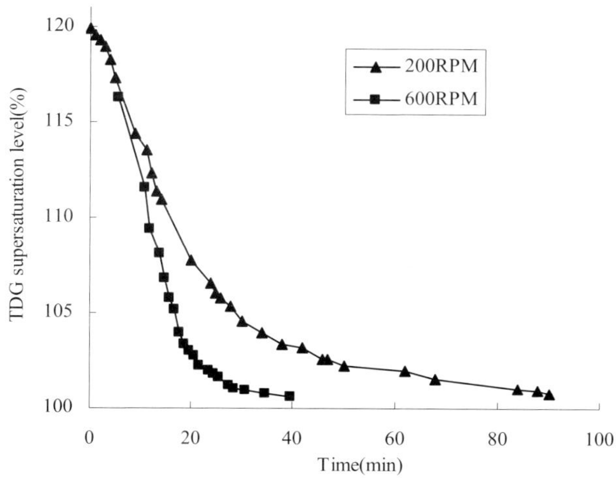

Fig. 3 Various TDG levels with time under different rotation speed.

Under the two experiment conditions, it took about 95 minutes and 35 minutes for the supersaturated water in the beaker to recover to normal-saturated sate. The time is significantly faster than the natural river. It was shown that water depth and hydrostatic pressure in the beaker is much smaller than in the natural river, resulting in the dissolved gases releasing more easily back into the atmosphere. It also shows that the water depth is an important factor affecting the dissipation process of the supersaturated TDG.

CONCLUSION

Based on the analysis of the research in China and elsewhere, the TDG field observation of Ertan high dam is summarised in this paper. It shows that the supersaturation level is not only associated with the flood rate, but also directly associated with the discharge mode, i.e. the flow distribution between the discharge structures. It indicates that the study on the supersaturated TDG must be combined with the specific needs of the project and its flood characteristics. The dissipated process in Yalongjiang River is very slow and the convergence with branches is helpful for the dissipation of the supersaturated TDG. The TDG variation during the branches convergence follows the flow-weighted well-mixed formula. The dissipation rate of the supersaturated TDG is closely related to the turbulence characteristics and water depth.

All the results in the paper are based on the Ertan field observations. As the study on TDG supersaturated problem of high dams in China has just started in recent years, the prototype observation on supersaturated TDG is of great value. The results are helpful for us to accumulate work experience and to improve the knowledge of the TDG supersaturation problem of the high dam project in China. It can also provide basis data and reference for further study.

Acknowledgements The project was sponsored by the National Natural Science Foundation of China (Grant no. 50579043) and the Open Foundation of the State Key Laboratory of Hydraulics and Mountain River Engineering.

REFERENCES

Ebel, W. J., Raymond, H. L., Monan, G. E., *et al.* (1975) Effects of atmospheric gas supersaturation caused by dams on Salmon and steelhead trout of the Snake and Columbia Rivers, National Marine Fisheries Service, Northwest Fisheries center, Seattle Washington, USA.

Heqing Huang (2002). Computational model of total dissolved gas downstream of a spillway. PhD Thesis, University of Iowa, USA.

Jiang Liang (2008) Study on Total dissolved gas supersaturation downstream of High Dam. PhD Thesis, Sichuan University, China.

Jiang Liang, Li Jia, Li Ran, Li Ke-feng (2008) Study on dissolved gas supersaturation downstream of Zipingpu dam. *Adv. in Water Sci.* **19**(3), 367–371.

Politano, M., Carrica, P. & Turan, C. (2005) Multidimensional simulation of the two-phase flow downstream of spillways for total dissolved gas prediction. EWRI, Virginia, USA.

Tan, De-cai (2006) Research on the lethal effect of the dissovled gas supersaturation result from Three Gorges project to fish. [D] Chongqing: Southwest University, China.

United States Environmental Protection Agency (2006) National Recommended Water Quality Criteria.

US Army Corps of Engineers (2005) Technical Analysis of TDG Processes. US Army Corps of Engineers – Northwest Division, Environmental Resources and Fish Planning Offices, USA.

Water Resources Protection Bureau of Yangtze River (1983) Investigation of gas bubble disease and spill of Gezhouba dam.

Weitkamp, D. E. & Katz, M. (1980). A review of dissolved gas supersaturation literature. *Trans. of the Am. Fisheries Soc.* **109**(6), 659–702.

Wu, Cheng-gen (1994) Gas bubble disease of rainbow trout. *Fishery of China* **10**.

Study on ecological treatment of sewage at Xinzhen River

HONGBING LUO[1,4], XIAOMING WU[2], GU HUANG[3], XIAOXUE HUANG[4] & RUI XU[4]

1 *College of Urban and Rural Construction, Sichuan Agricultural University, Du Jiangyan 611830, China*
luohongbing66@163.com

2 *Pearl River Water Resources Committee of Water Resources Institute of Science, Guangzhou 510611, China*

3 *Shenzhen Branch of the Northwest Municipal Engineering Design and Research Institute Co. Ltd. Shenzhen 518048, China*

4 *College of Architecture and Environment, Sichuan University, Chengdu 610065, China*

Abstract Due to a serious lack of freshwater and serious river water pollution in Shenzhen city, China, studies on sewage treatment for the Xinzhen River in Shenzhen city were conducted. A 1-dimensional model (MIKE-11) was used to simulate the tidewater trace in order to find the furthest tidewater trace distance in Xinzhen River. A new connected channel and wide water gates should be built between Xinzhen River and Bi-boundary River so that tide hydrodynamics can be exploited. According to appropriate operating programs, the tidal energy can effectively cause the exchange of water body and improve water quality of Xinzhen River. The calculated result of ecological water requirement was 11×10^4 m^3/day at Xinzhen River and 12.5 × 104 m^3/day at XiXiang River. The total conducted pipe line is 11.17 km and the highest pump head is 23.99 m at Xinzhen River. The perfect aquatic plant system made up of emerging plants, floating leaf plants, submerged plants and floating plants such as Lotus, Calamus, Water Chestnut, *Hydrilla verticillata*, etc. can be used in water purification in Xinzhen River. Configuration of ecological landscape engineering must give prominence to indigenous plants and appropriately introduced outside plant species. In coastal cities some plants of constructed wetland such as *Canna indica*, *Cyperus*, *Vetiveria zizanioide*, etc., especially the mangrove, can improve the river water environment.

Key words water scarcity; hydrodynamic control; ecological treatment; countermeasures

INTRODUCTION

It is important to use limited water resources to improve water environments, e.g. some coastal cities have a serious lack of freshwater and terrible environmental pollution; water pollution of tidal rivers have become more serious in recent decades in China. River ecosystems have become severely disrupted by scrambled weeds, clogged silt beds, and toxic pollutants in recent years. Ecological re-engineering, including wet vegetation and ecological bank rehabilitation is an effective measure to restore and improve river water functions, and it has been widely applied to improve water quality and restore the ecosystem in many places (Hefting *et al.*, 2005; Kovacic *et al.*, 2006).

Hydraulic characters play an important role among technologies (Shields *et al.*, 2003). By using hydraulic dynamics of a river or tidal river, the stagnant sewage was flushed to other places in order to reduce the pollution (Hong Li-jian, 2000; Hu Benxiong, 2004; Zhen Hao-zhi, 2006). Numerical models of tides and surges are essential tools for coastal management, and the most popular methods are finite differencing (Maddock & Pingree, 1978; Couch & Copeland, 2003), finite elements (Navon, 1988), finite volume (Toro, 2001; Montazeri-Namin *et al.*, 2004) and the method of characteristics (Lai, 1977). Engineering countermeasures, such as reducing pollutant loads at each pollution source, are very important for effective water quality management in coastal areas; methods for reducing pollutant loads need to be identified and implemented to restore water quality and environmental capacity (Cerco, 1995; Lee & Park, 2002).

The coastal urban river should be understood and regulated from the viewpoint of ecology to maintain and improve the ecological structure and the ecological function of the urban river. Meanwhile, it is regarded as reaching a win–win relationship between environmental protection and river regulation. The objectives of this study are: (1) under water scarcity conditions, to determine whether tidal river sewage can be exchanged to the sea by using tide hydraulic dynamics. (2) Whether some ecological engineering can improve the water environment for Xinzhen River.

MATERIALS AND METHODS

Description of the study area

This river study is the Xinzhen River, which belongs to the Pear River Estuary and is located in the centre of Baoan district in Shenzhen coastal city in China (Fig. 1(a)). Shenzhen city has a serious lack of water. Xinzhen River is the main channel of flood discharge for Xin'an Street community and has been regulated, with 8.15 km length, 8–35 m river bottom width and 2.0–4.0 m depth. This river has three low water retaining dams with 1.0–1.5 m water-head; the sediment was very full, very near to these dams, so that rising riverbed and flood discharge was greatly influenced. However, tidal segment of this river only extends from national highway 107 to Big Shovel Bay, with 2.8 km length. Xinzhen River covers an area of 14.18 km^2 and it is severely polluted by a large amount of sewage and urban surface runoff mixed with garbage being directly discharged into the Xinzhen River. It has not reached the V standard of environmental quality standards of surface water because of incomplete interception of sewage, new pollution discharge ports, runoff, etc.

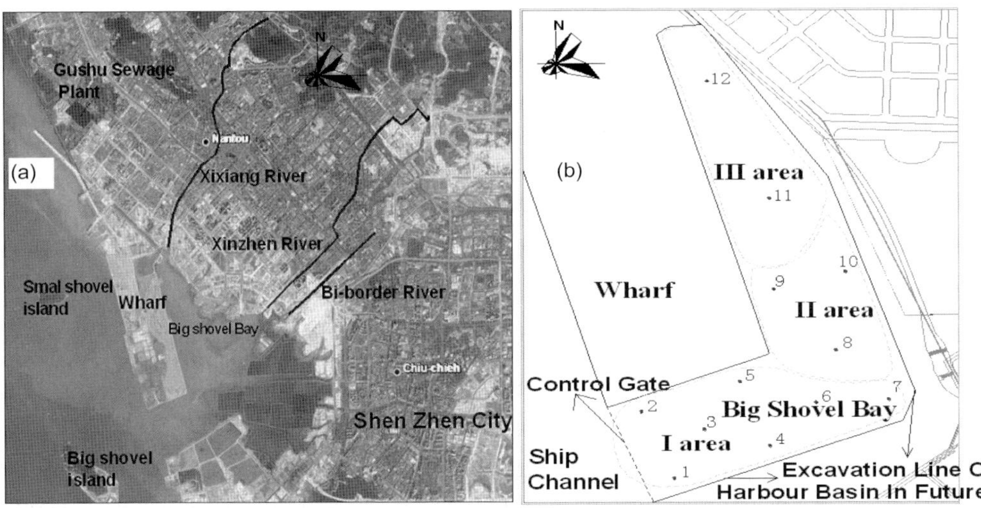

Fig. 1 (a) Study area map and (b) physical model of Big Shovel Bay.

Analyses of hydraulic dynamics

Mike-11 is a widely used 1-D flow routing modelling tool developed by the Danish Hydrologic Institute (DHI, 2003). This modelling system is used extensively by the engineering industry and numerous hydrological consulting groups. The base control equations of 1-D model (MIKE-11) are as follows:

Equation of continuity (Conservation of Mass) is shown in equation (1), and momentum equation (Newton's second law) is shown in equation (2).

$$\frac{\partial Q}{\partial x} = \frac{\partial A}{\partial t} \tag{1}$$

$$\frac{\partial Q}{\partial t} + \frac{\partial(\alpha \frac{Q^2}{A})}{\partial x} + g \cdot A \cdot \frac{\partial h}{\partial x} + \frac{gQ|Q|}{C^2 A \cdot R} = 0 \tag{2}$$

where h is the tidal river water level at time t; Q is the volumetric flow rate of water at time t; A is flow cross-section area of tidal river at time t; R is the hydrodynamic radius at time t; C is the

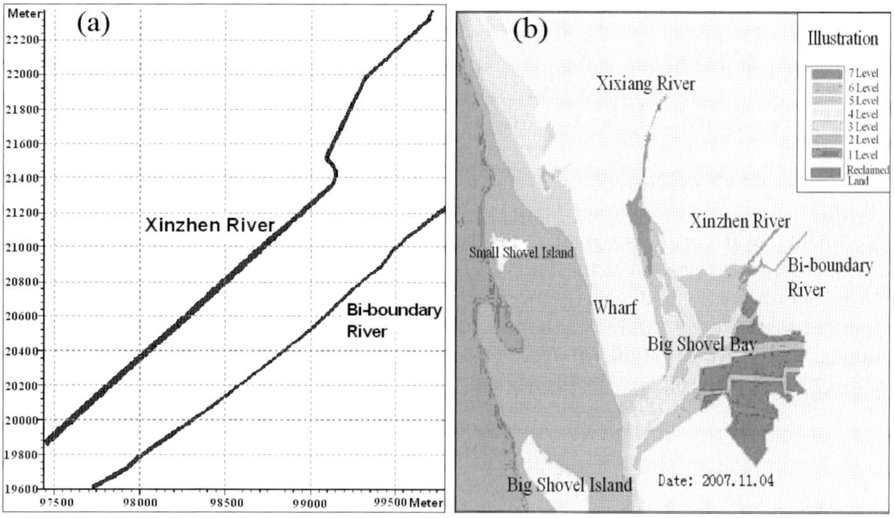

Fig. 2 (a) The physical model, and (b) pollution distribution.

coefficient of Chezy; g is the acceleration of gravity; x is the position coordinates; t is the time. The solution of these equations was described in DHI Water and Environment (DHI, 2003). Figures 1(b) and 2(a) show the physical model of Big Shovel Bay, Xinzhen River and Bi-boundary River, the observation number was summed up in 12 points in Fig. 1(b).

River ecological treatment

In order to improve the water environment, to re-use water resources, and to treat river sewage, many multidisciplinary analyses such as ecology, hydraulic, dynamics, environmental science, landscape science, etc. ought to be integrated for coastal rivers. Eco-biological technology is a new and rapid development in recent years, which uses life action of cultivated and vaccinated micro-organisms or plants to treat wastewater, in which contamination from water can be transported, changed and degraded during water purification. These technologies were mainly composed of strengthened micro-organisms, aquatic plants purification, constructed wetland technology, stabilization lagoon technology, biological membrane technology, eco-concrete technology, multi-natural riverway technology, mixed biotechnology, etc.

RESULT AND DISCUSSION

Hydrodynamics control

Simulation result of river hydrodynamics A 1-D model (MIKE-11) was used to simulate the tidewater trace in order to find the furthest tidewater trace distance in Xinzhen River. The discrete solution of equation (1) and equation (2) is a six-point implicit scheme, the solution of discrete linear equation is the chase-after method. The boundary conditions of the physical model are: (1) upper boundary is flow volume of inlet cross-section at upstream, and the upstream flow volume is the average volume for many years, (2) low boundary is water level of the outlet cross-section downstream, the downstream water level is the typical spring tide process, and (3) if the water gate in this physical model is open, then this water gate was considered as the inner-boundary (flow volume point). The Xinzhen River flow volume ranges from –10.5 m^3/s to 6.5 m^3/s, and the Bi-boundary River's ranges from –9.5 m^3/s to 5.5 m^3/s in three days. As for model verification, since the river condition was not actually measured, roughness factor can be made an initial draft base on current river condition, planning situation and the hydraulic calculation manual. Two times of the tidewater trace distance every day is an important index, which is particularly important in influencing exchange between river water body and tidewater body. A new connected channel and wide water gates with six holes and 5 m length (in Fig. 3(b)) can be built between Xinzhen River and Bi-boundary River so that tide hydrodynamic can be preferably used.

The simulation result of current tidewater trace is shown in Fig. 3(a), when the spring tide rises, the furthest distance of tidewater trace is about 2300 m, the tidehead is about 2500 m from the estuary in Xinzhen River. It is obvious that the tidewater influence is of evident subsection and very different during two natural tides every day at Xinzhen River and Bi-boundary River. The colour in Fig. 3 is blacker, and sewage accumulation in the river is more serious. Figure 3(b) showed new tidewater trace after the connected channel was established. By using the tide hydrodynamics, the tidewater will be clockwise flow so that the sewage in the river can be effectively discharged from the estuary to the sea.

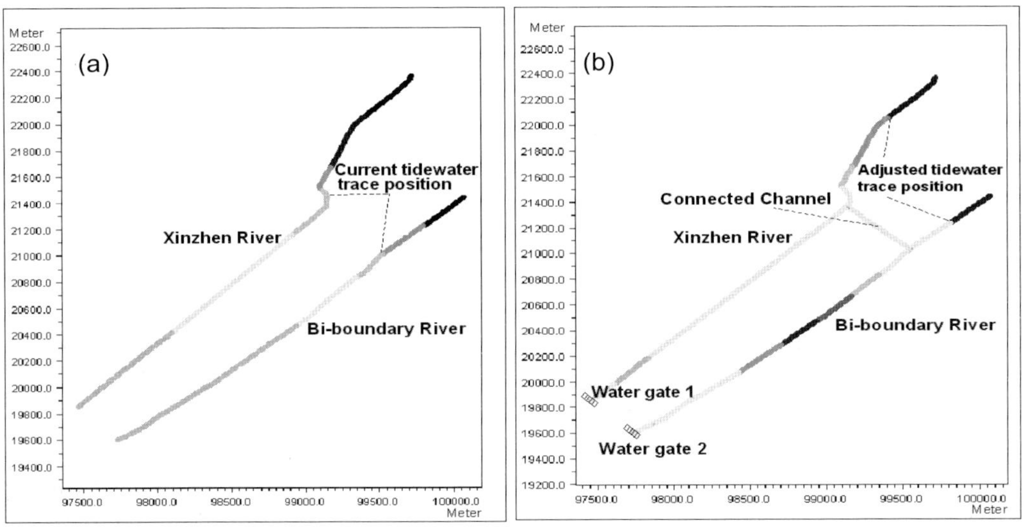

Fig. 3 (a) Current tidewater trace, and (b) the simulation result of tidewater trace after adding the connected channel and water gates in Xinzhen River and Bi-boundary River.

River hydrodynamic control program The relative concentration of 12 points in Fig. 1(b) was measured in the light for Rhodamine B. Figure 2(b) shows the pollution distribution in the study area; it is obvious that the pollution of rivers and estuaries was very serious, especially the tidal riverway, where most pollutants remained. Figure 1(b) showed there were three types of pollution situation, the first (I area) is a strong exchange water area where the concentration was almost exponentially degraded to zero after one and a half days from the first to seventh observation point. The second (II area) is a medium exchange water area, with a concentration around 0.4–0.5 mg/L after one and a half days from the eighth to the tenth observation point. The third area (III area) is a weak exchange water area with a concentration around 0.8–0.9 mg/L after two days from the eleventh to the twelfth observation point, the velocity of this water area was comparatively weak. On the basis of the simulation result in Fig. 3, the detailed hydrodynamics control project programs are as follows:

(1) When the spring tide comes with better water quality, after opening the water gate at Xinzhen River, the tide water flows into Xinzhen River and then into Bi-boundary River via the connecting channel. When the spring tide reaches the highest tide level, the water gate must

be immediately shut down until the following spring tide arrives so that the higher water level can be kept in Xinzhen River. This process can cause the full exchange between the tidewater and polluted river water. When the ebb tide comes, the water gate at Bi-boundary River must be opened and the water gate at Xinzhen River must be closed; the tide can flush sludge at the bottom of the riverbed and flow into the estuary so that the river water quality can improve because pollutants and sediment are carried into the sea.

(2) By dynamically controlling the water level for every river, the water gate can separately operate in order to make better use of tide hydrodynamics. If the floating litter and sediment before every water dam is taken away, the tidal river way must strengthen daily management.

Ecological treatment

It is significant to use ecological technology to fight pollution; re-using sewage is advantageous as Shenzhen city has a serious lack of fresh water.

Ecological water requirement engineering The ecological water requirement of the river for diluting sewage has two calculation methods, one is the 1-D quantity and quality of the water model, and the other is water consumption of the landscape. In general, the calculation result of dilution water for according with water standards was 10×10^4 m^3/day, the supplemental water for landscape was 1.04×10^4 m^3/day, so the total ecological water requirement of ecological engineering may be considered as 11×10^4 m^3/day for Xinzhen River.

After designing the ecological water requirement engineering, the pipe line is 11.171 km in total length, with the beginning of GuShu sewage treatment plant and the end of the upstream of Xinzhen River and Xixiang River. The diameter of the main pipe is 1000 mm and the branch pipe is 800 mm. The total ecological water requirement is 22.5×10^4 m^3/day, including 11×10^4 m^3/day of Xinzhen River and 12.5×10^4 m^3/day of Xixiang River. The highest pump head is 23.99 m at Xinzhen River and 10.1 m at Xixiang River. This engineering also explained that the ecological treatment of urban rivers should be considered as a whole.

River water ecological purification engineering The purification engineering of water quality in Xinzhen River will use the biotechnology which includes the bio-floating island and the biological multi-island, etc. Establishing the perfect aquatic plant system made up of emerging plants, floating leaf plants, submerged plants and floating plants, can effectively cut down the algae in urban rivers and degrade the concentration of organic pollutants, nitrogen and phosphorus. These spatial plants can show the beautiful landscape with surrounding circumstances, especially with the river water body. In this engineering, the emerging plants are *Lotus*, *Calamus*, *Scirpus validus* Vahl and water lilies; the floating plants are *Eichhornia crassipes*, Gordon Euryale seed and water chestnut; and the submerged plants are *Potamogeton crispus*, *Ceratophyllum demersum* and *Hydrilla verticillata.*

Constructed wetland engineering The residual river sewage will be conducted in constructed wetland by a new channel. The residual sewage will be treated by intercepting the wastewater in the Xinzhen River and Bi-boundary River. To protect the water environment of Big Shovel Bay, the constructed wetland can be planted downstream, which will work well in contributing to wastewater treatment. The tidewater can flow naturally into constructed wetland in order to improve the water ecological environment in Xinzhen River. Based on the original plant system in this area, it will have a good configuration in controlling the following plants: *Typha orientalis* Presl, *Phragmites australis*, *Cyperus alternifolius*, *Canna indica*, *Cyperus papyrus*, *Vetiveria zizanioides*, *Thalia dealbata*, white ginger lily, *Iris pseudoacorus, Coix puellarum* Balansa and *Cyperus.*

Ecological landscape engineering To embody the rule of sustainable development and economics, some methods such as saving operation costs, strengthening management, choosing fine plants, etc. can be carried out to persist in the organically unification of economic benefit, ecological benefit and social benefit. Configuration plants for Xinzhen River are core to the nature of the landscape, and some base countermeasures are as follows:

(1) To give prominence to indigenous plants, and to stop use of foreign invasive plants. The indigenous plants of these urban rivers in Shenzhen city can embody the main ecological landscape function and these indigenous plants are *Metasequoia*, *Castanopsis hystrix*, *Sapium discolor*, *Ilex rotunda*, *Manglietia glauca*, *Machilus nanmu*, *Rudbeckia laciniata* and *Melastoma candidum*. Of course, some outside plants with stronger purification function, resistant to pollution and with landscape effects, can be introduced as species.

(2) To emphasize the biodiversity protection and restoration, the currently existing plants should be protected and plant oriented. The river bank plants can be selected from the following plants: *Ficus nervosa*, *Caryota ochlandra*, water banyan, *Syzygium jambos* (L.), Elaeocarpaceae, *Taxodium distichum*, water pine, *Taxodium ascendens* Brongn, *Litsea glutinosa* and *Ficus simplicissima*, etc.

CONCLUSION

Major conclusions drawn from this study are as follows:

(1) It is certainly significant that controlling the tidal hydrodynamics can flush stagnant pollutants to sea in order to reduce the pollution of tidal rivers at coastal cities. By establishing the water conservancy facilities such as the connected channel between the Xinzhen River and Bi-boundary River, water gates with a single directional flap valve, etc. and by appropriately operating programs of these water conservancy facilities, the tidal energy can effectively cause the exchange of water body and improve water quality of Xinzhen River.

(2) It is necessary to calculate the ecological water requirement for urban rivers, the output water of sewage treatment plant can be considered as reclaimed water re-use and be conducted to the upstream. After detail engineering design in this study, the conducted pipe lines are 11.171 km, the total ecological water requirement is 22.5×10^4 m^3/day, including 10×10^4 m^3/day of Xinzhen River and 12.5×10^4 m^3/day of Xixiang River.

(3) River water purification can use the perfect aquatic plant system made up of emerging plants, floating leaf plants, submerged plants and floating plants. Configuration of ecological landscape engineering must give prominence to indigenous plants and appropriately introduce foreign plant species. Constructed wetland, especially the mangrove in coastal cities, can improve the river water environment.

Acknowledgements The authors are grateful to staff in the Pearl River Water Resources Committee of Water Resources Institute of Science, especially, Heqing Du, Zhihui Dong, San Guo, Fenghua Yu, Xiaoying Ding, Jianping Wang, Yong Lei and Rongli Chen. Thanks to all staff at Shenzhen Branch of the Northwest Municipal Engineering Design and Research Institute Co. Ltd. Thanks also to staff at Baoan Municipal Bureau of Water Affairs in Shenzhen city and in the Baoan Water Environment Comprehensive Improvement Office for their attention and assistance.

REFERENCES

Avid, A., Kovacic, R. M. & Twait, M. P. (2006) Use of created wetlands to improve water quality in the midwest – lake Bloomington case study. *J. Ecol. Engng* **28**(3), 258–270.

Cerco, C. F. (1995) Response of Chesapeake Bay to nutrient load reductions. *J. Environ. Engng* **121**, 549–557.

Couch, S. J. & Copeland, G. J. M. (2003) Tidal straining, mixing and Lagrangian flow residuals around headlands. *J. Marine Environ. Engng* **7**, 25–45.

DHI (2003) MIKE11 Reference Manual. Horsholm, Denmark, Danish Hydraulic Institute, Denmark.

Hefting, M. M., Clement, J.-C., Bienkowski, P. *et al.* (2005) The role of vegetation and litter in the nitrogen dynamics of riparian buffer zones in Europe. *Ecol. Engng* **24**(5), 465–482.

Hong Li-jian (2000) Study on the technique of diverting water to flush out pollutants in urban rivers of Fuzhou City. *Pearl River* **1**, 38–42.

Hu, Ben-xiong (2004) Optimization research for the sewage flushing of the Shenzhen River. *Water Resour. Hydropower Engng* **35**(9), 20–23.

Lai, C. (1977) Computer simulation of two-dimensional unsteady flows in estuaries and embayments by the method of characteristics—basic theory and the formulation of the numerical method. Water Resources Investigations. US Geological Survey, pp. 77–85.

Lee, D. I. & Park, C. K. (2002) The modelling of winter water quality in the estuary of the Nakdong River. *J. Korea Soc. Water Quality*. **18**, 349–360 (in Korean).

Maddock, L. & Pingree, R. D. (1978) Numerical simulation of the portland tidal eddies. *Estuarine Coastal Marine Sci.* **6**, 353–363.

Montazeri-Namin, M., Lin, B. & Falconer, R.A. (2004) Modelling estuarine and coastal flows using an unstructured triangular finite volume algorithm. *Adv. Water Resour.* **27**, 1179–1197.

Navon, I. M. (1988) A review of finite-element methods for solving the shallowwater equations. In: *Computer Modelling in Ocean Engineering* (ed. by B. A. Schrefler & O. C. Zienkiewicz), 273–279. AA Balkema Publishers, Rotterdam, The Netherlands.

Shields F. D., Cooper C. M., Jr., Knight, S. S. *et al.* (2003) Stream corridor restoration research: a long and winding road. *Ecol. Engng.* **20**(5), 441–454.

Toro, E. F. (2001) *Shock-Capturing Methods for Free-Surface Shallow Flows*. John Wiley and Sons, Inc., New York, USA.

Wu Linlin, Huang Minsheng & Deng Hong (2006) Advances and application of bio-ecological abatement for urban polluted river. *Water Purification Technol*. **25**(6), 11–15.

Zhen Hao-zhi (2006) Discussion on sewage flushing of Shiqi River in Zhongshan city. *Guangdong Water Resour. Hydropower* 2, 66–68.

Water environmental capacity at watershed scale: a case study in the Yong River system

BAOLIN SU[1], CHENGWEN WANG[2] & SHENGTONG CHENG[2]

1 *College of Water Sciences, Beijing Normal University, Beijing 100875, China*
subl@bnu.edu.cn

2 *Department of Environmental Science and Engineering, Tsinghua University, Beijing 100084, China*

Abstract Water environmental capacity (WEC) is important to environmental protection and management of a river system. Many relevant studies were mainly focused on point source pollution, such as effluents from industrial sites and centralized wastewater treatment plants (WTPs), and non-point source (NPS) pollution was usually not taken into account while calculating WECs. In this study, a new modelling approach was proposed which considered NPS pollutant losses and different hydrological conditions, such as dry, normal and wet seasons. The results showed that WECs varied with hydrological conditions and NPS pollutant losses into the river system. During the dry season, WECs of chemical oxygen demand (COD) and ammonium nitrogen (NH_3-N) are small owing to low streamflow and low dilution capacity, and only a small part of NPS pollutant losses into the river. However, the WECs of COD and NH_3-N increased with increase of streamflow during the wet season, and the NPS pollutant losses into the river also increased. This study also showed that water resources development level in the upstream area, such as reservoir construction and water diversion from upstream directly to urban area in the downstream area, had a great influence on the streamflow and subsequent WECs of the river reaches in the downstream. A reasonable development level of water resources in a watershed is very important to sustainable utilization of water resources, not only in quantity, but also in quality.

Key words water environmental capacity (WEC); water resources development; non-point source (NPS); watershed; river

INTRODUCTION

Water environmental capacity (WEC) is defined as a threshold of pollutant loadings discharged into a river system under certain hydrological, drainage conditions and water environmental objectives. It is important to water environmental management of a river system and one of the bases to total amount control of pollutant discharges into a river system. There are many research activities in China on WEC calculation depending on the river water quality model (Huang *et al.*, 2000a,b; Wan & Wu, 2000; Lu & Liu, 2001; Sun *et al.*, 2001; Hu *et al.*, 2002; Cheng, 2003; Mao & Wang, 2003; Xu *et al.*, 2003; Ji *et al.*, 2004; Lu *et al.*, 2004; Luo *et al.*, 2004; Kang, 2005; Sun & Feng, 2005; Xiong & Luo, 2005). However, such kinds of research were predominantly focused on point source pollution, such as effluents from industrial sites and centralized wastewater treatment plants (WTPs), and non-point source (NPS) pollution was not taken into account while calculating WEC. Because the hydrological condition used in WEC calculation is generally confined to a relatively dry hydrological condition, for example $7Q_{10}$ streamflows, WEC calculation based on such a dry hydrological condition is obviously safe to river water environmental protection, but it is too strict and luxurious, because WEC is one kind of resource (self-purification capability) of a river system and we can utilize it in a reasonable manner. For example, we can increase discharge loadings during the wet season because streamflow is large and usually the river system has a stronger self-purification capability when streamflow becomes larger. In the meanwhile, pollutant loadings from non-point sources also increase with surface runoff during normal and wet seasons. Therefore it is necessary to research seasonal distribution of WEC during different hydrological conditions so that we can develop and utilize it in a better way from the viewpoint of watershed environmental management.

The objectives of this paper are to research water environmental capacity on a watershed scale by focusing not only on point source pollution, but also on non-point source pollution, and to discuss the influences of water resources development in the upper stream areas on environmental capacity of the downstream areas.

STUDY AREA

The Yong River is located in Ningbo City, Zhejiang Province of China. It includes two main rivers named the Fenghua River and the Yao River (Fig. 1), which play different roles in water supplies and water environment functions in the city. According to their use functions and objectives, the Yao River was zoned as III Class water environmental function zone in its main reaches from Shenwang section to the floodgate of the Yao River and supply water for agricultural and industrial use; and the main reaches of the Fenghua River were zoned as water environmental function zone of III and IV Classes from the upper stream at the Xikou section to the Sanjiangkou section. The downstream of the confluence of the Yao River and Fenghua River is called the Yong River, i.e. from the Sanjiangkou section to the Youshan section, which flow into Hangzhou Bay. The main reaches of the Fenghua River and the Yong River act as the main receiving water area for industrial and domestic pollution discharges of Ningbo City.

There are some reservoirs constructed in the upstream areas of the Yong River watershed, especially of the Fenghua River watershed. The main reservoirs are the Jiaokou Reservoir, Tingxia Reservoir and Hengshan Reservoir in the Fenghua River sub-watershed, and Siming Lake Reservoir in the Yao River sub-watershed. The first three reservoirs are the main water resources for drinking and industrial water supplies of the city. Besides, the Yao River is a channel reservoir which acts as an important water resource for industrial use in Ningbo City and partially of Zhoushan City.

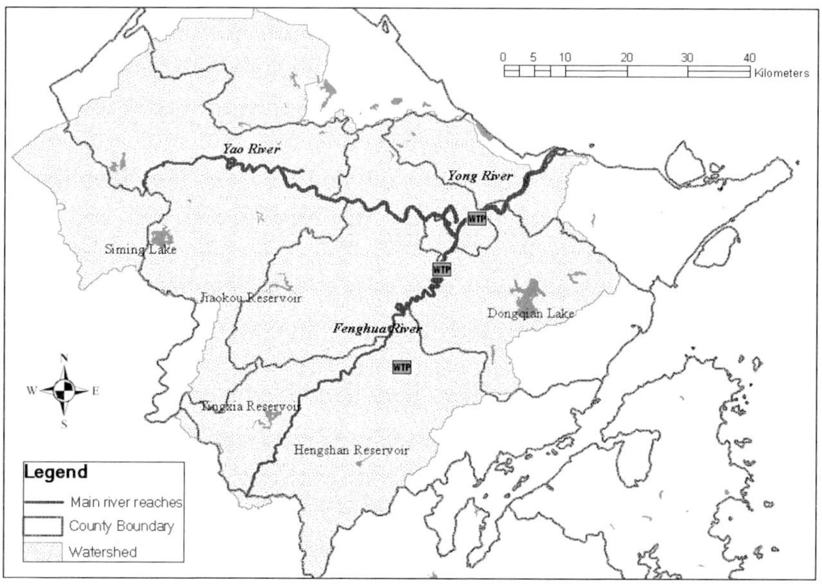

Fig. 1 The study area – the Yong River watershed.

Table 1 shows the main river reaches and monitored river cross-sections and current industrial point source drainage outlets and their basic data. Table 1 also gives present water quality levels of these main river reaches. The data show that the water quality is relatively good in the Yao River, because current water quality can meet water quality objective of III Class of Chinese Environmental Quality Standard for Surface Water (GB8383-2002) (SEPA & AQSIQ, 2002). However, the water quality of main reaches of the Fenghua River can not meet their specific water quality standards because of a lot of discharges of point source pollution from industrial sites and residential areas, and non-point source pollution from agricultural residential areas, livestock activities and agricultural fertilizer losses, as confirmed by recent studies in a river system and emphasized in this paper.

Table 1 Basic facts of main reaches.

River	Reach	Start-End section	Streamflow (m^3/s) 75%	50%	25%	u_a(m/s) 75%	Control Section	CWQ	OWQ	psCOD (t/a)	psNH$_3$-N (t/a)	npsCOD (kg/(m·a))	npsNH$_3$-N (kg/(m·a))
(1)	(2)	(3)	(4)	(5)	(6)	(7)	(8)	(9)	(10)	(11)	(12)	(13)	(14)
Yao	1047	Shenwang-Kaifengqiao	4.91	14.39	27.76	0.40	Qingjiangdu362	III	III	9.90	1.04	57.31	4.37
	1048	Kaifengqiao-Yuyao	4.91	14.41	27.78	0.30	Dipuwang753	III	II	18.00	1.20	11.01	0.99
			7.08	20.78	40.08	0.30	Nianciqiao754	III	III				
	1050	Yuyao-Jiezhizha	9.23	27.08	52.23	0.20	Xiachendu442	III	III	238.98	17.23	7.95	0.90
	1051	Jiezhizha-Chejiu	9.44	27.69	53.41	0.20	Pukouzha755	III	III	64.50	9.50	15.52	1.08
			11.68	34.27	66.09	0.20	Chengshan	III	III				
	1052	Chejiu-YaojiangDazha	12.28	36.03	69.48	0.20	Qinglindu606	III	III	117.20	7.86	15.52	1.08
			11.14	34.94	68.47	0.20	YaojiangDazha	III	III				
Fenghua	996	Xikou-Xiaozheng	2.22	6.51	12.55	0.50	Xiaozhen444II	II	II	67.86	11.70	60.46	2.75
	997	Xiaozheng-Fangqiao	2.30	6.74	13.00	0.40	Jiangkou751	III	III	45.07	6.51	32.31	1.60
	998	Fangqiao-Fangshidu	7.05	20.68	39.88	0.30	Fangshidu2	IV	III	165.94	21.38	13.89	1.33
	999	Fangshidu-Tongpenpu	7.78	22.84	44.05	0.25	Tongpenpu772	V	IV	276.64	26.69	53.96	5.19
	1186	Tongpenpu-Sanjiangkou	8.90	26.11	50.37	0.20	Chenglanyan440	V	IV	1611.51	113.33	6.06	0.67
						0.10	Sanjiangkou53	V	IV				
Yong	1187	Sanjiangkou-Youshan	19.84	58.22	112.23	0.05	Zhangjiangang54	IV	IV	4454.56	857.34	10.73	0.67
							Youshan252	IV	IV				

Note: (4), (5) and (6) are streamflows for each river reach under different hydrological frequency of 75% (dry season), 50% (normal season) and 25% (wet season), respectively; (7) u_a-esitmated average velocities of the reach; (9) CWQ-current Chinese water quality classes according to GB3838-2002 (SEPA & AQSIQ, 2002); (10) OWQ-objective of water quality; (11) and (12) are point source loadings of chemical oxygen demand (COD) and ammonia-nitrogen (NH_3-N); (13) and (14) are initial non-point source loadings of COD and NH_3-N into a river from reach-specific catchments during dry season, which were estimated by IESN (2005) according to an arbitrary ratio of 0.06, i.e. only 6% of estimated NPS pollutant yields could lose into a reach from its catchment area.

METHODS AND MATERIALS

Water quality models

One-dimensional river water quality model and tidal river water quality model are applied to estimate the water environmental capacity of COD and NH_3-N, respectively, for the Yao River and the tidal-affected main river reach of the Fenghua River and the Yong River.

One-dimensional river water quality model The one-dimensional (1-D) river water quality model assumes that pollutant concentration only varies along the longitudinal direction of a river, and is applicable to following conditions at the same time: (a) narrow and shallow river reach; (b) pollutants can be mixed completely in a short time; and (c) pollutant concentrations do not vary so much on the cross-section, and concentration gradient in vertical and traverse direction can be neglected.

In this study, the 1-D model is applied to the Yao River and upstream reaches of the Fenghua River, and its basic formula is as in equation (1):

$$C(x) = C_0 e^{-Kx/u_x} \tag{1}$$

where u_x is average velocity of a river reach; K is decay constant of specific pollutant; x is longitudinal distance from a river cross-section to a drainage outlet; C_0 is pollutant concentration at $x = 0$; and $C(x)$ is pollutant concentration at $x = x$.

One-dimensional tidal river water quality model Because the main reaches of the Fenghua River and Yong River are influenced by tide from the Hangzhou Bay, 1-D tidal river water quality model is applied, and its basic formula is as follows (China SEPA, 2005):

$$\frac{\partial C}{\partial t} + u_x \frac{\partial C}{\partial x} = \frac{\partial}{\partial x}\left(E_x \frac{\partial C}{\partial x}\right) - KC \tag{2}$$

If we adopt average values of hydraulic parameters in a half tidal cycle, then equation (2) can be solved under steady conditions. That is, the drainage is steady, and the term $\frac{\partial C}{\partial t} = 0$ and the solutions are as:

To the upstream location of drainage outlet during flood tide period ($x < 0$):

$$C(x) = \frac{C_0^{'}}{M} \exp\left(\frac{u_x}{2E_x}(1+M)x\right) \tag{3}$$

To the downstream location of drainage outlet during the ebb tide period ($x > 0$):

$$C(x) = \frac{C_0^{'}}{M} \exp\left(\frac{u_x}{2E_x}(1-M)x\right) \tag{4}$$

and $C_0^{'} = W/(Q_h + Q_r)$, and $C_0^{'} = W/Q_r$ while $Q_h << Q_r$. $M = \sqrt{1 + 4KE_x/u_x^2}$, where, E_x is longitudinal dispersive coefficient (m^2/s); K is decay constant; Q_r is average flow of freshwater (m^3/s); Q_h is average flow of wastewater effluents (m^3/s); u_x is average velocity of freshwater (m/s); and W is pollutant loadings of calculated reach (g/s).

Consideration of non-point source pollution In the Guideline of Water Environmental Capacity Calculation (China SEPA, 2005), the point source loadings are the only discharges in consideration while calculating WEC, and non-point source pollutant loadings are simply subtracted from calculated WEC. As is known, the contribution of point source discharge is accorded with the theory of linear superposition, and non-point source pollutant discharge can be regarded as many tiny point source discharge outlets along a river reach. Therefore, non-point source pollutant loadings are divided as definite tiny point source loadings along a river reach unit (for example 100 m in this study), and pollutant loadings are assumed to be discharged into this reach unit from the end point of unit itself, and its contribution to specific cross-section is calculated with 1-D river water quality model.

The pollutant loading discharged into a river reach at upstream location x of a control section can be calculated as:

$$dW_l = Pdx \tag{5}$$

and its contribution to the pollutant concentration of the control section is as following:

$$(Pdx)\exp(-Kx/u_x) \tag{6}$$

where P is non-point source pollutant loadings discharged into a river reach from reach-specific catchments (kg/(m·a)); and dx-length of divided reach unit (100 m in this study).

Conceptualization of river reaches

River reach conceptualization is an important step before the water quality model can be applied. Fig. 2 showed the conceptualization diagram of a river reach, which indicates that there are n point-source pollution discharge outlets and tributaries (Q_1,C_1, Q_2,C_2,…Q_n,C_n), and m water uses from the river reach q_j (j = 1, 2, …m). Here, Q_0,C_0 is streamflow rate and pollutant concentration at the start section of the reach, and Q_s,C_s is streamflow rate and concentration at the river section concerned (such as control section). Generally, C_s can be set as objectives of the control section according to their water environmental function level (Ningbo EPB, 2003) and surface water standard (GB3838-2002). X_i is the distance from the ith drainage outlet to the control section. Non-point source loss into the reach is considered as linear source pollution and divided as many tiny point sources in the water quality model.

The conceptualization of the Yong River is shown briefly in Fig. 3, which shows the main generalized tributary branches of the Yao River, Fenghua River and Yong River. Table 2 lists the water quality data of 2004 and estimated streamflow rate of main branches.

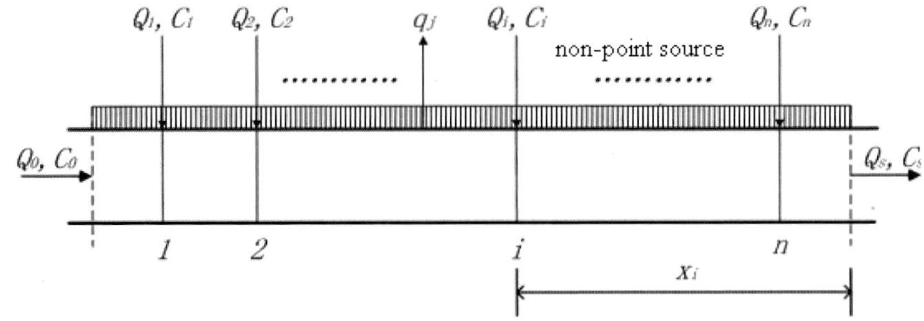

Fig. 2 Schematic diagram of river reach conceptualization.

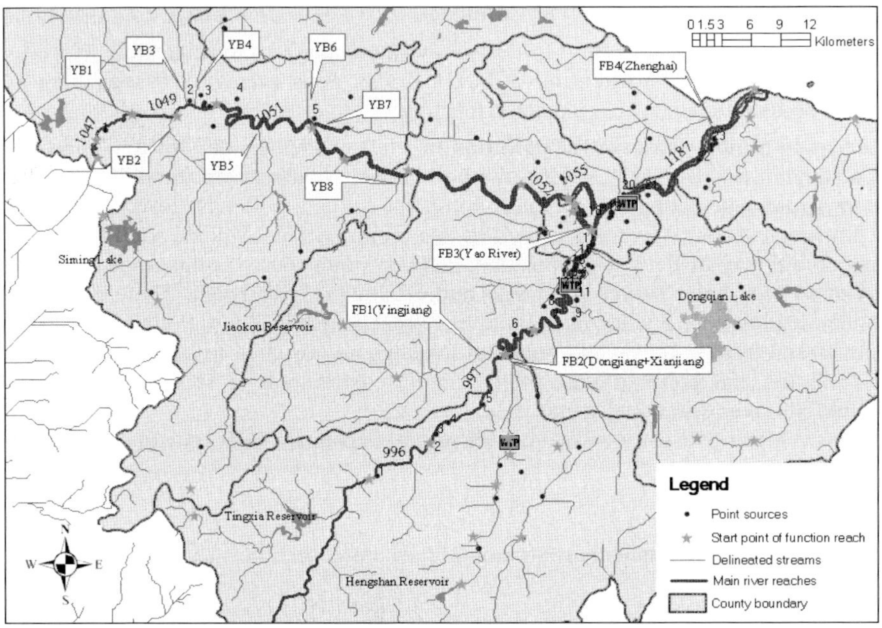

Fig. 3 Conceptualization of the Yong River system.

Table 2 Basic data of main branches of the Yong River.

River	Branch	CWQ (mg/L)		Streamflow (m^3/s)			OWQ
		COD	NH_3-N	75%	50%	25%	
Yao	YB1	5.28	0.60	2.50	7.33	14.14	III
	YB2	2.86	0.26	0.66	1.94	3.75	III
	YB3	5.41	0.79	1.41	4.13	7.97	III
	YB4	6.00	1.00	2.08	6.11	11.78	III
	YB5	2.96	0.30	0.81	2.38	4.58	III
	YB6	5.31	0.63	0.10	0.28	0.54	III
	YB7	5.31	0.63	0.43	1.27	2.45	III
	YB8	5.49	0.62	0.34	1.00	1.94	III
Fenghua	FB1	6.01	1.11	2.23	6.54	12.61	III
	FB2	3.87	1.34	2.19	6.42	12.38	III
	FB3	5.59	0.50	11.22	35.16	68.90	IV
Yong	FB4	10.55	1.52	1.45	4.25	8.19	IV

Note: CWQ (mg/L) – average water quality of year 2004.

There are 140 point pollution sources in the Yong River system, including three wastewater treatment plants (WTPs) (IESN, 2005). These point pollution sources are predominantly located along the main reaches of the Fenghua River and the Yong River, which are the main receiving water areas for pollution discharge. In the modelling, the point discharge outlets are conceptualized as 23 outlets along the main reaches of the Fenghua River and the Yong River, and 6 outlets along the main reaches of the Yao River. The pollutant loadings of each main reach are listed in Table 1.

SIMULATION RESULTS

Model calibrations

The model calibrations are conducted using the trial-and-error method, and water quality standards of different water quality levels are used to confine and finalize the trials. In the calibration, the NPS pollutant loadings from the catchment area of each river reach were estimated considering NPS contributions from agricultural fertilization, dispersive rural residential areas, and/or impermeable urban area by using the exporting coefficient method (IESN, 2005). And the estimated NPS loadings were generalized as a kind of linear source pollution and were evenly discharged into river reach, and a ratio was used (as a calibrated parameter) to represent actual NPS loss part from the catchment into the river reach in different hydrological conditions, and it was also calibrated using observed hydrological and water quality datasets during different seasons. Here, dry, normal and wet seasons are defined according to hydrological frequency at 75%, 50% and 25%, and are assumed to correspond to the observed data in periods of: (a) January, February, November and December; (b) March to May, and October; and (c) June to September, respectively. The water quality data of year 2004 in dry, normal and wet seasons are used in calibrations for control river sections and main branches.

For the sake of simplification of modelling, decay constants are not differentiated among different water quality levels and different hydrological conditions in the study, but are differentiated among different rivers. The calibrated parameters are listed in Table 3, which shows the average decay constants of COD and NH_3-N, and the average ratios of NPS losses into the rivers in different hydrological conditions.

Observed COD and NH_3-N are plotted against calculated ones and linear regressions are conducted using a function of $y = x$ and relative coefficients (r-square value) are obtained (Table 4). The r-square values showed that the simulations are well agreeable to observed data, and established water quality model can be applicable to the river system of this study.

Table 3 Calibrated decay constant K and ratios of NPS losses into reaches.

River reaches	Decay constant K (1/d)		Averaged ratios of NPS losses into river reaches		
	COD	NH_3-N	75%	50%	25%
Yao River	0.20	0.10	10.9%	43.7%	45.4%
Fenghua River	0.20	0.15	11.4%	36.9%	51.7%
Yong River	0.20	0.15	11.4%	36.9%	51.7%

Table 4 Model calibrations.

Item	r-square values		
	75%	50%	25%
COD	0.965	0.912	0.785
NH_3-N	0.694	0.642	0.612

Calculation of water environmental capacity

Based on the calibrated parameters and the water quality model, the water environmental capacity of COD and NH_3-N can be calculated through the following procedures:

(a) adjust (reduce or increase) pollutant loadings of each conceptualized outlet from upstream to downstream;
(b) run the water quality model and output calculated concentrations of control sections concerned;
(c) compare the calculated COD and NH_3-N at specific control sections with water quality standards. If the calculated concentration is higher than corresponding water quality standards, then reduce pollutant loadings; if the calculated concentration is lower than standards, then increase pollutant loadings correspondingly; and go back to step (b);
(d) once the calculated concentrations at all control sections are as close as possible to their water quality standards, stop trials and sum pollutant loadings of all outlets. And output this summation as one set of point source water environmental capacity.
(e) according to above procedures, water environmental capacities of the studied rivers are determined, which include both point source and non-point source water environmental capacities (Table 5).

Table 5 Calculation results of water environmental capacity (WEC) (unit: t/a).

River	Hydrological Condition	Point Sources		Non-pint Sources		Total WECs	
		COD	NH3-N	COD	NH3-N	COD	NH3-N
Yao	75%	1344.06	110.25	4246.83	90.96	5590.89	201.21
	50%	2240.10	183.75	9964.40	364.93	12204.50	548.68
	25%	6859.80	746.20	12764.08	379.58	19623.88	1125.78
Fenghua	75%	22573.01	857.34	5850.72	444.74	28423.73	1302.08
	50%	29884.13	1124.34	19008.63	1244.45	48892.76	2368.79
	25%	42592.79	1432.51	26602.20	1454.21	69194.99	2886.72

DISCUSSIONS

The ratio of NPS losses into reaches

As discussed earlier, NPS pollution considered in this study includes pollutant loading contributions from agricultural fertilization, dispersive rural residential areas, and/or impermeable urban area. Generally, the ratios of NPS pollutant losses into reaches are arbitrarily estimated by the researchers, depending on characteristics of the study area, such as topographic features, land use and tillage patterns for agricultural areas, and popularization of drainage system in a reach-specific catchment for residential areas. It is very hard to quantify this coefficient at most times because the researchers always lack some important observed data and references of drainage system of a studied city; even the technicians of municipal departments usually failed to know an overview of the drainage system in their cities. All of the above lead to the fact that estimated ratios of NPS pollutant losses (both for agricultural areas and residential areas) are arbitrary and not approvable using observed data.

However, these important ratios can be determined accurately in a relative manner based on methods of this study. Once we obtain the actual point source pollution discharges from industrial sites and some residential areas (through centralized WTPs of a city), we can estimate these ratios depending on water quality modelling and observed water quality data. As shown in Table 3, the averaged ratios of NPS losses into river reaches during different hydrological conditions (seasons) were obtained through model calibrations. These results are close to levels of streamflow analyses of referenced hydrological station, which showed that the streamflows during dry (January, February, November and December), normal (from March to May, and October) and wet (from

June to September) seasons amount to 12.7%, 28.7% and 58.6%, respectively. They are verified indirectly to be reasonable for the Yong River system.

Seasonal variation of water environmental capacity

Because hydrological condition used in WEC calculation is usually confined to a relatively strict hydrological condition, such as dry season, with a hydrological frequency of 75% or 90%, WEC calculation based on such dry hydrological condition is obviously safe to river water environmental protection and management, but to some extent it is too strict and luxurious. As we know, WEC is one kind of resource (self-purification capability) of a river system and we could utilize it in a reasonable manner. For example, we could increase discharge loadings during the wet season because streamflow is large and usually the river system has a stronger self-purification capability when streamflow becomes larger. This is useful for optimization of WEC utilization in a watershed.

The simulation results (Table 5) show that water environmental capacity varied with hydrological conditions and corresponding non-point source pollutant loadings into the river system. During the dry season (75% hydrological frequency), water environmental capacities of chemical oxygen demand (COD) and ammonium nitrogen (NH_3-N) are small due to low streamflow and low dilution capacity, and only a small part of NPS pollutant losses into the river due to less surface runoff generated during dry period (this part of NPS pollutant losses is contributed predominantly from dispersive residential areas and urban areas without perfect drainage system). However, the COD and NH_3-N capacities increased with increase of streamflow during the wet season. In the meantime, the NPS pollutant losses into the river also increased due to more surface runoff generated, which showed that NPS contribution was also important and water environmental protection and management of the Yong River system should not be neglected. It is recommended that the NPS pollution issues should be taken into account in the watershed management of the Yong River in the future.

Influences of water resources development in the upstream area

As shown in Fig. 3, there are three main reservoirs constructed in the upstream area of the Fenghua River are the Jiaokou Reservoir, Tingxia Reservoir and Hengshan Reservoir, which are the main water resources for drinking and industrial water supplies of the city, and the current level of water resources development is about 33.7%. Water diversion, from upstream directly to urban area in the downstream area, had a great influence on the streamflows of the Fenghua River in the downstream area, especially largely decreased the river base flow in the down stream area, which subsequently decreased largely water environmental capacity of the river.

According to current development planning and overall water resources planning of the Ningbo City (PDIHEH, Ningbo, 2004), water demand will increase with increase of local economy and water resources development in the upstream of the Yong River will be enhanced further. Because the current development level (33.7%) is close to the upper limit of reasonable level of water resources development in a natural river system, which is known as 35–40%, influences of water resources development on water environment and eco-system in the downstream area should be emphasized. In this study it is assumed that development level of water resources in the upstream of the Fenghua River will increase to about 50% in the future, and new estimates of water environmental capacity under new hydrological conditions are listed in Table 6.

Table 6 Calculation results of water environmental capacity (WEC) by increasing water resources development level to 50% (unit: t/a).

River	Hydrological Condition	Point Sources		Non-pint Sources		Total WEC	
		COD	NH_3-N	COD	NH_3-N	COD	NH_3-N
Fenghua	75%	17918.31	728.74	3333.79	157.47	21252.10	886.21
	50%	26740.87	907.29	10072.90	493.96	36813.77	1401.26
	25%	35795.53	970.67	16817.71	783.29	52613.24	1753.96

By comparing with Table 5, the total COD and NH_3-N WEC of the Fenghua River would decrease about 24.0–25.2% and 31.9–40.8%, respectively while water resources development level increasing from current level of 33.7% to about 50% in the future. It is shown that the influences are serious, especially because the current water quality has already exceeded their function objectives under present pollutant discharges and water resources development.

CONCLUSIONS

Both point source and non-point source pollution are important to water environmental protection and water environmental capacity calculation in a river system. A new modelling approach was proposed considering non-point source pollution loadings and different hydrological conditions. The results showed that water environmental capacity varied greatly with hydrological conditions. During the dry season, water environmental capacities of COD and NH_3-N are small owing to low streamflow and low diluted capacity, and only a small part of NPS pollutant losses into the river due to less surface runoff. However, the COD and NH_3-N capacities increased with increase of streamflow during the wet season, and in the meantime, the NPS pollutant losses into the river also increased due to more surface runoff. This study also showed that water resources development level in the upstream area, such as reservoir construction and water diversion from upstream directly to industrial and domestic uses in the downstream, had great influences on the streamflow and subsequent water environmental capacity of the river reaches in the downstream area. It indicates that a reasonable development level of water resources in the watershed is very important to sustainable utilizations of water resources not only in quantity but also in quality, and this should be kept in mind when drawing up regional water resources planning.

REFERENCES

Cheng, S. T. (2003) Environmental capacity and permissible discharge of rivers. *Water Resour. Protection* **2**, 8–10 (in Chinese).

China SEPA (2005) Guidelines for water environmental calculation in China, National Environmental Protection Agency (in Chinese).

Hu, G. H., Zhao, P. L. & Wang, R. X. (2002) Study on water environmental capacity for Mengjin-Huayuankou section of Yellow River. *Water Resour. Protection.* **1**, 26–28, 68 (in Chinese).

Huang, Z. L., Li, Y. L., Li, J. X. & Chen, Y. C. (2004) Water environmental capacity for the reservoir of Three Gorge Project. *ShuiLi XueBao.* **3**, 7–14 (in Chinese).

Institute of Environmental Science of Ningbo City (IESN) (2005) Water pollution prevention planning of the Yong River system. Study report by IESN (in Chinese).

Ji, L., Li, J. & Wang, J. Q. (2004) Research on water environmental capacity in Nanfeihe River in Hefei City. *J Anhui University (Natural Science Edition)* **28**(6), 71–75 (in Chinese).

Kang, Y. F. (2005) Calculation of water environmental capacity and countermeasures for the water pollution control of Zhuzhanghe River's Changzhi section. *SCI/Tech Information Development & Economy* **15**(15), 115–116 (in Chinese).

Lu, J., Ding, J. C., Shi, H. C. & Wang, F. S. (2004) Current water quality status and water environmental capacity in upper stream of Zihe River basin. *J. Shandong Univ. Technol. (Sci & Tech).* **18**(1), 42–45 (in Chinese).

Lu, Q. L. & Liu, D. M. (2001) Preliminary approach on the environmental capacity and total amount control of Inner Mongolia section of Yellow River. *Environmental Protection Inner Mongolia* **13**(1), 19–22 (in Chinese).

Luo, J., Pang, Y. Luo, Q. J. & Lin, Y. (2004) Study on water environmental capacity for reversing current channels in plain river network region in Taihu Lake basin. *J. Hohai Univ. (Natural Science)* **32**(2), 144–146 (in Chinese).

Mao, Y. Y. & Wang, X. D. (2003) Pollutants bearing capacity of Huaihe River basin in Jiangsu Province. *Water Resour. Protection* **19**(3), 13–15 (in Chinese).

SEPA, AQSIG (2002) Chinese Environmental quality standards for surface water (GB8383-2002), National standards of People's Republic of China (in Chinese).

Ningbo Environmental Protection Bureau (Ningbo EPB) (2003) Zoning of water environmental function in the Ningbo City. Technical report (in Chinese).

Planning and Designing Institute of Hydraulic Engineering and Hydropower of Ningobo City (PDIHEH, Ningbo) (2004) Overall water resource planning of the Ningbo City. Technical report (in Chinese).

Sun, W. H., Yao G. J. & Pang, Y. (2001) Discussion of asymmetrical coefficient-based calculation method for water environmental capacity. *Water Resour. Protection* **2**, 25–26, 44 (in Chinese).

Sun, X. X. & Feng, Y. Q. (2005) Analysis and research on influence factors of environmental capacity of water. *Ground Water* **27**(5), 389–391 (in Chinese).

Wan, B. & Wu, Y. M. (2000) Study on calculation and allocation method of river environmental capacity. *J Wuhan Univ. Hydrol. & Elec Engng* **33**(1), 75–76 (in Chinese).

Xiong, F. & Luo, J. (2005) Calculation model analysis of river water environmental capacity. *China Measurement Technology*. **31**(1), 116–118 (in Chinese).

Xu, G. Q., Chu, J. D., Wu, Z. Y. & Chen, Q. J. (2000a) Numerical computation of aquatic environmental capacity for tidal river network. *ACTA J. Environ. Sci.* **20**(3), 263–268 (in Chinese).

Xu, G. Q., Chu, J. D., Wu, Z. Y. & Chen, Q. J. (2000b) Study on effect factors water environmental capacity for tidal river network. *Adv. Water Sci.* **11**(4), 375–380 (in Chinese).

Xu, Z. X., Lu, S. Q. & Lin, W. Q. (2003) Calculating analysis on water environmental capacity of tidal river networks. *Shanghai Environ. Sci.* **22**(4), 254–257 (in Chinese).

The potential of a hydraulic method to estimate minimum ecological instream flow for data scarce basins

SUXIA LIU[1], XINGGUO MO[1], ZHONGHUI LIN[1] & LINA JI[1,2]

1 *Key Laboratory of Water Cycle & Related Land Surface Processes, Institute of Geographical Sciences and Natural Resources Research, Chinese Academy of Sciences, Beijing 100101, China*
liusx@igsnrr.ac.cn

2 *Beijing Management Division of North Grand Canal, Beijing 101100, China*

Abstract By deducing a general expression between discharge and each of such hydraulic variables as wetted perimeter, mean cross-sectional depth of flow, mean cross-sectional flow velocity, cross-sectional area, channel width, and hydraulic radius for a channel, a simple hydraulic method to analytically estimate minimum ecological instream flow requirements (MEIFR) is proposed. The method can quickly give an estimate of MEIFR based only on observed data of mean cross-sectional depth of flow and channel width, so has potential for applications in ungauged basins. By applying the method to the rivers in Western Route South-to-North Water Transfer project in China, where data for MEIFR are scarce, the MEIFR values derived by the new method are shown to be in agreement with those by popular, conventional and well-developed MEIFR techniques, such as the Tennant method and wetted perimeter method, with small gaps in between.

Key words ecological instream flow requirements; hydraulic variable, ungauged basins; Western Route South-to-North Water Transfer Project in China

INTRODUCTION

With water resources around the world coming under increasing pressure, allocating water between abstractors and the environment needs to be based progressively more on sound science (Acreman, 2005). The questions as how much the MEIFR should be kept to guarantee the realization of its ecological functions must be answered before making any water resources development plan. For headwater areas, the challenge is even bigger as data are always scarce there.

Karim *et al.* (1995) classified the variety of previously developed MEIFR assessment techniques into: (i) historic flow methods, such as the Tennant method (Tennant, 1976); (ii) hydraulic rating methods, such as the wetted perimeter method (Annear & Conder, 1984); and (iii) habitat rating methods, such as Instream Flow Incremental Methodology (IFIM) (Stalnaker *et al.*, 1995) and its primary component – Physical Habitat Simulation System (PHABSIM) (Milhous *et al.*, 1989), the Australian "Holistic Approach" (Arthington *et al.*, 1992) and the South African Building Block Methodology (BBM, King & Tharme, 1994).

The historic flow method requires seasonal flows for fish and wildlife as percentages of mean annual flow, the simplicity meaning it has found wide applications in the world. However, it is not directly related to ecological indexes. Habitat rating methods consider lots of ecological indexes but are very complex and require considerable time, money and technical expertise.

Hydraulic rating methods estimate a desired low-flow value from a habitat index that incorporates stream channel characteristics. People have recognized that hydraulic biotope classes can be considered as significantly different in terms of flow hydraulics (Wadeson & Rowntree, 1998). Based on the relationship between the discharge and habitat expressed by hydraulic variables, with relatively fewer requirements for ecological data, hydraulic rating methods should have found prosperous applications in setting MEFIR. However, comparing with the development of historic flow methods and habitat rating methods, the development of the hydraulic rating methods has been slow.

There are perhaps two reasons to hinder the application of the hydraulic rating methods. Firstly, so far, in hydraulic rating methods only wetted perimeter is used as a hydraulic variable to represent the habitat. Only the wetted perimeter method is relatively matured for seeking MEIFR in the category of hydraulic rating methods. Using the relationship between channel wetted perimeter (an indicator of habitat availability) and discharge has found applications mainly in fishery (Gippel & Stewardson, 1998). In China, for example, there has been much research on

MEIFR (as shown in the review by Liu *et al.*, 2004), but most of them use the Tennant method, with only a few dealing with the wetted perimeter method. There are so many hydraulic variables, e.g. the mean cross-sectional depth of flow, the mean cross-sectional flow velocity, cross-sectional area, channel width, and hydraulic radius among others; it is not unusual to ask why just the wetted perimeter so far is used in hydraulic rating methods to seek MEIFR. To answer this question, a general theoretic framework between discharge and habitat via each of all the hydraulic variables is desperately needed, which has not yet been established.

Secondly, the advantage of hydraulics over other scientific disciplines is its physical and mathematical background. However, so far the research methods used in most hydraulic rating methods for setting MEIFR are mainly data fitting and visual comparison. If hydraulic rating methods contain more mathematical deduction than data fitting, they will have a more promising future.

Although hydraulic rating methods have less ecological data requirements, they still have large data requirements on discharge matched with channel characteristic data. How to set MEIFR for ungauged basins poses a big challenge to all the MEIFR evaluation techniques, including hydraulic rating methods. Ungauged basins exist almost everywhere in the world. They can be due to sparse observational network in the headwater area, with very high altitude above sea level where it is very difficult to take measurements, or the change of hydrological situation due to climatic change or land use/cover change, which makes the hydrological data available so far lose the values (Sivapalan *et al.*, 2004). To establish a method suitable in ungauged basins needs more theory to know the system. By containing more mathematical deduction other than just via data fitting, hydraulic rating methods will have stronger potential to be able to seek MEIFR in ungauged basins.

Based on the above analysis, this paper proposed a new hydraulic rating method for setting MEIFR. It is then applied in the rivers of the Western Route South-to-North Water Transfer project (WRSNWT) in China to test the new method.

THEORY

As in Liu *et al.* (2006), based on the definition of the discharge, assuming the channel is uniform, via the Manning formula and Chezy equation, for a simple triangular cross section channel, the general analytical expression between discharge Q and a hydraulic variable Y can be written as in a power law relationship:

$$Q = aY^{b} \tag{1}$$

or

$$Y = a^{-\frac{1}{b}} Q^{\frac{1}{b}} \tag{2}$$

Y can be either wetted perimeter (P), or the cross-sectional area of the flow (A), or the water width (B), or the mean cross-sectional depth of flow, simply noted as a water depth (h), *or* velocity (V), or hydraulic radius (R). Parameters a and b are shown in Table 1. 2θ is the angle between the left and right bank, which can be estimated by:

$$B = 2h\tan\theta \tag{3}$$

By using curvature technique (Gippel & Stewardson, 1998), the MEIFR based on the point with maximum curvature is deduced (Liu *et al.* 2006) as:

$$\text{MEIFR} = a^{\frac{1}{1-b}} b^{\frac{b}{1-b}} \left(\frac{2-b}{1-2b}\right)^{\frac{b}{2b-2}} \tag{4}$$

The above is the basic principle of the new hydraulic rating method, which is similar to that of wetted perimeter method (Gippel & Stewardson, 1998), but has been extended to any of the hydraulic variables instead of only the wetted perimeter, with more analytical deduction.

Table 1 The general expression between discharge (Q) and the hydraulic variable (Y) for triangular cross-sectional channels (P, wetted perimeter; A, area; R, hydraulic radius; h, water depth; B, channel width; V, flow velocity).

Y	Relation with Q	Coefficient a_y	Coefficient b_y
P	$Q = a_P P^{b_P}$	$a_p = \frac{1}{32}\frac{\sqrt{S}}{n}(\sin 2\theta)^{5/3}$	$b_p = \frac{8}{3}$
A	$Q = a_A A^{b_A}$	$a_A = \frac{1}{2}\frac{\sqrt{S}}{n}(\sin 2\theta)^{1/3}$	$b_A = \frac{4}{3}$
R	$Q = a_R R^{b_R}$	$a_R = 8\frac{\sqrt{S}}{n}(\sin 2\theta)^{-1}$	$b_R = \frac{8}{3}$
h	$Q = a_h h^{b_h}$	$a_h = 2^{1/3}\frac{\sqrt{S}}{n}(\sin 2\theta)^{-1}(\sin\theta)^{b_h}$	$b_h = \frac{8}{3}$
B	$Q = a_B B^{b_B}$	$a_B = 2^{-7/3}\frac{\sqrt{S}}{n}(\sin 2\theta)^{-1}(\cos\theta)^{b_B}$	$b_B = \frac{8}{3}$
V	$Q = a_v v^{b_v}$	$a_v = 8(\frac{\sqrt{S}}{n})^{-3}(\sin 2\theta)^{-1}$	$b_v = 4$

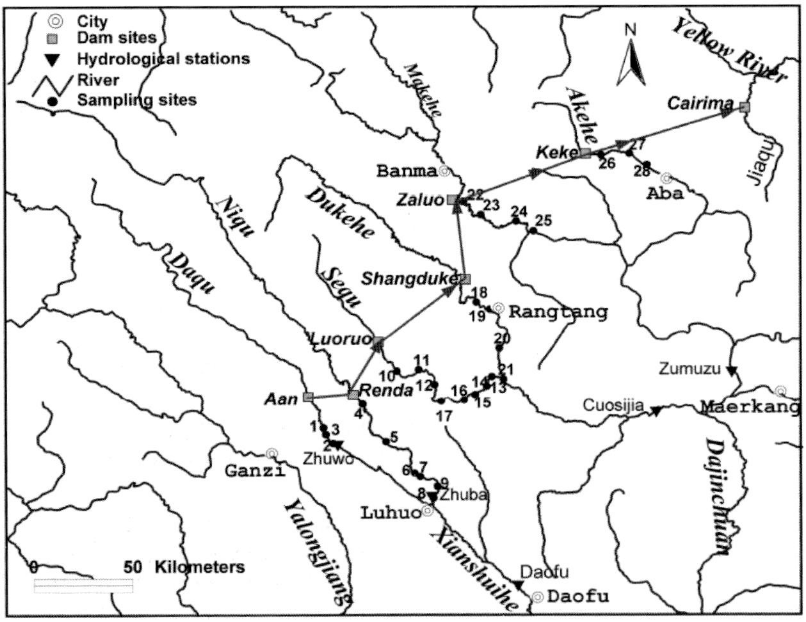

Fig. 1 The study area with the arrows showing the water transfer route to the Yellow River in the Western Route South-to-North Water Transfer Project in China (Liu *et al.*, 2006) (the numbers in the figure show the sites for water depth and channel width observation).

The detail steps of the new method are as follows: (1) Collecting the data of water depth h and water width B. They can be collected at one site with several samplings or at several sites with one sample each. (2) Calculating the angle between the left and right bank of the river based on equation (3). (3) Collecting the data of roughness n and hydraulic slope S. If without data, they can be assigned the values according to experiences or from the look-up table of hydraulic handbooks. (4) The point with maximum curvature as shown in equation (4) is the MEIFR.

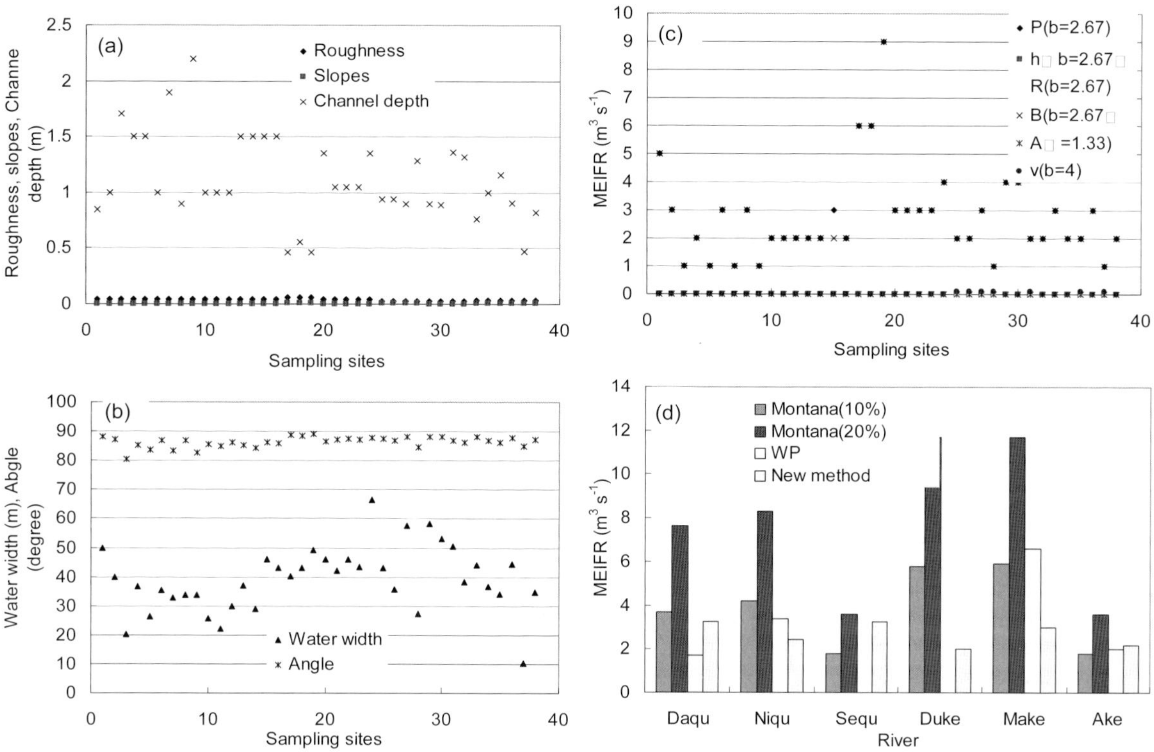

Fig. 2 (a) and (b) The data of water depth, channel width, roughness slopes on 38 sites with the angle derived by equation (3). (c) The result of MEIFR ($m^3 s^{-1}$) for all the sampling sites in the six rivers and for all the hydraulic variables. (d) The comparison of the results of MEIFR among the new method, Tennant (10%), Tennant (20%), wetted perimeter method.

The key of the method is the data of water depth and water width, which are relatively easy to take measurements to compare with discharge. The distinguishing feature of the method is that it is not dependent on discharge series data.

The prerequisite of this method is that the cross-section is triangular. Basically it is suitable in headwater areas where triangular cross sections are very common (Liu *et al.*, 2006). Because the headwater area is also the area where data are often in short supply because of poor conditions for taking measurements, a new method with marginal observed data is needed, and is especially meaningful for these regions.

STUDY REGION AND DATA

The Western Route South-to-North Water Transfer project in China (WRSNWT) project is located in one of the few regions in the world where the primitive habitat has been undisturbed, which is a unique home for some very rare fish, e.g. *Hush bleekeri* Kimura and *Schizothorax (Racoma) prenanti* (Tchang). For a detailed introduction of the project please refer to Liu *et al.* (2006).

In order to test the models, the water depth and channel width data were measured by the authors along six rivers involved in the project from July to August 2004 by simple leather scales on a local bridge or near a tributary. Combining with the survey data by the Yellow River Survey Agency in previous years, there are 38 samplings in total, the location of which is shown in Fig. 1. The roughness and hydraulic slopes data are obtained from the hydraulic handbooks and local experience references.

RESULTS

MEIFR derived by the new method

Based on the observed data of width and depth, the angles are calculated as shown in Fig. 2(a),(b). The angles over six rivers are in the same order. MEIFR for each of the hydraulic variables over 38 samplings is calculated based on equation (4) and Table 1, as shown in Fig. 2(c). The results of MEIFR corresponding to the hydraulic variables A, V, R and B are very small, all being very close to zero. The results of MEIFR corresponding to the hydraulic variable P and h are very close to each other. Thus there are only two result points for each sampling site on Fig. 2(c). As it is hard to accept that MEIFR is almost zero, only the MEIFR corresponding to P and h is finally adopted. After making an average over the MEIFR results corresponding to P and h, which is then averaged over all the sites (shown in Fig. 1) along each river, we get the MEIFR for each river, as shown in Fig. 2(d).

The comparison of the results between the new method, the Tennant method and the wetted perimeter method

As there are no observed MEIFR data available to check the new method, the comparison of the results among different methods has to be used to test the new method.

The first method for comparison is the Tennant method. Within the study area, the only stations with time series data of discharge are at Daofu, Zhuba, Cuosijia and Zumuzu stations. By using analogy with catchment area, the time series data of discharge at the planning dam sites for six rivers can be obtained. By taking the ratio of 10% of average annual discharge and 20% of average discharge as a standard, the results of MEIFR determined by the Tennant method are shown in Fig. 2(d).

Another method for the comparison is the wetted perimeter method, which needs discharge data matched with cross-section data. Within the study area, the only stations with such data are Daofu, Zhuba and Zumuzu stations from 1965 to 1987. By fitting data of wetted perimeter and discharge into a logarithmic relationship, it is easy to get the transition point, and hence MEIFR, for these stations by the curvature method. Still using the analogy with catchment area, the result of MEIFR from the wetted perimeter method for the six rivers is shown in Fig. 2(d).

Obviously, the results of the new method are close to the results by the Tennant method (10%) and wetted perimeter method. The results of the Tennant method (20%) are larger than all other results.

CONCLUSIONS

A new method is proposed to determine MEIFR for ungauged basins. There are two motives to developing the new method. One is from water managers' view to provide a simple method to obtain MEIFR for ungauged basins. Another is driven by the innovation to try to establish a theoretical platform for hydraulic rating methods.

The theory this new method is based on is similar to that of the wetted perimeter method. However, under the purely analytical frame, it seeks MEIFR analytically, not via data fitting. This feature provides a great possibility to apply this method in ungauged basins.

Acknowledgements The Natural Sciences Foundation of China (40671032, 40671033) and the Lt. Co. of Yellow River Survey are greatly thankful. Profs Changming Liu and Jun Xia, Mr Xiaobo Xuan, Hongan Zhang, Zhiping Shi and Baohui Men are specially appreciated for their cooperation in the project.

REFERENCES

Acreman, M. (2005) Linking science and decision-making: features and experience from environmental river flow setting. *Environ. Modelling & Software* **20**, 99–109.

Annear, T. C. & Conder, A. L. (1984) Relative bias of several fisheries instream flow methods. *N. Am. J. Fish. Mgmt*, **4**, 531–539.

Arthington, A. H., Bunn, S. E., Pusey, B. J., Blodorn, D. R., King, J. M., Day, J. A., Tharme, R. E. & O'Keeffe. J. H. (1992) Development of an holistic approach for assessing environmental flow requirements of riverine ecosystems. In: *Proceedings of an International Seminar and Workshop on Water Allocation for the Environment* (ed. by J. J. Pigram, & B. P. Hooper) (November 1991). The Centre for Water Policy Research, University of New England, Armidale, Australia.

Gippel, G. J. & Stewardson, M. J. (1998) Use of wetted perimeter in defining minimum environmental flows. *Regulated Rivers: Res. Manage.* **14**(1), 53–67.

Karim, K., Gubbels, M. E. & Goulter, I. C. (1995) Review of determination of instream flow requirements with special application to Australia. *Water Resour. Bull.* **31**(6), 1063–1077.

King, J. M. & Tharme, R. E. (1994) Assessment of the Instream Flow Incremental methodology and initial development of alternative Instream Flow methodologies for South Africa. Water Research Commission Report no. 295/1/94.

Liu, S., Mo, X., Zhu, Y. & Huang, H. (2004) A new method for ecological water consumption calculation and application in Haihe River basin. *J. Nat. Resour.* **19**, 662–671 (in Chinese).

Liu, S., Mo, X., Xia, J., Liu, C., Lin, Z., Men, B. & Ji, L. (2006) Estimating the minimum in-stream flow requirements via wetted perimeter method based on curvature and slope techniques. *J. Geogr. Sci.* **16**(2), 242–250.

Milhous, R. T., Updike, M. A. & Schneider, D. M. (1989) Physical habitat simulation system reference manual-version II, Instream Flow Information Paper no. 26. US Fish and Wildlife Service, Fort Collins, USA.

Sivapalan, M., Takeuchi, K., Franks, S. W., Gupta, V. K., Karambiri, H., Lakshmi, V., Liang, X., McDonnell, J. J., Mendiondo, E. M., O'Connell, P. E., Oki, T., Pomeroy, J. W., Schertzer, D., Uhlenbrook, S. & Zehe, E. (2003) IAHS decade on Predictions in Ungauged Basins (PUB), 2003–2012: shaping an exciting future for the hydrological sciences. *Hydrol. Sci. J.* **48**(6), 857–880.

Stalnaker, C. B., Bovee, K. D. & Walddle, T. J. (1995) Importance of the temporal aspects of habitat hydraulics to fish populations studies. *Regul. River: Res. Manage.* **12**, 145–153.

Tennant, D. L. (1976) Instream flow regimens for fish, wildlife, recreation and related environmental resources. *Fisheries* **4**, 6–10.

Wadeson, R. A. & Rowntree, K. M. (1998) Application of the hydraulic biotope concept to the classification of instream habitats. *Aquatic Ecosystem Health and Management* **1**, 143–157.

Preliminary study on estuarine ecosystem health evaluation in Zhangweixin River

ZHI-HE CHEN, XIAO-HONG CHEN, TAO PENG & KAI-RONG LIN
Department of Water Resources and Environment, Sun Yat-sen University, Guangzhou 510275, China
chzhhe@mail.sysu.edu.cn

Abstract Evaluation study on the estuarine ecosystem is the foundation and basis for ecological studies in the estuary area. In this paper an index system for estuarine eco-health evaluation of Zhangweixin River is established, with 15 indexes in three major types of indicators: such as environmental, ecological and socio-economic, by using the analytic hierarchy process (AHP). Multi-criterion analysis is used to evaluate the estuarine eco-health of Zhanweixin River. The results indicated that the period of river flow cut-off and phytoplankton diversity were important indexes for evaluating estuarine eco-health of Zhangweixin River. The main cause of the "unhealthy" status of the estuarine ecosystem of Zhangweixin River is related to the significant reduction of water flowing into the sea. As the basis of study on estuarine eco-environmental water demand, evaluation results provide basic support for the research on improving the estuarine eco-environment by utilizing flood resources.

Key words estuarine ecosystem; ecosystem health; analytic hierarchy process (AHP); Zhangweixin River

INTRODUCTION

The estuary link connecting river and marine ecosystems with each other is an important channel for them to exchange and transport substance and energy. Interaction between the sea and land enables the estuarine ecosystem to possess unique environmental characteristics and important functions of ecological service, as well as becoming a relatively fragile ecosystem. The estuarine ecosystem is mainly influenced by hydrology and environment from the basin, as well as social development and change. For example, in Haihe basin in recent years, a remarkable increase of off-channel water demand, massive discharge of channel water division, disorderly emission of domestic sewage and industrial wastewater, etc. have caused the estuarine ecosystem to degenerate rapidly, and a large quantity of species have disappeared or are endangered (Zhang *et al.*, 2007). There are numerous factors affecting estuarine eco-health, which can only be recovered by finding suitable and specific ways to solve problems after searching out the main cause of deterioration of estuarine eco-health in combination with control and restoration measures. This paper will provide basic support for scientific management, rational allocation and efficient utilization by establishing an evaluation system for estuarine eco-health, seeking out the main indexes of estuarine eco-health, combining the previous research to analysis and description. The evaluation results as the prophase research for the research on water demand of estuarine eco-environment in flood season.

STUDY REGION

Zhangweixin River is located in the common border of Shangdong and Hebei Provinces, with overall channel length of 246.6 km and controlled catchment area of 37 040 km^2. From the Sinusi Water Control Project, Dezhou, Shangdong Province, downward it flows through Shangdong and Hebei Provinces, and flows into Bohai Bay from Dakou River. There are seven sluice gates built along the river, with storage capacity of 0.1 billion m^3, playing an important role in allocating industrial and agricultural water. The Xinji Sluice Gate situated in the lowest downstream, which is 37.6 km from the estuary, is a tide resistance sluice. The flood control standard for this reach is the 50-year flood, design flood discharge is 3500 m^3/s and design drainage discharge at the Xinji Sluice below is 1250 m^3/s.

The estuary of Zhangweixin River is located between northern latitude 38°03′57″–38°20′24″ and eastern longitude 117°35′02″–117°56′59″. Up from the Xinji Sluice Gate and down to the

Dakou River (including partial extension area of the shallow sea), the total estuary area is 131.02 km^2, where the land area is 85.97 km^2 and the sea area is 45.05 km^2. The estuary location of Zhangweixin River is shown in Fig. 1.

The runoff of Zhangweixin River is decreasing by the change of natural and human activities, while in a dry year there is scarcely any runoff discharge. Annual change of runoff shows great disparity. As shown in Fig. 2, from 1973 to 2008 the total runoff discharge of Zhangweixin River was 7.3662 billion m^3 and average annual runoff discharge was 0.2046 billion m^3, where peak of runoff discharge were 1.97 and 1.49 billion m^3 in 1977 and 1996, respectively, and there was zero discharge in 15 years. There has been a larger amount of runoff discharge in four years since 2003, due to abundant inflow in the upstream Wei River.

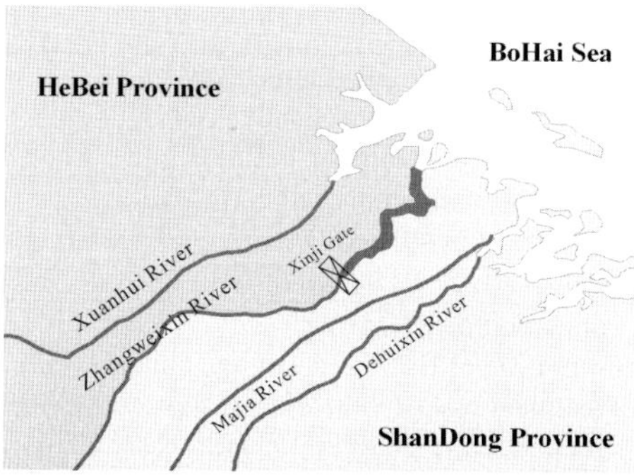

Fig. 1 Estuary location of Zhangweixin River.

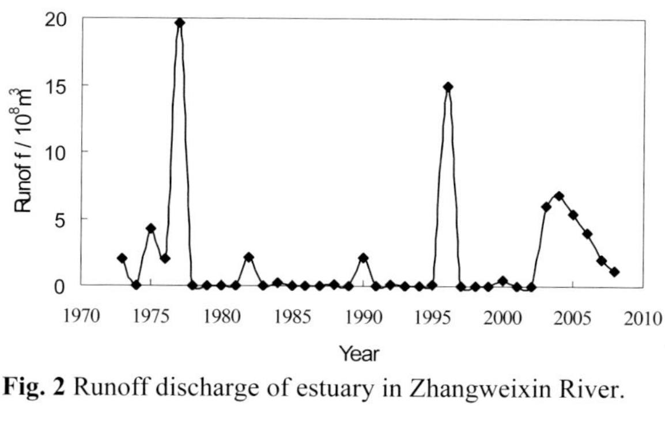

Fig. 2 Runoff discharge of estuary in Zhangweixin River.

EVALUATION METHOD FOR ESTUARINE ECOSYSTEM HEALTH

On the basis of analysing the scientific connotation of estuarine ecosystem health, an index system for evaluation of estuarine ecosystem health is established with analytical hierarchy process (AHP) (Vaidyaa, 2006) and a set-pair analysis model for evaluation of estuarine ecosystem health is proposed, providing a scientific basis for the recovery of the estuarine ecosystem and calculation of estuarine ecological water demand during the flood season.

Establishment of evaluation index system

Considering the principle of science, integrity, hierarchy and operability, an evaluation index system for the hierarchy of estuarine ecosystem health is established after analysing and selecting, as shown in Table 1. Evaluation index system divided into three layers: the first layer is target layer (A), which is estuarine ecosystem health; the second layer is criterion layer (B), in which there are all elements for realizing the general goal, including environmental, ecological and socio-economic elements; the third layer is index layer (C), in which there are all indexes for reaching target and criterion layers, 15 indexes in total.

Table 1 Evaluation Index System for estuarine ecosystem health.

Target Layer (A)	Criterion layer (B)	Index layer (C)
Estuarine ecosystem health	Environmental elements (*B*1)	Dissolved oxygen (*C*11)
		Chemical oxygen demand (*C*12)
		Eutrophication index (*C*13)
		Proportion of average annual runoff (*C*14)
		Proportion of river flow cutoff in the whole year (*C*15)
	Ecological elements (*B*2)	Biomass of phytoplankton (*C*21)
		Phytoplankton diversity index (*C*22)
		Biomass of zooplankton (*C*23)
		Zooplankton diversity index (*C*24)
		Biomass of benthos (*C*25)
		Benthos diversity index (*C*26)
	Socio-economic elements (*B*3)	Population density (*C*31)
		Proportion of fishing in average annual (*C*32)
		Treatment rate of industrial wastewater (*C*33)
		Treatment rate of domestic wastewater (*C*34)

Environmental indexes

Environment indexes directly reflect conditions of the estuarine abiotic environment. Dissolved oxygen and chemical oxygen demand indexes are used to signify water quality conditions at the estuary.

Runoff discharge reflects the magnitude of water circulation volume at the estuary. The scale of runoff discharge is related to the scale of estuarine ecosystem and the service function of the ecosystem. The proportion of average annual runoff can, to a certain extent, suggest the degeneration degree of the estuarine ecosystem.

The period of river flow cutoff is selected to signify vitality of the estuarine system. When the river flow cutoff period exceeds certain limits, the estuarine ecosystem will be seriously damaged, resulting in overall degeneration of the system function.

Ecological indexes

Ecological indexes reflect the conditions of biological components and function of the estuarine ecosystem. The biomasses of phytoplankton, zooplankton and benthos are the best indexes for production capacity of the estuarine ecosystem, most direct expression of the structure and function of estuarine ecosystem and comprehensive embodiment of environmental quality of the estuarine ecosystem.

Socio-economic indexes

Health conditions of the estuarine ecosystem are closely related to human activities. Population density, amount of fishing and treatment rate of wastewater which have a great influence on the estuarine ecosystem health are selected in evaluation.

Determination of index weight

Decompose the estuarine health into composition elements according to data information and evaluation basis, group these elements into orderly hierarchical structure by dominance relationship, quantify the judgment by pairwise comparison to form a judgment matrix, and ultimately resolve the system analysis into the determination of weight of relative importance of lowest layer (decision scheme) to the top layer (general goal) or the ranking problem of relatively optimal order, thereby providing a basis for decision scheme selection. Procedures of determination of the index weight by AHP include: establishment of judgment matrix, calculation of hierarchy single ranking, uniformity inspection and hierarchy general ranking.

Establishment of judgment matrix

To determine the relative importance of each element to the upper layer, establish judgment matrix according to the calibration method of 1–9 and their reciprocals proposed by Saaty (2003). Establish judgment matrix of relative importance for each layer according to the expert scoring results, as shown in Tables 2–5.

Table 2 Judgment matrix A–B and Eigenvalue.

A–B	A	B1	B2	B3	W_B
	B1	1	1	3	0.4286
	B2	1	1	3	0.4286
	B3	1/3	1/3	1	0.1429

λ_{max} = 3.00, CI = 0, RI = 0.58, CR = 0<0.1.

Table 3 Judgment matrix B1–C1 and Eigenvalue.

B1–C1	*B1*	*C11*	*C12*	*C13*	*C14*	*C15*	W_{C1}
	C11	1	1	1	1/3	1/5	0.0922
	C12	1	1	1	1/3	1/5	0.0922
	C13	1	1	1	1/3	1/5	0.0922
	C14	3	3	3	1	1	0.3072
	C15	5	5	5	1	1	0.4163

λ_{max} = 3.00, CI = 0.0079, RI = 1.12, CR = 0.0070<0.1.

Table 4 Judgment Matrix B2-C2 and Eigenvalue.

B2~C2	*B2*	*C21*	*C22*	*C23*	*C24*	*C25*	*C26*	W_{C2}
	C21	1	1	3	3	5	5	0.32397
	C22	1	1	3	3	5	5	0.32397
	C23	1/3	1/3	1	1	2	2	0.11494
	C24	1/3	1/3	1	1	2	2	0.11494
	C25	1/5	1/5	1/2	1/2	1	1	0.06109
	C26	1/5	1/5	1/2	1/2	1	1	0.06109

λ_{max} = 6.0074, CI = 0.0015, RI = 1.24, CR = 0.0012<0.1.

Table 5 Judgment Matrix B3-C3 and Eigenvalue.

B3~C3	*B3*	*C31*	*C32*	*C33*	*C34*	W_{C3}
	C31	1	1	1/7	1/7	0.0625
	C32	1	1	1/7	1/7	0.0625
	C33	7	7	1	1	0.4375
	C34	7	7	1	1	0.4375

λ_{max} = 4.00, CI = 0, RI = 0.94, CR = 0<0.1.

Hierarchy single ranking and uniformity inspection

Methods for calculating weight value in AHP are mainly power method, sum method, root method, minimum square weight method, etc. The most simple sum method is used in this study to calculate maximum eigenvalue and eigenvector.

Normalize the judgment matrix by column, and obtain matrix q_{ij}, where:

$$q_{ij} = \frac{\delta_{ij}}{\sum_{k=1}^{n} \delta_{ki}} \quad (i,\ j = 1, 2, \ldots, n) \tag{1}$$

Sum the matrix q_{ij} by row, and obtain vector $c = (c_1, c_2, \ldots, c_n)^T$, where:

$$c_i = \sum_{j=1}^{n} q_{ij} \quad (i = 1, 2, \ldots, n) \tag{2}$$

Normalize $c = (c_1, c_2, \ldots, c_n)^T$ and obtain eigenvector corresponding to the maximum eigenvalue:

$$w_j = \frac{c_j}{\sum_{k=1}^{n} c_k} \quad (j = 1, 2, \ldots, n) \tag{3}$$

Calculate maximum eigenvalue of the judgment matrix λ_{max}:

$$\lambda_{max} = \frac{c_j}{\sum_{i=1}^{n} c_k} \quad (j == 1, 2, \ldots, n) \tag{4}$$

Check the judgment matrix with random consistency ratio (CR), i.e.:

$$CR = \frac{CI}{RI} \tag{5}$$

where, CI is the consistency index of the judgment matrix and RI is the random consistency index. RI is available in Table 6. CI can be calculated with the following equation:

$$CI = \frac{\lambda_{\max} - n}{n - 1} \tag{6}$$

where λ_{max} is the maximum characteristic root of the judgment matrix and n is the order.

Table 6 Average Random Consistency Index (*RI*) (Li *et al.*, 2006).

n	3	4	5	6	7	8	9
RI	0.58	0.94	1.12	1.24	1.32	1.14	1.45

When CR is less than 0.1, consider the judgment matrix having satisfying consistency, or else adjust the judgment matrix until obtaining satisfying uniformity.

Hierarchy general ranking

According to the single sequencing for each layer, calculate the sequencing weight of relative importance of the bottom layer to the top layer. Check the random consistency ratio (CR) of hierarchy general ranking, with calculation formula as follows:

$$CR = \frac{\sum_{j=1}^{n} w_j CI_j}{\sum_{j=1}^{n} w_j RI_j} \tag{7}$$

where w_j is the ranking coefficient for n elements at A layer, CI_j is the consistency index of single ranking for some elements at B layer to a certain element A_j at A layer; RI_j is the corresponding average random consistency index.

Evaluation model for estuarine ecosystem

During evaluating the conditions of estuarine ecosystem health, in addition to the establishment of evaluation index system and determination of weight, another important work is to establish evaluation standard system, which is the foundation of estuarine ecosystem health evaluation. Determine the evaluation standard by field survey, using national standard and relevant research results for reference, expert consultation, public participation and other methods (Leppard & Munawar, 1992; Ladson & White, 1999; Baina *et al*, 2000; MEPA of PRC, 2002), as shown in Table 7, which can be divided into very healthy, healthy, sub-healthy, unhealthy and morbid according to the health level.

Table 7 Evaluation grade for estuarine ecosystem health.

Evaluation Index	Very healthy	Healthy	Sub-healthy	Unhealthy	Morbid
*C*11 (mg/L)	>6	6–5	5–3	3–2	<2
*C*12 (mg/L)	<15	15–20	20–30	30–40	>40
*C*13 (mg/L)	<0.15	0.15–1	1.5–1	1.5–2	>2
*C*14 (%)	>90	90–70	70–40	40–10	<10
*C*15 (%)	<10	10–30	30–60	60–90	>90
C21 (mg/m^3)	>3	3–2	2–1	1–0.5	<0.5
C22	>3	3–2	2–1.5	1.5–1	<1
*C*23 (mg/m^3)	>200	200–150	150–100	100–50	<50
C24	>4	4–3	3–2	2–1	<1
*C*25 (mg/m^2)	>20	20–15	15–10	10–5	<5
*C*26	>3	3–2	2–1.5	1.5–1	<1
*C*31 (person/km^2)	<100	100–300	300–500	500–700	>700
*C*32 (dollar)	>8000	8000–3000	3000–1000	1000–500	<500
*C*33 (%)	>95	95–80	80–60	60–40	<40
*C*34 (%)	>95	95–80	80–60	60–40	<40

The types of indexes selected are complicated, with different units and dimensions, including positive indexes (the larger the index is, the higher the health level will be) and negative indexes (the smaller the index is, the higher the health level will be). Therefore, all indexes lack comparability and they need to be standardized prior to comprehensive evaluation. In order to be simple, definite and easy to calculate, give each level the standardized score values which are set between 0 and 1, see Table 8 for details.

Table 8 Quantitative standard for estuarine ecosystem indexes.

Evaluation grade	Very healthy	Healthy	Sub-healthy	Unhealthy	Morbid
Standardized Score Value	>0.8	0.6–0.8	0.4–0.6	0.2–0.4	<0.2

There are many evaluation methods aimed at multiple levels, multiple indexes and large-scale systems, mainly multi-criterion analytical method, fuzzy decision analysis method, support vector machine (SVM), etc. Point count model in multi-criterion analytical method is adopted to evaluate the estuarine ecosystem.

The multi-criterion analytical method is the one commonly used in system evaluation. In this method, qualitative index is quantified by grading scoring to enable the evaluation results of different evaluation units to be comparable. Point count model evaluation is expressed with the following formula:

$$E = \sum_{i=1}^{n} W_i P_i \tag{8}$$

where, E is comprehensive index for estuarine ecosystem health, W_i is the weight of index i, P_i is the standardized value of index i.

RESULTS AND DISCUSSION

Indexes for estuarine eco-health

Tables 2–5 refer to the calculation for single ranking for layer elements, uniformity inspection and results of hierarchy general ranking, which has shown that index layer to goal layer is meeting the requirement of uniformity inspection. The *CR* value calculation is less than 0.1, showing that the general ranking has satisfactory uniformity. Table 9 gives the calculation results for the general ranking weight of index layer. Of these 15 indexes in the index layer, the four indexes: proportion of average annual runoff (*C*14), proportion of river flow cutoff in the whole year (*C*15), biomass of phytoplankton (*C*21) and phytoplankton diversity index (*C*22) are larger than 0.1, where the maximum value is 0.1784 in *C*15, indicating that runoff discharge of Zhangweixin River is an important index affecting the estuarine eco-health. The runoff of Zhangweixin River reduced to 0.04 billion m^3 and 0.17 billion m^3, respectively, in the 1980s and 1990s from 0.87 billion m^3 in the 1950s. At present the runoff discharge is mainly sewage and flood during the flood season (Zheng *et al*, 2005). According to the distribution of runoff discharge within the year, the 80.9% is concentrated in flood season and only 19.1% is distributed in the non-flood season (Zhang *et al.*, 2007). Since the 1980s, due to a shortage of water resources, there has been basically no freshwater flowing into the sea from April to June, causing inshore salinity increase, which is not

Table 9 Calculation for hierarchy general ranking.

Upper Layer	*B*1	*B*2	*B*3	General ranking weight for layer C
Lower Layer	0.4286	0.4286	0.1429	
*C*11	0.0922	0	0	0.0395
*C*12	0.0922	0	0	0.0395
*C*13	0.0922	0	0	0.0395
*C*14	0.3072	0	0	0.1317
*C*15	0.4163	0	0	0.1784
*C*21	0	0.32397	0	0.1389
*C*22	0	0.32397	0	0.1389
*C*23	0	0.11494	0	0.0493
*C*24	0	0.11494	0	0.0493
*C*25	0	0.06109	0	0.0262
*C*26	0	0.06109	0	0.0262
*C*31	0	0	0.0491	0.0070
*C*32	0	0	0.2052	0.0293
*C*33	0	0	0.3728	0.0533
*C*34	0	0	0.3728	0.0533

suitable for fish to spawn and hatch, and for larva to grow, while large amounts of nutrients brought by freshwater flowing into the sea cause the frequent occurrence of red tide (Zhang *et al.*, 2007). The estuarine biodiversity of the Zhangweixin River is reducing gradually, mostly affecting hydrophytes and phytoplankton. The calculation results of general ranking weight of index layer also indicate that the biomass of phytoplankton (*C*21) and phytoplankton diversity index (*C*22) are important indexes for estuarine eco-health of Zhangweixin River.

Estuarine eco-health evaluation

Evaluation results show that comprehensive index (*E*) of Zhangweixin River is 0.3410, meaning an unhealthy condition, shown in Table 10, where the environment health index is 0.1438/0.4286 = 0.3355, indicating that conditions of abiotic environment of estuarine ecosystem in Zhangweixin River are poor. Among the evaluation indexes, dissolved oxygen (*C*11), chemical oxygen demand (*C*12) and eutrophication index (*C*13) are less than 0.2, presenting a prominent issue of water quality at the estuary of Zhangweixin River. Liu *et al.* (2007) have compared the pollutant content of water and mud samples from Haihe estuary and indicated that heavy metal pollution in the water is mainly from industrial wastewater; nitrogen and phosphorous are mainly from domestic sewage and agricultural fertilization. Ecological health index is 0.1450/0.4286 = 0.3382, signifying that the structure and function of biological components in estuarine ecosystem of Zhangweixin River are unreasonable and biological productivity level is low. According to the recent survey data, freshwater and semi-seawater fishes in the estuarine and offshore area of Bohai Bay have shown signs of disappearance, the migratory fish – Japanese eel and *Acipenser dabryanus* are basically extinct, the distribution of some important economic fish such as sea bream and hairtail are substantially reduced and they have become rare species (Zheng *et al.*, 2005). Socio-economic heath index is 0.0522/0.1429 = 0.3653, signifying pressure of socio-economic activities of humans that the estuarine ecosystem of Zhangweixin River is undertaking is comparatively large.

Table 10 Comprehensive indexes for estuarine ecosystem.

Evaluation index	Status value of index	Standardized value (P_i))	Weight of index (W_i))	Single index
*C*11	1.2	0.120	0.0395	0.0047
*C*12	159	0.006	0.0395	0.0002
*C*13	12.7	0.120	0.0395	0.0047
*C*14	43.39	0.2226	0.1317	0.0293
*C*15	58.08	0.5872	0.1784	0.1048
*C*21	0.79	0.316	0.1389	0.0439
*C*22	1.5	0.4	0.1389	0.0555
*C*23	105	0.42	0.0493	0.0207
*C*24	1.3	0.26	0.0493	0.0128
*C*25	6	0.24	0.0262	0.0063
*C*26	1.05	0.22	0.0262	0.0058
*C*31	224	0.7235	0.0070	0.0051
*C*32	7738	0.7895	0.0293	0.0232
*C*33	50	0.3	0.0533	0.0160
*C*34	30	0.15	0.0533	0.0080
Comprehensive index for estuarine ecosystem health (*E*)	0.3410 (unhealthy)			

CONCLUSION

Estuarine eco-health is most sensitive to the change of natural and human activities, making the estuary a relatively fragile ecosystem. These 15 indexes in environmental, ecological, socio-economical and other influencing factors are adopted in this study to evaluate estuarine eco-health of Zhangweixin River. The results indicated that reduction of runoff discharge is a main factor of rapid deterioration of estuarine ecology in Zhangweixin River. At present, a flood control engineering system has formed in the Zhangweixin River basin, playing roles of flood control and disaster reduction. In addition, it also provides favourable engineering conditions conducive to utilizing flood resources. Therefore, developing and utilizing flood resources in the basin can be considered to improve the estuarine eco-health. This paper is the preliminary study on estuarine eco-environmental water demand. Estuarine ecological water demand and utilization of flood resources will be discussed in further investigations.

Acknowledgements This study was supported by the Key Project of the Natural Science Foundation of China (no. 50839005, no. 50809078 and no. 50909107) and National Key Technology Support Program (no. 2006BAB14B07).

REFERENCES

Baina, M. B., Harigb, A. L. & Loucksc, D. P. (2000) Aquatic ecosystem protection and restoration: advances in methods for assessment and evaluation. *Environ. Sci. & Policy* **3**, 89–98.

Costaa, C. B. & Vansnick, J. C. (2008) A critical analysis of the eigenvalue method used to derive priorities in AHP. *European J. Operational Res.* **187**(3), 1422–1428.

Ladson, A. R. & White, L. J. (1999) *An Index of Stream Condition: Reference Manual* (second edition). Department of Natural Resources and Environment, Melbourne, Australia.

Leppard, G. G. & Munawar, M. (1992) The ultrastructural indicators of aquatic ecosystem health. *J. Aquatic Ecosystem Health* **1**(4), 309–317.

Li, S., Qiu, W. & Zhao, Q. L. (2006) Applying analytical hierarchy process to assess eco-environment quality of Heilongjiang Province. *Environ. Sci.* **27**(5), 1031–1034.

Liu, C., Wang, Z. Y. & Huang, W. D. (2007) Water and sediment pollution at river mouths of Haihe River basin. *J. Hydraul. Engng* **38**(8), 920–925.

MEPA of PRC (2002) *Environmental Quality Standards for Surface Water (GB3838 -2002)*. Environmental Science Press, Beijing, China.

Saaty, T. (2003) Decision-making with the AHP: why is the principal eigenvector necessary. *European J. Operational Res.* **145**(1), 85–91.

Vaidyaa, O. & Kumar, S. (2006) Analytic hierarchy process: an overview of applications. *European J. Operational Res.* **169**(1), 1–29.

Zhang, Z. Y., Sun, Y. C. & Han, J. (2007) Feasibility analysis of ecological restoration in Haihe River Estuary. *Haihe Water Resour.* **8**, 14–18.

Zheng, J. P., Wang, F. & Hua, Z. L. (2005) Research on ecological water requirement of Haihe River Estuary. *J. Hohai Univ. (Natural Sciences)* **33**(5), 518–521.

Analysis of climate change impact on precipitation in Danjiangkou reservoir basin by using statistical downscaling method

HUA CHEN[1], JING GUO[1], SHENG-LIAN GUO[1], C-Y XU[2] & VIJAY P. SINGH[3]

1 *State Key Laboratory of Water Resources and Hydropower Engineering Science, Wuhan University 430072, China* chua@whu.edu.cn

2 *Department of Geosciences, University of Oslo, NO-0316 Oslo, Norway*

3 *Department of Biological and Agricultural Engineering, Texas A & M University, Scoates Hall, 2117 TAMU, College Station, Texas 77843-2117, USA*

Abstract General circulation models (GCMs) are often used to assess the impact of climate change. As they operate on coarse scales, the simulation results obtained from GCMs are not particularly in accordance with comparatively smaller river basin scales. A statistical downscaling method based on a least square support vector machine (LS-SVM) is proposed for downscaling daily total precipitation series from GCMs outputs for the Danjiangkou reservoir basin in China. The downscaling method is developed and validated using large-scale predictor variables derived from the National Center for Environmental Prediction – National Center for Atmospheric Research (NCEP/NCAR) reanalysis data set. The performance of the LS-SVM downscaling method is also compared to a well-known statistical downscaling model (SDSM). The downscaling results suggest that the LS-SVM is an efficient method for downscaling daily precipitation series in the study region. The statistical relationship obtained is used to project future precipitation from CGCM2 for IPCC SRES A2 and B2 scenarios. The downscaling results, corresponding to both scenarios, show that there is an increase in the average annual precipitation downscaled from CGCM2 by about 277.51 mm for the IPCC SRES A2 scenario and by about 157.65 mm for the IPCC SRES B2 scenario, by the 2080s.

Key words climate change; statistical downscaling; least square support vector machine (LS-SVM); Danjiangkou reservoir basin; SDSM

INTRODUCTION

New problems, arising from impact of climate changes in ungauged basins, reveal that past hydrological records may no longer be adequate to make predictions and forecasts about the future. It is desirable to know the necessary information for reliable prediction under climate change. To cope with this, it is necessary to combine downscaled General Circulation Models (GCMs) climate change scenarios with PUB to assess the potential long-term change of the water resources in ungauged basins. GCMs, which describe the atmospheric process by mathematical equations, are one of the most important tools in the study of climate variability and climate change impact on hydrology. Although GCMs are able to simulate reliably the most important mean features of the global climate at large scales, they perform poorly at smaller spatial and temporal scales relevant to the regional impact analyses (Grotch & MacCracken, 1991; Wilby *et al.*, 2007). The main reason is that the spatial resolution of GCM grids is too coarse to resolve many important sub-grid scale processes and GCM output is often unreliable at individual grid and sub-grid scales (Wilby *et al.*, 1999; Xu, 1999).

In recent years numerous statistical downscaling methods have been applied to deal with this problem of mismatch of spatial and temporal scales between GCMs and watershed hydrological models. Statistical downscaling concerns how to draw empirical relationships that transform large-scale features of the GCM (Predictors) to regional-scale variables (Predictands), such as precipitation and temperature (Tripathi *et al.*, 2006; Fowler *et al.*, 2007). A diverse range of empirical downscaling techniques have been developed over the past two decades and each method generally lies in one of the three major categories, including weather pattern schemes (Bárdossy *et al.*, 2005), weather generators (WGs) (Kilsby *et al.*, 2007) and regression models (Fowler *et al.*, 2007). Among these methods, regression models which are also termed transfer functions may be the most popular approaches and are directly used to quantify a relationship between the predictand and a set of predictors, such as multiple regression models (MLP) (Wilby *et al.*, 1999), artificial neural networks (ANNs) (Zorita & von Storch, 1999), canonical correlation analysis (CCA) (Busuioc *et al.*, 2001), and singular value decomposition (SVD) (Huth, 1999). One of the well-recognized statistical downscaling tools that implements a regression and stochastic method is the Statistical Downscaling Model (SDSM; Wilby *et al.*, 2002). SDSM is used in this study as a benchmark model because it appears to be one of the most widely used models for precipitation and temperature downscaling.

Recently, support vector machines (SVM), which are novel machine learning algorithms proposed by Vapnik (1995, 1998) have been widely applied in the fields of classification and regression analysis (Tripathi *et al.*, 2006; Ghosh & Mujumdar, 2007; Yu & Liong, 2007). Although SVM has extensive applications in various fields, it has some drawbacks in dealing with large sample data, such as slow training speed, low implementation efficiency, and inadaptability to noise and outliers. As an improved algorithm based on SVM, the least square support vector machine (LS-SVM) developed by Suykens & Vandewalle (1999), which is basically a ridge-type learning machine, has received considerable attention and has been verified to deal efficiently with large sample data sets. In this study, LS-SVM was applied to establish a statistical relationship between large-scale GCM simulations and precipitation in the Danjiangkou reservoir basin in China.

The paper is organized as follows: after the introduction, a description of the study area and the data used in this study are provided; the methods are then briefly introduced, including the LS-SVM and SDSM; in the next section the comparative results are discussed and analysed from the two downscaling methods and the impact of climate change on precipitation is assessed; finally, some concluding remarks are made in the last section.

STUDY AREA AND DATA

The study area selected in this study for the application of downscaling models is the Danjiangkou reservoir basin, which is the source of water for the middle route of the well-known South-to-North Water Diversion Project (SNWDP) in China (Fig. 1; Chen *et al.*, 2007). The basin lies in a subtropical monsoon zone and its annual precipitation is about 700–1000 mm. Daily precipitation was selected as predictand in this study from 1961 to 2000 and its data were provided by the National Climatic Centre of China for eight National Meteorological Observatory (NMO) stations within the Danjiangkou reservoir basin as showed in Fig. 1. The altitudes and longitudes of the stations are listed in Table 1. One of the most important steps in downscaling is to select appropriate predictors or characteristics from GCMs, and there are some main factors to constrain the choice of predictors proposed (Wilby *et al.*, 1999). Precipitation is a consequence of mean sea level pressure (MSLP), geopotential height (GH), and specific humidity (SH) (Wilby *et al.*, 1999; Wetterhall *et al.*, 2005; Ghosh & Mujumdar, 2007). Therefore, MSLP, surface air temperature (2 m), 500 hPa GH and SH, 850h Pa GH and SH are considered as the predictors for predicting precipitation in the present study. The downscaling methods are developed and validated using large-scale NCEP/NCAR reanalysis predictor variables which are available free on the internet (http://dss.ucar.edu/pub/reanalysis/). Climate variables corresponding to the future climate change scenario for the study area are extracted from the CGCM2. The future climate change considered in this study corresponds to the so-called business-as-usual scenario. Thus the CGCM2 output used for this study is the result of the IPCC SRES A2 and B2 scenarios. The A2 scenario envisions population growth to 15 billion by the year 2100 and rather slow economic and technological development, and the B2 scenario envisions slower population growth (10.4 billion by 2100) with a more rapidly evolving economy and more emphasis on environmental protection (IPCC, 2001).

As the Danjiangkou reservoir basin has distinct seasonal variation of the atmospheric circulation and precipitation as the result of the subtropical monsoon circulation, the downscaling models will be calibrated in four seasons: winter (December, January, and February), spring (March, April, and May), summer (June, July, and August) and autumn (September, October, and

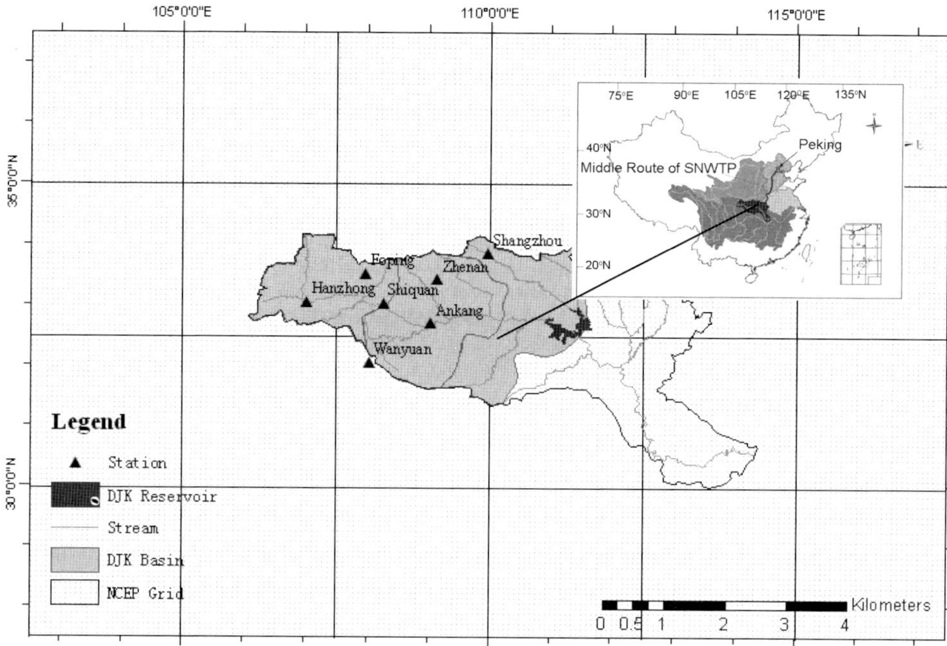

Fig. 1 Location of the Danjiangkou reservoir basin in China and the meteorological stations used in the study.

Table 1 The latitude and longitude of the meteorological stations in the Danjiangkou reservoir basin.

No.	Station name	Latitude(°N)	Longitude(°E)	Elevation(m a.s.l)
1	Hanzhong	33.04	107.02	508.4
2	Foping	33.32	107.59	1087.7
3	Shangzhou	33.52	109.58	742.2
4	Xixia	33.18	111.30	250.3
5	Wanyuan	32.06	108.03	129.2
6	Shiquan	33.03	108.16	484.9
7	Ankang	32.43	109.02	290.8
8	Yunxian	32.51	110.49	201.9

November), respectively. In order to evaluate the performance of each method, the entire data set was split into two parts, the training set, which is taken as 1961–1990, and the testing set as the remaining 10 years. For assessing the climate change impact on precipitation in the basin, the data series are divided into four distinct periods, namely, the current (1961–2000), the 2020s (2010–2039), the 2050s (2040–2069), and the 2080s (2070–2099) to facilitate trend analysis. Prior to downscaling the observations, NCEP/NCAR reanalysis and CGCM2 outputs data are standardized to reduce systematic biases in mean and variances. The geographical extent, 102.5°E–115°E, 27.5°N–37.5°N, is chosen to include all areas with noticeable influence on the circulation patterns that govern weather in the Danjiangkou reservoir basin. Figure 1 shows the NCEP grid points (2.5° latitude × 2.5° longitude) superposed on the map of Danjiangkou reservoir basin. CGCM2 grids (3.75° latitude × 3.75° longitude) were spatially interpolated into the NCEP grids by using the inverse distance weighting method.

There are six predictor variables at 24 NCEP grid points with a dimensionality of 144 for statistical downscaling methods, which will be a dimension disaster for personal computers.

Principal Component Analysis (PCA) was used to reduce dimensions and compress data while retaining most of the information content of the original data set. It can be observed that the first eight Principal Components have represented 90.1% of the information content of the original predictors, and therefore they are used as input to the statistical downscaling methods in this study.

METHODS

The least square support vector machine

As an improved algorithm based on SVM, the least square support vector machine (LS-SVM) developed by Suykens & Vandewalle (1999) has received considerable attention. LS-SVM is a kind of SVM under a quadratic loss function. In LS-SVM, the non-sensitive loss function is replaced by a quadratic loss function and the inequality constraints are replaced by equality constraints. The standard framework for LS-SVM estimation is based on a primal-dual formulation.

Given a training data set $\{x_i, y_i\}_{i=1}^{N}$, the goal is to estimate a model of the form:

$$y = \omega^T \varphi(x) + b \tag{1}$$

where $x \in R^n$, $y \in R^n$, $\varphi(x)$: $R^n \rightarrow R^m$ is the mapping to a high dimensional (and possibly infinite dimensional) feature space and weight $w \in R^n$ and basis $b \in R$ are parameters to be determined. The following optimization problem is formulated:

$$\min_{\omega,b,e} \quad \frac{1}{2}\omega^T \omega + \frac{C}{2}\sum_{i=1}^{N} e_i^2 \tag{2}$$

$$s.t. \quad y_i = \omega^T \varphi(x_i) + b + e_i, i = 1, 2, \cdots, N$$

where C is the penalized parameter which controls the trade off between the fitting accuracy and generalized ability. With the application of Mercer's theorem on the kernel matrix Ω as $\Omega_{ij} = K(x_i, y_j) = \varphi(x_i)^T \varphi(y_j), (i, j = 1, 2, \cdots, N)$ it is not required to compute explicitly the nonlinear mapping $\varphi(\cdot)$ as this is done implicitly through the use of positive definite kernel functions K. There are several possibilities for the choice of a kernel function, including linear, polynomial, sigmoid, splines, and radial basis functions (RBF). An RBF kernel can map the training into a possibly infinite-dimensional space and is computationally simple. Moreover, RBF can effectively handle the nonlinear relationship between predictors and predictands. The Lagrangian was used to solve the optimization problem. The resulting LS-SVM model in the dual space becomes:

$$y(x) = \sum_{i=1}^{N} \alpha_i K(x, x_i) + b \tag{7}$$

where $\alpha_i \in R, (i = 1, 2, \cdots, N)$ are the Lagrange multipliers.

The ridge regression formulation is presented in the cost function, and its regularization parameter γ avoids ill conditioning due to possible multicollinearity among the m dimensions of φ. Usually the training of the LS-SVM model involves an optimal selection of the kernel parameter σ and γ, which can be done using cross-validation techniques. A robust cross-validation score function (De Brabanter *et al.*, 2002) is an effective method for dealing with outliers and non-Gaussian noise distribution on the data, which is very difficult for the traditional cross-validation criterion, and it is introduced to select the optimal learning parameter of LS-SVM.

Statistical downscaling model (SDSM)

The statistical downscaling model (SDSM) (Wilby *et al.*, 2002) is a hybrid between a multilinear regression method and a stochastic weather generator. The model, which has been applied in many catchments in the world (Wilby *et al.*, 2002; Coulibaly *et al.*, 2005; Wetterhall *et al.*, 2006) uses large-scale predictors to linearly condition local-scale weather generator parameters. SDSM 4.2 decision support tool was designed for assessing local climate change impacts using a robust statistical downscaling technique (Wilby & Dawson, 2007). It facilitates the rapid development of multiple, low-cost, single-site scenarios of daily surface weather variables under present and future climate forcing, and implements bias correction and variance inflation techniques to reduce the standard error of the estimate and to increase the amount of variance explained by the models to achieve the best possible downscaling performance.

RESULTS AND DISCUSSION

To assess the accuracy of the downscaling methods in producing precipitation input to hydrological models, comparison between observed and simulated mean daily precipitation and standard deviation (SD) is shown in Table 2. It can be observed that the two methods produce small differences in the mean daily precipitation values, as shown in Table 2. In the testing period LS-SVM performs better than SDSM to simulate mean daily precipitation, except in winter, and SDSM is better in terms of SD, except in summer. However, it is also found that SD of the generated series is consistently smaller than that of the observed series for both methods. The underestimation of the observed variance of precipitation has also been found in previous studies (Srikanthan & McMahon, 2001), which is a common problem to be improved. The reason may be that regression based statistical downscaling models often cannot explain the entire variance of the downscaled variable (Wilby *et al.*, 2004) and cannot mimic the extreme precipitation observed in the record. To check the precipitation dynamics simulated by the two methods, the monthly values, averaged over all stations in the Danjiangkou reservoir basin, are reasonably well captured by the two methods, as shown in Fig. 2. The long-term average intra-annual variation is well simulated (Fig. 3) except in autumn when simulated precipitation values are lower than observed values.

In the present study, LS-SVM method is proposed for statistical downscaling of daily precipitation from simulations of GCMs. LS-SVM is a reformulation of standard SVMs and is closely related to regularization networks and Gaussian processes but additionally emphasize and exploit primal-dual interpretations. Compared with SDSM, the proposed LS-SVM has been shown to be suitable for downscaling GCMs simulations to study the impact of climate change on hydrology in the region. The statistics of the mean and variance of precipitation are more often expected to be downscaled with sufficient accuracy when averaged over longer time periods in hydrologic modelling, because precipitation is inherently stochastic, strongly intermittent, and a nonlinear process (Deidda, 1999) that has a great impact on local climate and water balance. The presented downscaling results are typical examples that provide useful regional scale precipitation scenarios for hydrological modelling.

Table 2 Comparison of observed and downscaled precipitation statistics for the Danjiangkou reservoir basin (mm).

		Training					Testing				
		Obs.	LS–SVM		SDSM		Obs.	LS–SVM		SDSM	
			Values	Bias	Values	Bias		Values	Bias	Values	Bias
Spring	Mean	2.12	2.22	0.1	2.09	–0.03	1.78	1.88	0.1	2.02	0.24
	SD	4.7	3.23	–1.47	3.16	–1.54	3.94	2.79	–1.15	3.1	–0.84
Summer	Mean	4.52	4.58	0.06	4.91	0.39	4.47	4.37	–0.1	4.81	0.34
	SD	7.95	5.17	–2.78	4.52	–3.43	8.18	5.26	–2.92	4.49	–3.69
Autumn	Mean	2.84	2.84	0	2.83	1.54	2.14	2.07	–0.07	2.29	0.15
	SD	6.17	3.72	–2.45	4.38	–3.34	5.1	2.84	–2.26	4.03	–1.07
Winter	Mean	0.28	0.29	0.01	0.29	0.01	0.26	0.21	–0.05	0.24	–0.02
	SD	0.92	0.58	–0.34	0.51	–0.41	0.93	0.51	–0.42	0.49	–0.44

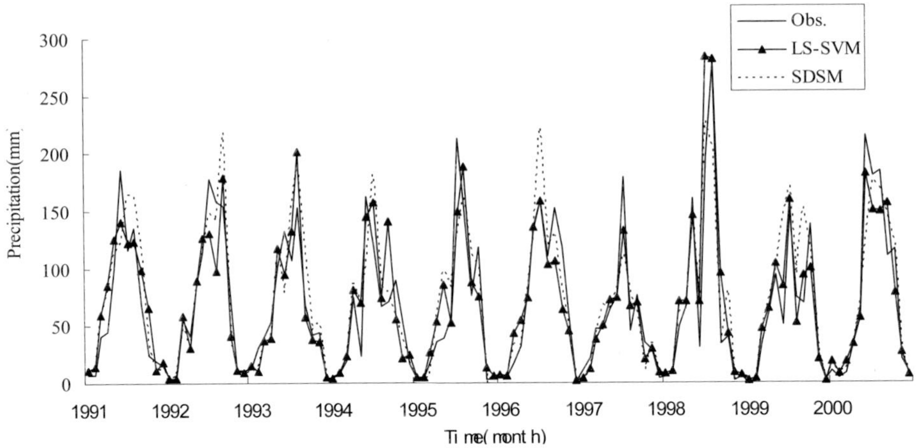

Fig. 2 Monthly mean observed and downscaled precipitation in the Danjiangkou reservoir basin during the testing period.

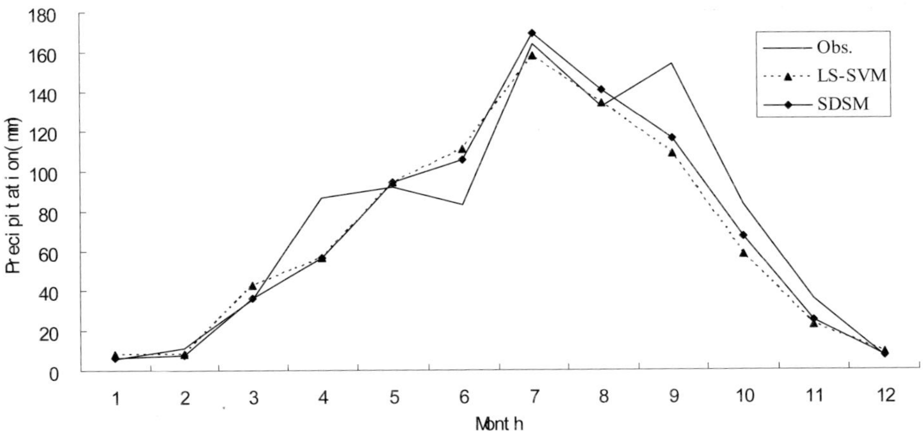

Fig. 3 Mean monthly observed and downscaled precipitation during 1991–2000 in the Danjiangkou reservoir basin.

The analysis in the above section has shown the LS-SVM method as a feasible potential alternative for climate impact studies in hydrology in the catchment. Therefore, the LS-SVM downscaling model is selected to downscale the CGCM2 output based on the IPCC SRES A2 and B2 scenarios to obtain future regional simulations of precipitation. Comparisons of observed precipitation and simulated precipitation from CGCM2 based on the IPCC SRES A2 and B2 scenarios downscaled by the LS-SVM (mm) are listed in Table 3. The results show the performance of the downscaling models on a seasonal basis. A significant increase in precipitation based on both GCMs scenarios can be observed in the 2050s and 2080s. A decrease in projected precipitation is found for the A2 scenario in all seasons and the B2 scenario except in the autumn in the 2020s.

The monthly statistics of actual observed values and the current and future CGCM2 simulations downscaled with LS-SVM are summarized and plotted in Fig. 4. This figure shows that the monthly mean values of observed precipitation are quite close to that of the LS-SVM downscaled from the NCEP/NCAR reanalysis data of the current time period (1961–2000). Figure 4(a) and (b) also show an increase in the mean daily precipitation for both scenarios between the current and the future time periods in May, June, September, and December, while a decrease in February, March, April, and October. There is no significant change for the future downscaled precipitation for both scenarios in January and November. The results show that there is a discordant for the future downscaled precipitation from both scenarios in some months of the year. However, the seasonal variation in the future downscaled precipitation from both scenarios is concordant in almost all seasons in different future periods.

Table 3 Comparison of observed and downscaled precipitation by LS-SVM from CGCM2 based on the IPCC SRES A2 and B2 scenario (mm).

				A2					
	Obs	Current Value	Bias	2020s Value	Bias	2050s Value	Bias	2080s Value	Bias
Winter	24.27	26.29	2.02	20.83	−3.44	27.04	2.77	30.40	6.13
Spring	186.33	193.61	7.28	171.57	−14.75	208.90	22.58	241.99	55.67
Summer	415.21	416.85	1.64	382.28	−32.93	419.42	4.21	514.39	99.18
Autumn	242.38	241.14	−1.24	240.58	−1.80	286.24	43.86	358.51	116.13
Annual	868.80	878.54	9.74	815.96	−52.84	942.51	73.71	1146.31	277.51
				B2					
	Obs	Current Value	Bias	2020s Value	Bias	2050s Value	Bias	2080s Value	Bias
Winter	24.27	26.29	2.02	21.88	−2.39	26.81	2.54	29.51	5.24
Spring	186.33	193.61	7.28	181.90	−4.43	203.04	16.71	241.15	54.82
Summer	415.21	416.85	1.64	384.93	−30.29	433.02	17.81	455.55	40.34
Autumn	242.38	241.14	−1.24	271.63	29.25	298.55	56.17	299.25	56.87
Annual	868.80	878.54	9.74	861.07	−7.73	962.32	93.52	1026.45	157.65

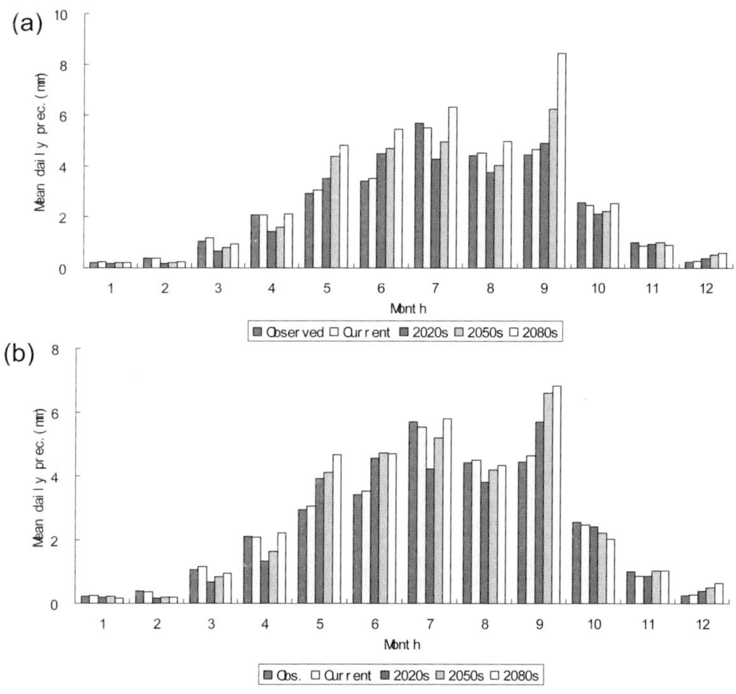

Fig. 4 Observed and downscaled precipitation with LS-SVM from the CGCM2 based on the IPCC SRES A2 (a) and B2 scenarios (b).

CONCLUSIONS

This study investigates the applicability of the LS-SVM as the statistical downscaling method for generating daily precipitation at the Danjiangkou reservoir basin in China, and its capability in downscaling daily precipitation amount and other characteristics were compared with SDSM, which is the most widely used regression/based statistical downscaling method. LS-SVM is also used to investigate the possible impact of climate change on precipitation. The results will be helpful to cope with new problems arising from climate changes in ungauged basins.

The study results show that LS-SVM is effective as a statistical downscaling technique as compared to the commonly used statistical downscaling method. The main advantage of this downscaling method is its ability to solve problems in nonlinear function estimation and is closely related to regularization networks and Gaussian processes. The downscaling results, corresponding to both scenarios, show that LS-SVM estimates an increase in the average annual precipitation downscaled from CGCM2 for the IPCC SRES A2 scenario by about 277.51 mm by the 2080s and an increase by about 157.65 mm for the IPCC SRES B2 scenario during the same time period. The results also demonstrate that emphasis should be given in identifying appropriate downscaling tools for impact studies by showing how different future climate scenarios could result in significantly different hydrological impact simulation results for the same watershed.

Acknowledgements The study is financially supported by the National Natural Science Fund of China (50679063; 50809049), International Cooperation Research Fund of China (2005DFA20520) and Research Fund for the Doctoral Program of Higher Education (200804861062). The authors are very grateful to the Chinese National Climate Centre, NOAA Earth System Research Laboratory and Canadian Centre for Climate Modelling and Analysis for providing valuable climatic data.

REFERENCES

Busuioc, A., Chen, D. & Hellström, C. (2001) Performance of statistical downscaling models in GCM validation and regional climate change estimates: application for Swedish precipitation. *Int. J. Climatol.* **21**, 557–578.

Bárdossy, A., Bogardi, I. & Matyasovszky, I. (2005) Fuzzy rule-baseddownscaling of precipitation. *Theor. Appl. Climatol.* **82**, 119–129.

Chen, H., Guo, S.L., Xu, C-Y. & Singh, V. P. (2007) Historical temporal trends of hydro-climatic variables and runoff response to climate variability and their relevance in water resource management in the Hanjiang basin. *J. Hydrol.* **334**, 171–184.

Coulibaly, P., Dibike, Y. B. & Ancitil, F. (2005) Downscaling precipitation and temperature with temporal neural networks. *J. Hydrometeorol.* **6**(4), 483–496.

Deidda, R. (1999) Multifractal analysis and simulation of rainfall fields in space. *Phys. Chem. Earth Part B-Hydrol. Oceans & Atmos.* **24**(1–2), 73–78.

De Brabanter, J., Pelckmans, K., Suykens, J. A. K. & Vandewalle, J. (2002) Robust crossvalidation score function for LS-SVM non-linear function estimation. In: *International Conference on Artificial Neural Networks* (ICANN 2002, Madrid Spain), 713–719.

Fowler, H. J., Blenkinsopa, S. & Tebaldib, C. (2007) Linking climate change modelling to impacts studies: recent advances in downscaling techniques for hydrological modelling. *Int. J. Climatol.* **27**, 1547–1578.

Ghosh, S. & Mujumdar, P. P. (2007) Statistical downscaling of GCM simulations to streamflow using relevance vector machine. *Adv. Water Resour.* doi:10.1016 /j.advwatres. 2007.07.005.

Grotch, S. L. & MacCracken, M. C. (1991) The use of general circulation models to predict regional climatic change. *J. Climate* **4**, 286–303.

Huth, R. (1999) Statistical downscaling in central Europe: evaluation of methods and potential predictors. *Climate Res.* **13**, 91–101.

IPCC (2001) *Climate Change 2001: The Scientific Basis* (ed. by J. T. Houghton *et al.*). Cambridge University Press, Cambridge, UK.

Kilsby, C. G., Jones, P. D., Burton, A., Ford, A. C., Fowler, H. J., Harpham, C., James, P., Smith, A. & Wilby, R. L. (2007) A daily weather generator for use in climate change studies. *Environ. Modelling & Software* **22**, 1705–1719.

Srikanthan, R. & McMahon, T. A. (2001) Stochastic generation of annual, monthly and daily climate data: a review. *Hydrol. Earth System Sci.* **5**(4), 653–670.

Suykens, J. A. K. & Vandewalle, J. (1999) Least squares support vector machine classifiers. *Neural Process. Lett.* **9**(3), 293–300.

Tripathi, S., Srinivas, V. V. & Nanjundiah, R. S. (2006) Downscaling of precipitation for climate change scenarios: a support vector machine approach. *J. Hydrol.* **330**(3–4), 621–640.

Vapnik, V. N. (1995) *The Nature of Statistical Learning Theory*. Springer Verlag, New York, USA.

Vapnik, V. N. (1998) *Statistical Learning Theory*. Wiley, New York, USA.

Wetterhall, F., Halldin, S. & Xu, C-Y. (2005) Statistical precipitation downscaling in Central Sweden with the analogue method. *J. Hydrol.* **306**, 174–190.

Wetterhall, F., Bardossy, A., Chen, D., Halldin, S. & Xu, C.-Y. (2006) Daily precipitation-downscaling techniques in three Chinese regions. *Water Resour. Res.* **42** W11423, doi:10.1029/2005WR004573.

Wilby, R. L., Hay, L. E. & Leavesley, G. H. (1999) A comparison of downscaled and raw GCM output: implications for climate change scenarios in the San Juan River basin, Colorado. *J. Hydrol.* **225**(1–2), 67–91.

Wilby, R. L., Dawson, C. W. & Barrow, E. M. (2002) SDSM-a decision support tool for the assessment of regional climate change impacts. *Environ. Modelling & Software* **17**(2), 145–157.

Wilby, R. L., Charles, S. P., Zorita, E., Timbal, B., Whetton, P. & Mearns, L. O. (2004) The guidelines for use of climate scenarios developed from statistical downscaling methods. Supporting material of the Intergovernmental Panel on Climate Change (IPCC), prepared on behalf of Task Group on Data and Scenario Support for Impacts and Climate Analysis (TGICA). http:// ipcc-ddc.cru.uea.ac.uk/ guidelines/StatDown_Guide.pdf.

Wilby, R. L. & Dawson, C. W. (2007) User Manual of SDSM 4.2-A decision support tool for the assessment of regional climate change impacts. File Report of Department of Geography, Lancaster University, UK, 01-94.

Wilby, R. L., Beven, K. J. & Reynolds, N. (2007) Climate change and fluvial flood risk in the UK: More of the same? *Hydrol. Processes* **22**(14), 2511–2523.

Xu, C-Y. (1999) From GCMs to river flow: a review of downscaling methods and hydrologic modelling approaches. *Prog. Phys. Geog.* **23**, 229–249.

Yu, X. Y. & Liong, S. Y. (2007) Forecasting of hydrologic time series with ridge regression in feature space. *J. Hydrol.* **332**, 290–302.

Zorita, E. & von Storch, H. (1999) The analog method as a simple statistical downscaling technique: comparison with more complicated methods. *J. Climate* **12**, 2474–2489.

Impacts of industrial wastes on regional groundwater

XIN-HUA ZHANG[1], RUI-PING YAN[1], YU-CHENG XIAO[1], NENG-CHUAN LI[1] & HUA WANG[2]

1 *State Key Lab. of Hydraulics & Mountain River Engineering, Sichuan University, Chengdu 610065, China*
xhzhang@scu.edu.cn

2 *Sichuan Provincial Irrigation & Water Conservancy Bureau, Chengdu 610015, China*

Abstract The wastes left after mineral-resource development usually have a high concentration of chemical compounds. Environmental problems such as groundwater pollution will result from leakages of the waste-deposition-site if the site was not treated adequately during its early stage of construction. Due to the complicated hydro-geological conditions within the impact area of the waste site, it is difficult to reflect these complexities into a simulation model. Therefore, in this paper an unstructured finite volume model (FVM) is proposed. To overcome the difficulty in the selection of time step, a completely implicit numerical scheme of FVM was developed. A case study was conducted at a waste deposition site located in Qian-nan Prefecture, Guizhou Province and a satisfactory result has been achieved.

Key words groundwater; unstructured grid; pollution; simulation

INTRODUCTION

In the mining industry, the residues after beneficiation and production, referred to as "tailings", are usually stored in a designated deposition site, referred to in this paper as a "tailings stack". The tailings stack not only has a large quantity of fine tailings, but also a great deal of water because the ore-beneficiation needs lots of water. Therefore, the tailings stack usually has very high water content, called "process water". Process water normally contains a variety of chemical or pharmaceutical ingredients, which can be re-used by pumping them into the recycling system of a plant.

However, serious pollution will result if leakages from the tailings stack take place due to inappropriate treatment of the stack at its early stage of construction. The leakage impact of a tailings stack firstly reflects on the regional groundwater pollution, then on the surface water. Surface water is relatively easy to control, while the groundwater is difficult to manage. The focus of this study, therefore, is to investigate the impact of the tailings stack on regional groundwater by numerical simulation methods.

Currently, numerical methods used in groundwater can be classified as: (1) finite difference method (FDM), (2) finite element method (FEM), (3) boundary element method (BEM) and (4) finite analysis method (FAM) (Hao & Kang, 2006). Among them, the finite difference method and the finite element method are the most commonly used methods. The BEM is relatively quick and efficient for steady flow in homogeneous regions, but for non-homogeneous, non-steady flow problems, it becomes complex and difficult. The finite analysis method firstly divides the whole region into many small areas and obtains the analytical solutions in these small areas satisfying certain boundary conditions, and then acquires the solutions in the whole study region through establishing relationships with the regional analytical solutions. However, meshes generated in the FAM must be rectangular, which is difficult to meet if a study region is irregular. If adopting boundary fitting coordinates, we must carry out coordinate transformation, which will greatly increase the difficulty, as shown by Qian (1995). For a complicated study region with irregular boundaries, FEM has great flexibility in dealing with this situation because unstructured grids are used in FEM. But for the computational stability of FEM schemes, the selection of the time-step must be controlled strictly so that numerical oscillation or diffusion can be overcome as shown by Wood (1996), Xue *et al.* (2004), Zhang & Takeuchi (2004), and Zhang *et al.* (2007).

FVM has the merits of both FDM and FEM. Therefore, in this paper an unstructured FVM was proposed to study the impacts of a tailings stack on regional groundwater. Firstly, the governing equation of groundwater was discretized by integrating over the control volume of unstructured grids through the divergence theorem. In order to avoid the problems often encountered in the selection of spatial and temporal steps, then, a completely implicit FVM scheme was developed. A case study was conducted at a waste deposition site located in Qian-nan Prefecture, Guizhou Province. During its application, we firstly simulated the spatial distributions of the regional groundwater before the construction of the tailings stack. Then, based on the current operation height of the tailings stack and the quantities of leakages, further simulation was carried out. A comparison of groundwater distributions before and after the construction of the tailing stack was made, and finally the extent and scope of the influences of tailing stack on the regional groundwater were evaluated.

GOVERNING EQUATIONS

The differential equation of groundwater in porous media can be written as:

$$\mu\frac{\partial H}{\partial t}=\frac{\partial}{\partial x}\left(T_x\frac{\partial H}{\partial x}\right)+\frac{\partial}{\partial y}\left(T_y\frac{\partial H}{\partial y}\right)+\frac{\partial}{\partial z}\left(T_z\frac{\partial H}{\partial z}\right)+w_s \tag{1}$$

where H is the water head [L]; μ is the storage coefficient; $T(=K\cdot m)$ is the transmissivity $[L^2T^{-1}]$ in which, K is the hydraulic conductivity or permeability $[LT^{-1}]$ and m is the thickness of aquifers [L]; w_s is the source or sink $[LT^{-1}]$.

As the hydro-geological conditions in a study region are usually complex, it is difficult to accurately obtain the hydraulic conductivity in different directions. Furthermore, these hydraulic conductivities are not evenly distributed in spatial domains. Therefore, 2-D groundwater models have been adopted more often than the 3-D groundwater models for the study of groundwater related problems. The governing equation of 2-D groundwater can be expressed as:

$$\mu\frac{\partial H}{\partial t}=\frac{\partial}{\partial x}\left(T_x\frac{\partial H}{\partial X}\right)+\frac{\partial}{\partial y}\left(T_y\frac{\partial H}{\partial y}\right)+w_s \tag{2}$$

UNSTRUCTURED FVM SCHEMES OF GOVERNING EQUATIONS

In order to reflect the influences of irregular and complex boundaries in the simulation, in this study an unstructured grid, mainly based on an arbitrary triangular grid, was used. Unstructured FVM schemes can be obtained as proposed by Zhang *et al.* (2007). If H_i, H_j, H_m represent the values at the end of time step Δt, the completely implicit numerical schemes of unstructured FVM can be written as:

$$\sum_{t=1}^{m}\left[R_{ij}^{t}(H_j^{n+1}-H_i^{n+1})+R_{ik}^{t}(H_k^{n+1}-H_i^{n+1})\right]+A_iS_i=A_i\frac{H_i^{n+1}-H_i^{n}}{\Delta t} \tag{3}$$

Applying the iterative method of Gauss-Seidal to equation (3), water head H_i at the end of a time step Δt can be calculated based on the following iterative equation:

$$H_{i,n}^{(r+1)}=\frac{\sum_{t=1}^{m}\left[R_{ij}^{t}H_{j,k}^{r(r+1)}+R_{ik}^{t}H_{k,n}^{r(r+1)}\right]+\frac{A_i\mu_iH_{i,n}}{\Delta t}+A_iS_i}{\sum_{t=1}^{m}\left[R_{ij}^{t}+R_{ik}^{t}\right]+\frac{A_i\mu_i}{\Delta t}} \tag{4}$$

where r represents the number of iterations. Other coefficients in equation (4) can be calculated as:

$$\begin{aligned}&R_{ij}^{t}=-\frac{K_{ij}}{4A^{t}}\left(b_ib_j+c_ic_j\right),\ R_{ik}^{t}=-\frac{K_{ik}}{4A^{t}}\left(b_ib_k+c_ic_k\right)\\&A_i=-\sum_{t=1}^{m}\frac{1}{16A^{t}}\left[\left(b_k^2+c_k^2\right)\left(b_ib_j+c_ic_j\right)+\left(b_j^2+c_j^2\right)\left(b_ib_k+c_ic_k\right)\right]\end{aligned} \tag{5}$$

where t represents the tth triangle surrounding the node i, m represents the total number of the triangles surrounding the node i; b and c are the constants related to the coordinates of the triangle Δ_{ijk} and can be calculated by:

$$\begin{cases} b_i = y_j - y_k \\ b_j = y_k - y_i \\ b_k = y_i - y_j \end{cases}, \quad \begin{cases} c_i = x_k - x_j \\ c_j = x_i - x_k \\ c_k = x_j - x_i \end{cases} \tag{6}$$

the area of the triangle Δ_{ijk} can be calculated as follows:

$$A^{\Delta ijk} = A^t = \frac{1}{2}\left(b_j c_k - b_k c_j\right) \tag{7}$$

If the value-differences of H_i obtained from two successive iterations are very small, for instance, the relative value-difference is less than 10^{-5}, the value of water head H_i will be regarded as the final solution at the end of time-step Δt satisfying the requirements of convergence.

APPLICATIONS AND RESULTS ANALYSIS

In order to verify the applicability of the simulation model, a case study was conducted at the tailings stack of an enterprise for fertilizer production in the Qian-nan Pefecture, Guizhou Province. According to the regional topography, hydro-geological features, and the characteristics of the surrounding three rivers, the region that may possibly be impacted by the tailings stack was meshed into irregular-triangle-grid on the local map with a scale of 1:10 000. Elevations of each grid node in the study region can be directly obtained on the local map. Figure 1 displays the ranges and elevation distributions of the study region.

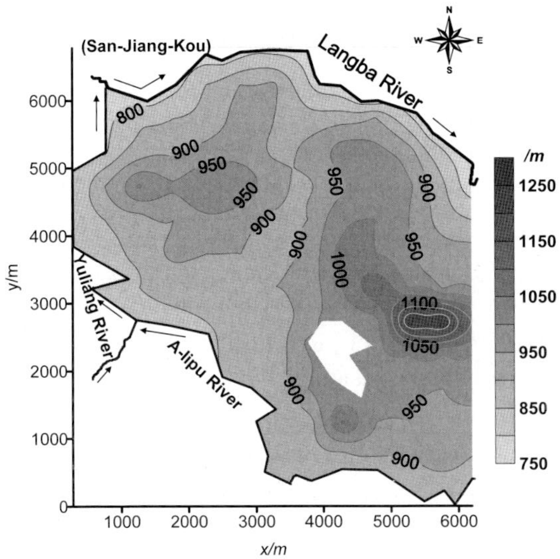

Fig. 1 Area and elevation distributions of the study region.

Determination of hydraulic conductivity

The parameter of hydraulic conductivity at each grid node was determined based on the location of the grid at the geological formations, lithology, hydro-geological characteristics, field pressure test

Table 1 Conductivities at different geological units.

Identifier of geological formations	Hydraulic conductivity K (m/day)
Q	7.5
T_2f^2	5.417
T_2f^{1-2}	0.086
T_2f^{1-1}	0.00864
T_2q^3	0.0000864
T_2q^2	5.4
T_2q^1	0.000078
T_2g	5.2
T_1a	0.07
T_1d	0.065
P_2c	0.06

data obtained by injection and pumping, etc. Details of the hydraulic conductivity obtained at each geological formation are listed in Table 1.

Boundary conditions

River boundary Three rivers adjacent to the tailings stack are the important boundaries for regional groundwater recharging and discharging. The multi-year averaged water levels of the three rivers were used as the boundary conditions of the simulation. Currently the tailings stack has been operated to an elevation of 936 m. Process water leaching from the stack has caused some environmental problem, especially the water pollution of the rivers. Therefore, the boundary conditions of the simulation can be classified into two categories, namely:

(1) Constant water head boundaries. This category includes two boundaries, one that directly contacts with the tailings stack and another is the three rivers. Because the tailing stack has now been accumulated to the elevation of 936 m and been fully filled by process water, water level of each grid immediately contacting with the stack can be set to 936 m. As for those grids located on the rivers, water levels of those grids will be set to the averaged annual water level of the river.

(2) Discharge boundaries. This kind of boundary refers to the leakage boundaries of the tailing stack, in which the seepage of process water is discharged into the regional groundwater system. As the boundary of the tailing stack lies in different geological units, the percolation properties of geological units are different. The total quantities of leaching from the tailing stack must be allocated to each grid or grid lines properly. Currently, the estimated quantity of daily leakage is approximately 6000 m^3. Considering the differences of hydraulic conductivities and lengths of the grid-lines, the production of hydraulic conductivity and the length of the grid-lines was used as a weight to allocate the quantity of leakage.

Initial conditions

In order to implement the numerical model for the adverse impact assessment of the tailings stack on regional groundwater, initial conditions must be determined in addition to the important boundary conditions. As the distribution of regional groundwater before the construction of the tailings stacks was unknown, we must at first simulate and get the initial groundwater distributions before the construction of the tailings stack. Then, the impact of the tailings stack on regional groundwater can be further evaluated on the basis of the initial groundwater distribution. For this end, two initial conditions are involved, i.e.:

– Initial condition before the construction of tailing stacks. River grids take the values of water level averaged in multi-years. Other grids within the region are supposed to have an initial water level of 765.00 m, which is the lowest elevation in the region.

- Initial conditions after the construction of tailings stack. Initial conditions at this situation are very important for the impact evaluation of the tailing stack. Currently the stack has been operated at the elevation of 936 m, results obtained through simulation under the condition of no tailings stack were used as the initial conditions after the construction of the tailings stack. Based on these initial conditions, regional groundwater distributions under the current operation height of the tailings stack can be further estimated.

RESULTS AND DISCUSSIONS

The multi-years averaged rainfall in the study area is 1176 mm and the multi-years averaged evaporation is 1147.9 mm. Rainfall and evaporation are approximately the same. In addition, the three rivers in the region belong to mountainous rivers, in which the rising and falling of water levels in the rivers during the flood season are quick and the time flooding lasted is short. Therefore, the impact of the rainfall and evaporation on the regional groundwater was ignored during the simulation.

Groundwater distribution at current operation height of the tailings stack

According to the boundary conditions and initial conditions, the spatial distribution of regional groundwater under the influence of existing tailings stack at the elevation of 936 m can be simulated. Results of the simulation were shown in Fig. 2, in which Fig. 2(a) represents the groundwater distribution and Fig. 2 (b) represents the cover thickness of the regional groundwater.

Accuracy of the results

In order to verify the accuracy and reliability of the results, three holes were drilled in the southeastern part of the study region at the geological section IV–IV′, as shown in Fig. 2(b). Depths of three holes drilled vary from 54.80 m to 71.80 m. Details of the boring holes are shown in Table 2. The averaged cover-thickness of groundwater in the three boring holes is 54.43 m. As indicated in Fig. 2(b), geological section IV–IV′ is just located at the cover-thickness contour line of 50 m. Bore hole ZK7 is adjacent to a grid node numbered as N-110, which is only

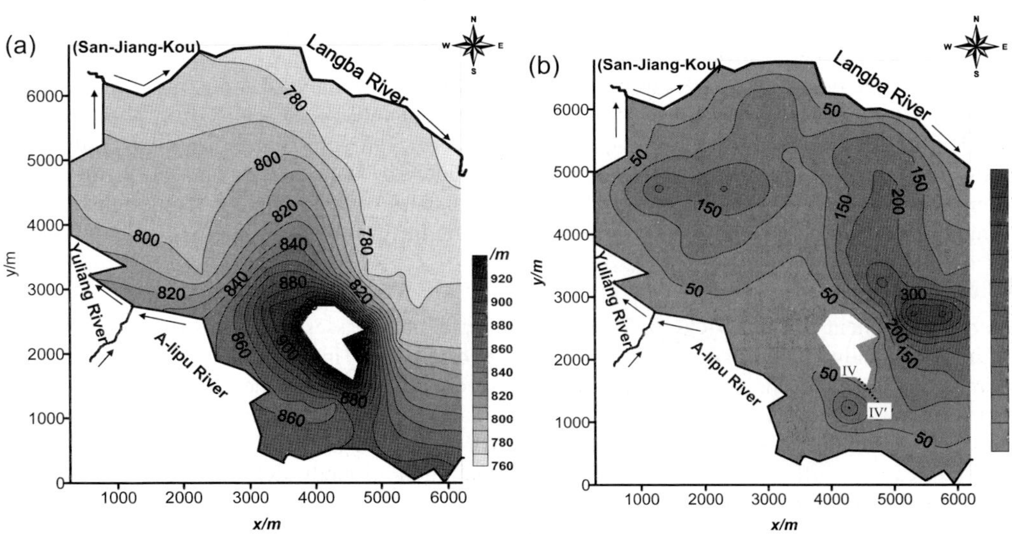

Fig. 2 Distributions of groundwater and groundwater-cover-thickness simulated when the tailings stack operates at current elevation of 936 m. (a) Spatial distribution of groundwater. (b) Cover thickness of regional groundwater.

Table 2 Bore data of the three holes drilled along Section IV–IV′.

No.	Ground elevation (m)	Hole depth (m)	Ground water level (m)	Cover-thickness (m)
ZK3	950.34	68.93	898.74	51.60
ZK7	935.86	54.80	882.56	53.30
ZK11	941.18	71.80	882.78	58.40

20 m apart. Water level in the bore hole ZK7 is 882.56 m, while the grid node N-110 has a water level of 882.18 m. The difference in water levels between them is only 0.38 m. Due to the complicated geological conditions in the region, the 20 m distance, and the inaccurate estimation of the leakage, –6000 m^3/day, it is reasonable to have some differences existing in the water levels between the bore hole and the grid node N-110. Moreover, this small difference implies that the numerical model, proposed in this study, has enough accuracy and reliability to estimate regional groundwater.

Uplift effect of the tailings stack on regional groundwater

As the tailing stack has operated to an elevation of 936 m and plenty of process water exists in the tailings stack, leakage of the process water will have a great impact on regional groundwater. In order to evaluate the influence, we can make a comparison between the groundwater levels simulated under conditions with and without construction of the tailings stack. Then, the uplift effects on water levels after the construction of the tailings stack, operating at current elevation of 936 m, can be obtained by subtracting the water level simulated from that without the tailings stack in each grid node. Results are shown in Fig. 3.

As shown in Fig. 3, the uplift effects of the tailings stack on regional groundwater are significant. More than 90% of the study region has been affected by the tailings stack with an increase of water level of more than one metre. The maximum increment is up to 150 m, which is adjacent to the tailings stack.

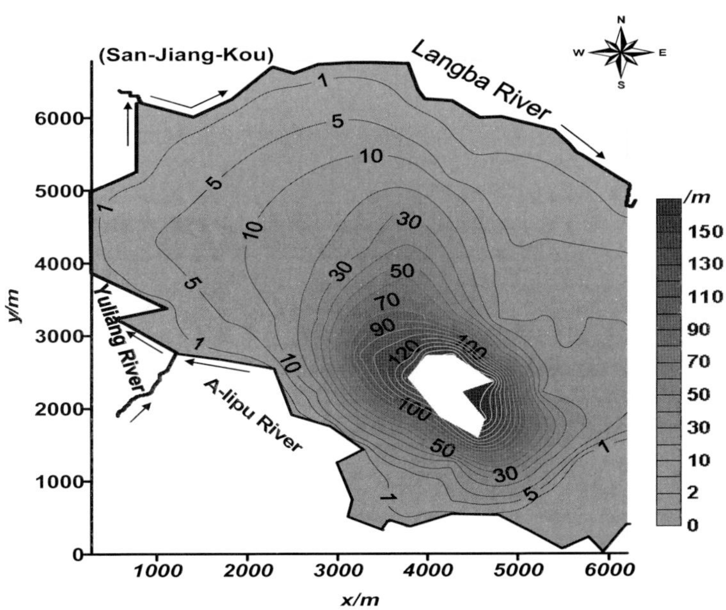

Fig. 3 Uplift effect of the tailings stack on regional groundwater.

CONCLUSIONS

This study presents an unstructured FVM model based on a triangular grid system for the impact evaluation of an industrial tailings stack on regional groundwater. A case study was conducted in Qian-nan Prefecture, Guizhou Province, China. Validation of the model has been made by comparing the results simulated with the data obtained by boring. Results indicate that:

(1) The unstructured FVM model is stable and accurate for the estimation of regional groundwater.
(2) Due to the numerical scheme of the unstructured FVM model is completely implicit, it will provide us with much more flexibility in the selection of the time step, which is a dilemma when using an explicit numerical scheme.
(3) As the unstructured FVM model is developed based on an irregular triangle system, it can integrate complicated boundary conditions, both inner and external, into the model and reflect their influences in the simulation. Therefore, this model can further improve its accuracy adequately and should have great potential for practical uses.

Acknowledgements This work has been supported in part by the State Key Laboratory of Hydraulics and Mountain River Engineering through grant no. 0805 and the National Natural Science Foundation of China with grant no. 50639100. The authors are grateful for the useful comments and suggestions provided by anonymous reviewers.

REFERENCES

Hao Zhi-Fu, Kang Shao-Zhong (2006) Current situation and development trend of numerical simulation of groundwater system. *Adv. Sci. Tech. Wat. Resour.* **26**(1), 77–81.

Qian Xiaoxing (1995) *Hydro-geological Computation*. China Water Power Press, China.

Wood, W. L. A (1996) Note on how to avoid spurious oscillation in the finite element solution of the unsaturated flow equation. *J. Hydrol.* **176**, 205–218.

Xue Yu-qun, Ye Shu-jun, Xie Chun-hong, *et al.* (2004) Application of multi-scale finite element method to simulation of groundwater flow. *Shuili Xuebao* **7**, 7–13.

Zhang Xiang-Wei & Takeuchi Kuniyoshi (2004) Methodology for modeling of groundwater flow in large area. *Shuili Xuebao* **6**, 7–13.

Zhang Xin-Hua, Zhang Xiang-Wei, Wang Hua, *et al.* (2007) Coupled simulation the effect of tailings stack on groundwater and groundwater quality. *J. Hydrodynamics, Series A*. **22**(5), 654–664.

Key word index

BENCHMARK PAPERS IN HYDROLOGY

The IAHS Series that collects together, by theme, the outstanding papers that provide the scientific foundations for hydrology today. Each thematic volume reprints 30+ papers selected by a leading expert, who comments on the significance and context of the individual papers.

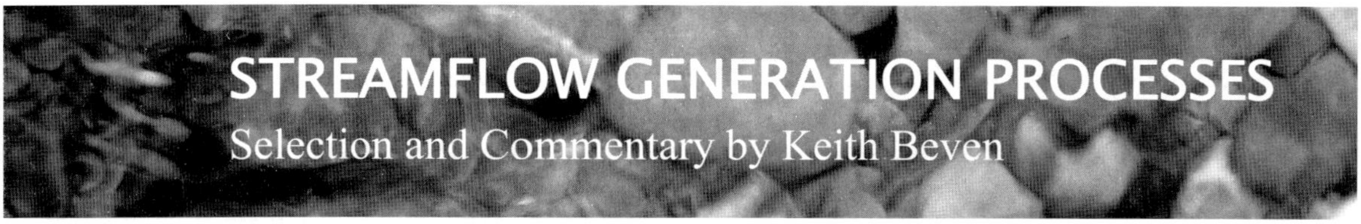

Keith Beven's selection of 31 papers on the theme of *Streamflow Generation Processes*, spans the period from 1933 to 1984, commencing with Horton's early papers on infiltration and on maximum groundwater levels. With the aid of the Introduction and Commentaries, they provide a stimulating insight to how this part of the field of hydrology developed.

IAHS BM1 ISBN 978-901502-53-4 (2006) A4 format, softback, 432 + x pp., £40.00, incl. postage

The measurement techniques section commence with the Wagon Wheel Gap catchment water balance (1921) paper, and Dalton's seminal essay *On Evaporation* (1802) starts the selection of papers on evaporation estimation, which also reprints the original Penman and Thornthwaite papers.

IAHS BM2 ISBN 978-901502-98-5 (2007) A4 format, softback, 526 + x pp., £40.00, incl. postage

All aspects of groundwater are considered, from a translation of Darcy's experimental results that led to Darcy's law, and classic papers by Meinzer, Theis and Hubbert, to Slichter's (1905) contribution identifying dispersion in the field, and are linked with more recent attempts to represent dispersion and heterogeneity with models.

IAHS BM3 ISBN 978-901502-74-9 (2008) A4 format, hardback, 626 + x pp., £55.00, incl. postage

Loague notes that hundreds, if not thousands, of hydrologic-response models have been developed, but that not all were created equal. This volume reprints 30 papers that exemplify the best in rainfall–runoff modelling. It charts developments from Mulvany's (1851) rational method for estimating peak flow, probably the first rainfall–runoff model, up to 1989. Papers on other empirical approaches, such as Sherman (1932) and Mockus (1949), are reprinted. As are Richards (1931) and Smith & Parlange (1978), the innovative contributions of Alan Freeze, and later Keith Beven, and the seminal papers of Moore & Clarke (1981) and Abbott *et al.* (1986).

IAHS BM4 ISBN 978-1-907161-06-3 (2010) A4 format, hardback, 506 + vi pp, £65.00, incl. postage

Groundwater and Climate in Africa

Edited by *Richard Taylor, C*
Mohammad Shamsudduha
IAHS Publ. 334 (*2009*) ISBN 978-

Current assessments of the impacts
variability and change on water
commonly exclude groundwater. Th
is of particular concern in Africa wh
water usage and future adaptations
to climate variability and change, to
population growth, place considerab
upon groundwater to meet
agricultural and industrial water
Groundwater and Climate in Africa
held in Kampala, Uganda, in June
the first such conference in Africa,
the first globally to focus on the
between groundwater and climate
and change.

This collection of papers selecte
conference includes the Kampala
and addresses five themes:

Impact of climate variability and cha
groundwater-based livelihoods

Impact of climate variability and cha
groundwater and groundwater-fed ec

Monitoring and modelling groundwa
replenishment

Estimation of groundwater resources
demand under a changing climate

Groundwater management in Africa

IAHS AISH Abstracts of the papers in this volume can be seen at: ***www.iahs.info***